Universal Energy Dynamics

Sandeep Chavan

Published by Gyrus Vision, 2026.

UNIVERSAL ENERGY DYNAMICS

First edition. March 5, 2026.

ISBN: 979-8233193606

Written by Sandeep Chavan.

Table of Contents

To those who question appearancesand seek the structure beneath events.

Understanding rarely arrives through certainty—it emerges through patient observationof how consequence shapes reality.

To my Students, my kids and my family

The universe does not operate through intention or punishment. It operates through consequence and correction.

— Sandeep J. Chavan

Author's Intent

This book exists because, after years of writing across physics, psychology, human behavior, society, and systems, one realization became unavoidable: everything I was saying was already governed by a single underlying logic. I did not invent that logic to write books; the books emerged because that logic kept insisting on being seen.

That logic is what I now call Universal Energy Dynamics (UED).

This is not a moment of declaration. It is a moment of alignment.

What UED Is — and What It Is Not

Universal Energy Dynamics is a physics-first framework that explains how reality behaves when energy exists away from equilibrium. It deals with delta, the pressure toward equilibrium, and the inevitable consequences that follow when imbalance is allowed, constrained, delayed, or forced.

UED is not a new force. It is not a new particle. It is not a speculative dimension. It does not replace equations, constants, or successful models.

UED corrects the level at which we believe explanations begin.

Modern physics works extraordinarily well within its designed limits. Classical mechanics, electromagnetism, relativity, quantum theory, and thermodynamics are not wrong. They are locally precise descriptions of behavior inside constrained delta windows. What they do not attempt—and were never meant—to do is explain *why* such windows exist, *why* repetition is possible, *why* time behaves differently across scales, or *why* resolution sometimes dominates over continuity.

UED lives exactly there.

It explains:

- why experiments are possible at all,
- why equations repeat in some regimes and fail in others,
- why time appears, stretches, collapses, or disappears,
- why motion feels continuous even when resolution is discrete,
- why nature relentlessly seeks equilibrium while life survives by delaying it.

UED does not compete with physics. It clarifies physics.

If modern physics is a collection of powerful tools, UED is the operating logic that explains why those tools work where they work—and why they must fail where they fail.

Why This Book Exists

This book exists as a reference spine.

Over the years, I have written extensively—sometimes technically, sometimes philosophically, sometimes through stories—about energy, consequence, awareness, intelligence, systems, leadership, markets, and human behavior. Readers often sensed a coherence beneath those works, even when the surface topics differed. That coherence was not thematic; it was structural.

Universal Energy Dynamics is that structure, written explicitly for the first time.

This book is not meant to persuade. It is not meant to provoke. It is not meant to replace anyone's education.

It is meant to orient.

A student of physics should be able to read this and understand *why equations behave the way they do*. A researcher should be able to locate the limits of their models without feeling threatened by them.

A thinker should be able to see how mind, matter, and society obey the same rules without collapsing into metaphor. And if someone like Feynman were to open this book, I would want him to smile—not because it flatters him, but because it respects clarity, restraint, and honesty.

This book exists because explanation must eventually reach a point where:

- nothing needs exaggeration,
- nothing needs mysticism,
- nothing needs denial.

UED is offered in that spirit.

It is not the final word on physics. It is the correct place to stand while doing physics.

If this book survives me, I hope it does so quietly—used not as a belief system, but as a lens. A lens that helps future readers see why reality behaves the way it does, and why understanding does not come from adding complexity, but from placing it at the right level.

That is the intent.

Preface

This book was not planned as a book.

Universal Energy Dynamics did not begin as a theory to be proposed, defended, or promoted. It emerged slowly, as a necessity, from repeated encounters with the same failure pattern appearing in unrelated domains. Whether the system was physical, psychological, organizational, or societal, the breakdowns shared a common structure: pressure accumulated silently, consequences were delayed, and when correction finally arrived, it appeared sudden, disproportionate, and unavoidable.

Existing frameworks explained parts of this behavior very well. None explained *why those explanations stopped working* when scale increased, coupling intensified, or feedback was delayed. The gap was not in data or intelligence. It was in the primitives being used to interpret what was already visible.

This book exists to address that gap—carefully and with restraint.

It does not attempt to unify science, replace mathematics, or offer a grand narrative of reality. It does not claim predictive power or moral authority. Wherever certainty would have been easy, it has been avoided. Wherever completeness would have required speculation, boundaries have been drawn instead.

The framework presented here treats *resolution*, not force or intent, as the primary driver of change. From that single shift, a consistent language becomes possible—one that can describe motion, stability, failure, behavior, leadership, and conflict without changing its core assumptions. The aim is not elegance, but durability.

This is not a book to be read for answers. It is a book to be read for *orientation*.

Some readers may find parts of it uncomfortable, not because they are complex, but because they remove familiar explanations without replacing them with comforting alternatives. That discomfort is not a flaw. It is a sign that the framework is doing what it was designed to do: reducing distortion rather than supplying closure.

You do not need to agree with everything written here. Agreement is irrelevant. What matters is whether the logic remains consistent when applied across domains, and whether it reduces the need for later correction.

If the book succeeds, it will not leave you with a new belief system. It will leave you intervening less often, noticing pressure earlier, and allowing resolution to complete without unnecessary interference.

If it fails, it will fail quietly—by being set aside.

That, too, would be an acceptable outcome.

— **Sandeep J. Chavan**

Thesis Overview — Universal Energy Dynamics

Modern science explains many phenomena with extraordinary precision, yet a persistent difficulty remains: systems that appear stable for long periods often change suddenly, sometimes catastrophically, even when no obvious force seems to cause the shift. This pattern appears not only in physics but also in biological systems, organizations, economies, and societies. The common thread across these domains is the delayed release of accumulated pressure under constraint.

Universal Energy Dynamics (UED) proposes that **systemic change is best understood as resolution of accumulated pressure rather than as the action of discrete forces between independent entities.** In this framework, what we observe as motion, transformation, or collapse is the consequence of unresolved gradients reaching structural limits.

UED introduces a simple canonical chain for diagnosing system behavior:

Constraint → Delta → Gradient → Threshold (ATT) → Coupling → Consequence

- **Constraint** defines the boundaries within which a system can resolve change.
- **Delta** represents unresolved pressure or imbalance accumulating within those boundaries.
- **Gradient** determines the direction in which resolution becomes possible.
- **Alignment Threshold Time (ATT)** marks the point beyond which gradual correction becomes impossible and forced adjustment occurs.

- **Coupling** describes how effects propagate across connected systems or scales.
- **Consequence** is the structural feedback that follows resolution.

Within this view, equilibrium is not static balance but **continuous resolution of emerging deltas within constraints**.

UED does not attempt to replace established scientific theories. Classical mechanics, relativity, quantum theory, and thermodynamics remain powerful descriptions within their respective domains. Instead, UED acts as a **diagnostic framework** that clarifies how pressure accumulates, why consequences appear delayed, and how stability can suddenly give way to correction.

The framework is intentionally restrained. It does **not predict specific events**, replace mathematical models, or offer philosophical meaning. Its purpose is orientation: to help observers identify where pressure is accumulating, which constraints are shaping resolution, and why consequences sometimes arrive later than expected.

If useful, UED should reduce explanation rather than multiply it. Its value lies not in expanding theory but in clarifying structure across systems that otherwise appear unrelated.

How to Read This Book

This book is written as one coherent whole, but it does not demand a single linear path. Reality itself does not unfold linearly; it resolves through layers, constraints, and contexts. Universal Energy Dynamics reflects that structure.

You can read this book cover to cover, slowly and sequentially—and it will hold. But you can also enter it from three distinct directions, depending on how your mind is trained and what questions brought you here.

There is no "correct" path. There is only alignment between reader and entry point.

What matters is not where you begin, but whether you allow the framework to settle before judging it.

1. Physics-First Reading Path

(For students, engineers, physicists, and technically trained minds)

If your instinct is to ask *"What exactly is being claimed, and where does it sit relative to existing physics?"*, this is your path.

Begin with the core kernel:

- the definitions of delta, equilibrium, and consequence,
- the reinterpretation of force, fields, motion, and geometry,
- the correction of time as repetition and resolution rather than flow.

Read the sections on:

- time and ATT,

- geometry and constraints,
- energy, dissipation, and thermodynamics.

This path treats UED as an advanced orientation layer for modern physics. It does not ask you to abandon mathematics or experiments. It asks you to recognize their domain of validity and their structural limits.

If you follow this path honestly, you will find that familiar equations do not collapse—they become *situated*. They stop pretending to be universal truths and start behaving like what they always were: precise tools operating inside carefully permitted delta windows.

Take this path if you value rigor, restraint, and conceptual cleanliness.

2. Mind-First Reading Path

(For psychologists, philosophers, educators, clinicians, and reflective readers)

If your instinct is to ask *"Why do humans behave this way, even when they know better?"*, this is your path.

Begin with the sections on:

- life as managed disequilibrium,
- the mind as a delta engine,
- awareness as resolution clarity,
- emotion as consequence feedback,
- anxiety, insight, trauma, and attention as alignment phenomena.

Then move backward into the physical foundations.

This path reveals something subtle but powerful: human psychology is not a special case. It obeys the same rules as energy systems, only at a more compressed and sensitive scale. Desire, fear, overthinking, calm, morality, and conflict are not moral failures or personality flaws—they are delta behaviors under constraint.

If you take this path, physics will stop feeling abstract. It will start feeling *familiar*.

Take this path if you are interested in understanding humans without reducing them, and in grounding psychology without turning it mechanical.

3. Systems-First Reading Path

(For leaders, policymakers, economists, strategists, and systems thinkers)

If your instinct is to ask *"Why do organizations, markets, and societies keep repeating the same failures?"*, this is your path.

Begin with the sections on:

- shared fields,
- power as delta control,
- institutions as alignment regulators,
- markets as consequence aggregators,
- geopolitics as multi-scale delta warfare.

Then trace these ideas downward into physics and upward into mind.

This path shows that large systems fail not because people are irrational, but because delta is mismanaged across scales. Policies fail when intent ignores field constraints. Revolutions occur when thresholds are crossed, not when opinions change. Stability emerges not from control, but from correct coupling.

If you follow this path, leadership will stop looking heroic or villainous. It will start looking structural.

Take this path if you want clarity without ideology, and realism without cynicism.

A Necessary Warning

Whichever path you choose, resist the urge to skim.

This book is not designed to impress quickly. It is designed to re-orient slowly. Many of its claims will feel obvious *after* you understand them and uncomfortable *before* you do. That discomfort is not confusion—it is your existing frameworks adjusting.

Read patiently. Allow definitions to stabilize. Let the same idea appear in multiple domains before concluding that it is repetition.

That repetition is intentional. It mirrors how reality itself reveals structure: not through novelty, but through consistent behavior across scale.

Final Note to the Reader

You do not need to agree with everything in this book for it to be useful. You only need to read it without forcing it into categories it does not belong to.

Universal Energy Dynamics is not asking for belief. It is offering orientation.

Choose your path. The framework will meet you there.

Scope, Language, and Discipline

This book is written with deliberate restraint. That restraint is not a limitation of ambition; it is a condition for clarity. Universal Energy Dynamics can only function as a reliable framework if its scope is defined, its language is disciplined, and its claims are placed at the correct level.

Without this discipline, any unifying idea—no matter how accurate—quickly collapses into exaggeration, ideology, or metaphorical comfort. This book avoids that trap intentionally.

No Replacement Claims

Universal Energy Dynamics does not replace existing physical theories.

It does not replace Newtonian mechanics, Maxwell's equations, relativity, quantum mechanics, thermodynamics, or any of their mathematical formalisms. Those frameworks work—and they work precisely because they operate within well-defined delta regimes. Their predictive power is real, earned, and indispensable.

What UED does is refuse to mistake effectiveness for fundamentality.

A theory can work extraordinarily well while still being context-bound. UED makes those contexts explicit. It explains *why* classical mechanics succeeds at macroscopic repetition, *why* quantum mechanics encounters resolution instead of continuity, and *why* relativity emerges when propagation constraints dominate behavior.

This distinction matters. Replacement destroys continuity. Correction preserves it.

UED is therefore not an overthrow of modern physics, but its next stabilizing layer—one that allows existing models to coexist without contradiction, competition, or forced unification.

Corrections of Ontology and Geometry

Where UED does intervene decisively is at the level where most confusion silently begins: ontology and geometry.

Modern physics often inherits its ontology unconsciously. Concepts such as "particles," "forces," "motion," "space," and "time" are treated as primitive entities rather than as descriptions of behavior under constraint. This works operationally, but it creates conceptual fractures when those primitives are pushed beyond their valid domains.

UED corrects this at the root.

- What exists are states and gradients, not independent objects.
- Motion is not traversal through space, but successive resolution of delta.
- Time is not a flowing dimension, but the duration of unresolved imbalance.
- Geometry is not a container, but a map of allowed resolution pathways.

These are not semantic games. They are ontological corrections that allow geometry to re-emerge cleanly, without paradox. Once ontology is clarified, geometry stops pretending to be absolute and starts behaving as what it always was: a constraint structure governing how resolution can proceed.

Nothing mathematical is discarded. What changes is what mathematics is understood to be describing.

Operational Definitions Over Metaphor

This book avoids metaphor wherever precision is possible.

Metaphors are powerful tools for intuition, but they become liabilities when they are mistaken for explanations. Phrases like "time flows," "forces act," "energy moves," or "systems want" may feel intuitive, but they quietly smuggle assumptions that later break under scrutiny.

Universal Energy Dynamics replaces metaphor with operational definitions:

- Delta is defined by measurable or inferable imbalance.
- Equilibrium is defined by minimum gradient states.
- Consequence is defined by redistribution, transformation, and diffusion.
- Alignment thresholds are defined by observable regime changes.

These definitions are intentionally reusable across domains. They apply to physics, biology, psychology, and social systems without changing meaning, only scale and sensitivity. That consistency is not poetic—it is structural.

When metaphor is unavoidable, it is clearly marked as such and never allowed to carry explanatory weight.

A Final Word on Discipline

This book does not promise answers to everything, and it does not attempt to close inquiry. Its discipline lies precisely in knowing where to stop.

Universal Energy Dynamics explains:

- why explanation itself is possible,
- why limits appear where they do,
- and why different domains obey the same structural rules.

Beyond that, experimentation, mathematics, and discovery must continue.

The goal here is not to end physics. It is to place physics on stable conceptual ground, so that future work—by others, long after this book—is not burdened by inherited confusion.

That is the scope. That is the language. That is the discipline.

Part I — The Kernel

Directional Overview

This part establishes the irreducible core of Universal Energy Dynamics. Everything that follows—whether it concerns time, motion, life, mind, or society—rests on the orientation set here. If this part is understood correctly, the rest of the book will feel less like explanation and more like recognition.

The starting point is deliberately simple, but not simplistic: equilibrium. Not as a philosophical ideal, not as a static end state, but as the default tendency of nature whenever imbalance is allowed to resolve. Equilibrium is not something systems "seek" with intent; it is what remains when no constraints prevent alignment. From this baseline, delta appears—not as an anomaly, but as a permitted deviation. Delta is not opposition to nature; it is the only way nature allows structure, motion, and experience to exist temporarily.

Within this framing, action is redefined. Action is not force applied to an object, nor intention imposed on matter. Action is always one of two things: the maintenance of delta against equilibrium pressure, or the resolution of delta back toward alignment. There is no third category. This distinction quietly dissolves many classical confusions about motion, work, effort, and change.

From here, consequence becomes unavoidable. Consequence is not introduced as ethics, morality, or reward–punishment logic. It is introduced as structural accounting. Whenever delta is created, constrained, delayed, or redirected, redistribution must occur. What we call "cause" is simply a shorthand label for pathways through which delta propagates and resolves. The Law of Consequence, in this context, is not belief-language or philosophical assertion; it is mechanism-language describing how imbalance cannot escape resolution, only transform its route.

With delta and consequence in place, ontology itself is corrected. The book intentionally steps away from "things" as primary entities. What exist are states—temporarily stabilized configurations of energy under constraint. Objects are not fundamental; they are bundles of resolved and unresolved delta that persist long enough to be named. Properties do not belong to objects; they describe how a state behaves when subjected to specific constraints. Force is no longer treated as a thing that acts, but as an effect that appears when delta resolution is obstructed.

Finally, the rule governing all of this is made explicit: nature contracts delta. Relaxation, dissipation, and alignment are not tendencies layered on top of physics; they are physics at its most basic level. Stability exists because some configurations minimize delta efficiently. Instability exists not as failure or noise, but as information—evidence of misalignment, constraint conflict, or pending resolution. Nothing is random in the sense of being meaningless; variability itself carries structural signal.

By the end of this part, the reader is not expected to memorize definitions or accept conclusions. What should shift is the *questioning stance*. The central question is no longer "What is real?" or "What exists fundamentally?" It becomes something quieter and far more powerful:

What is resolving, under what constraints, and with what consequences?

Once that shift occurs, the rest of Universal Energy Dynamics unfolds naturally.

Chapter 1 — The Only Starting Point: Equilibrium and Delta

Why We Must Start Here (and Nowhere Else)

Every explanation in physics, no matter how advanced or abstract, quietly assumes a starting point. Sometimes that starting point is hidden inside equations. Sometimes it is buried under language like *initial conditions*, *boundary constraints*, or *reference frames*. But it is always there. No explanation begins from nothing; it begins from what is already taken as given.

The difficulty is not that physics assumes a starting point. The difficulty is that we rarely stop to ask whether we have chosen the right one.

Historically, we have begun explanations with objects, forces, motion, space, or time. These choices felt natural because they matched experience: things move, forces act, time flows. But familiarity is not the same as fundamentality. Starting with motion already assumes imbalance. Starting with force already assumes resistance. Starting with objects already assumes stability. These are not origins; they are effects.

By beginning there, physics inherited a quiet confusion. We started by explaining *how things change* without first explaining *why change is possible at all.* As a result, many later paradoxes—about time, causality, measurement, and scale—were not discovered accidentally. They were planted at the beginning.

The only starting point that does not smuggle assumptions is equilibrium.

Equilibrium is not an outcome that systems strive toward. It is not a final state earned through dynamics. It is what remains when nothing prevents alignment. It requires no force, no motion, no agency, and no explanation. If nothing is disturbed, nothing needs to happen. That stillness is not emptiness; it is completeness.

This is the quiet idea that modern physics rarely states explicitly: nothing needs to be added for equilibrium to exist.

Once this is seen, imbalance stops looking like the natural state of the universe. It becomes what it truly is—a permission, not a default. Any deviation from equilibrium must be allowed, supported, constrained, or sustained. Without that permission, alignment resumes automatically.

This single shift changes how everything else is understood.

Action no longer means "something happening." It means something being allowed to remain unresolved. Motion no longer means travel through space; it means delta persisting across constraints. Stability no longer means rest; it means minimum unresolved difference. Even existence itself begins to look less like a collection of things and more like a landscape of temporarily held imbalances.

The intuition most of us carry—of a universe always in motion, always doing something—is backwards. What requires explanation is not why things settle, but why they ever don't.

That is why we must start here, and nowhere else.

1.1 — Equilibrium Is Not an End State, It Is the Default

Equilibrium is often described as something systems *reach*. The language itself is revealing. We speak of balance as if it were an achievement, a destination earned through motion, effort, or time. This framing quietly reverses the true order of things.

Equilibrium is not what happens after dynamics. It is what exists before dynamics are allowed.

When no constraints operate, when no boundaries hold imbalance in place, alignment does not need to be produced—it simply remains. No force pushes a system into equilibrium. No mechanism drives it there. In the absence of interference, there is nothing else it can do.

This is why equilibrium requires no explanation. It is the explanation that does not need one.

Deviation, on the other hand, always demands accounting. Any difference—of pressure, temperature, potential, concentration, attention, or intent—exists only because something prevents immediate alignment. Whether that prevention is structural, environmental, biological, or artificial, it must be present. Without it, the deviation collapses.

Physics often inverts this relationship by treating equilibrium as a special case and imbalance as the norm. That inversion is convenient for calculation, but misleading for understanding. What we usually call "normal behavior" in physics—motion, oscillation, interaction—already presumes held delta. The true baseline is not activity, but quiet alignment.

This distinction becomes clearer when we separate two commonly conflated ideas: static equilibrium and dynamic equilibrium.

Static equilibrium describes a condition where no observable change occurs because all gradients have resolved. It is the absence of internal pressure to change. Dynamic equilibrium, by contrast, involves continuous activity—flows, exchanges, oscillations—yet no net change in state. The system appears active, but its overall alignment remains stable.

Both are equilibrium. Neither requires intention, memory, or direction. The difference lies only in how resolution pathways are arranged. In static equilibrium, there are none left to traverse. In dynamic equilibrium, multiple pathways exist, but they cancel out at the level of the whole.

Understanding this prevents a common misunderstanding: that equilibrium means inactivity. It does not. It means no unresolved gradient at the scale of interest. Activity can exist inside equilibrium, but it cannot escape it.

This is why stability is far more common than chaos. Chaos is not the natural state of systems; it is a temporarily permitted condition that arises when multiple deltas interact under constraint. Without boundaries, feedback loops, or sustained gradients, chaotic behavior dissipates quickly. Alignment wins by default.

Chaos, turbulence, and instability attract attention precisely because they are rare and informative. They signal that delta is being actively maintained, obstructed, or misaligned. They are not evidence of disorder at the foundation of nature; they are evidence of constraint in operation.

Once this is recognized, equilibrium stops being something to explain or fear. It becomes the silent reference point against which all action, structure, and experience are measured. Nothing has to be done for equilibrium to appear. Something must always be done—or allowed—for it not to.

1.2 — Delta: The Only Reason Anything Happens

If equilibrium is the default, then the obvious next question is unavoidable: why does anything ever happen at all? The answer is not force, intention, motion, or time. The answer is *delta*.

Delta is not disorder. It is not chaos. It is not a violation of nature. Delta is simply permitted imbalance—a condition where alignment is delayed, constrained, or locally suspended. Nothing more, nothing less.

The mistake we often make is emotional rather than physical. We treat imbalance as something unnatural, something that must be explained away or corrected. In reality, imbalance is the only doorway through which structure, motion, and experience can appear. Without delta, nothing unfolds. With delta, everything that can happen, happens.

Delta can appear in many forms, but its essence is always the same. It may present itself as a gradient, a misalignment, or an asymmetry. Pressure exists because there is a difference between regions. Heat flows because temperature is uneven. Electrical current appears because potential is not uniform. These are not separate phenomena—they are different expressions of the same underlying condition: delta being held open.

Importantly, delta is not created in opposition to nature. Nature does not resist delta, nor does it celebrate it. Nature tolerates delta only under conditions. Those conditions may be boundaries, constraints, containers, feedback loops, or structural supports. When those conditions disappear, delta resolves without hesitation.

This is why delta should never be described as "against nature." It is not rebellion. It is temporary allowance.

The same logic extends seamlessly beyond physics.

A pressure difference allows air to move. A temperature gradient allows energy exchange. A difference between desire and reality allows motivation. A gap between attention and distraction allows thought.

In every case, action does not arise because something is added, but because alignment is delayed. When pressure equalizes, flow stops. When temperature balances, heat transfer ends. When desire is fulfilled, effort dissolves. When attention collapses into distraction or fixation, awareness fades.

Delta is therefore not a thing that acts. It is a condition that enables action. It does not push systems forward; it simply prevents them from settling immediately. That delay—whether brief or sustained—is where motion, behavior, and experience unfold.

This is why delta is universal across domains. It does not belong to physics alone. It does not change its meaning when applied to biology, psychology, or society. Only its scale, sensitivity, and resolution pathways change.

Once delta is understood this way, many long-standing confusions disappear. We stop asking what force caused motion and begin asking what imbalance was permitted. We stop asking why systems change and begin asking why alignment was postponed. We stop treating activity as the norm and stillness as the mystery.

Delta is the sole reason anything happens—not because it drives the universe forward, but because it allows the universe to hesitate before returning to equilibrium.

1.3 — Why Delta Must Be Permitted (Not Created)

It is tempting to say that systems *create* imbalance. We speak casually of forces generating motion, of humans creating change, of technology producing disruption. This language feels intuitive because it matches agency and effort. But it is structurally inaccurate.

Delta is never absolute freedom. It is never spontaneous rebellion against equilibrium. It is never created from nothing.

Delta exists only where resolution is prevented.

Nature does not manufacture imbalance; it allows imbalance to persist under specific conditions. The distinction is subtle but decisive. Creation implies authorship and intent. Permission implies constraint, context, and limits. Delta belongs to the second category.

To understand this, consider what happens in the absence of constraints. Any gradient—pressure, temperature, potential, concentration—collapses immediately. Alignment does not need instruction. It requires no memory, no rulebook, no external push. If nothing holds difference apart, difference disappears.

So whenever delta is observed, something must be holding it open.

That "something" is never abstract. It is always a boundary, a container, a structure, a rule, or a coupling. A pipe allows pressure difference to persist. A membrane allows chemical gradients to exist. A cell wall sustains metabolic imbalance. A social rule sustains power gradients. An attention frame sustains cognitive tension.

Remove the boundary, and delta resolves. Weaken the structure, and delta leaks. Change the rule, and the pathway of resolution shifts.

This is why delta cannot be described as freedom. It is the opposite. Delta is tightly constrained permission. It exists only inside limits, and those limits determine how long imbalance survives, how it behaves, and how it ultimately resolves.

This insight corrects a common misunderstanding about human action. Humans do not create imbalance in the universe. They engineer environments where imbalance can be sustained. Technology is not a generator of delta; it is a sophisticated system of constraints designed to delay equilibrium in useful ways.

An engine does not create motion out of nothing. It holds pressure, temperature, and chemical imbalance apart long enough to extract work. A computer does not create information. It stabilizes electrical and logical deltas inside rigid architectures. A city does not create activity. It concentrates and channels gradients—economic, social, energetic—through boundaries and rules.

Life itself operates on the same principle. A living organism does not generate energy in opposition to nature. It maintains internal gradients against an external equilibrium. Metabolism is not production; it is managed delay. Death is not failure. It is the moment when permission ends and equilibrium resumes.

The same applies to psychological and social systems. Motivation exists because desire and reality are held apart. Conflict exists because alignment is obstructed by identity, interest, or power. Institutions persist because rules prevent immediate resolution of competing gradients. When those rules fail, collapse follows—not because something new happened, but because permission expired.

Seeing delta this way removes unnecessary heroism and unnecessary blame. No force, leader, technology, or ideology creates imbalance in a fundamental sense. They design permission windows—some narrow and fragile, others wide and enduring. The quality of a system is not determined by how much delta it claims to create, but by how intelligently it manages the conditions under which delta is allowed to exist.

Once this is understood, action becomes less mystical and more precise. The real question is no longer *who caused change*, but what constraints allowed imbalance to persist, and for how long.

1.4 — Action Reframed: Maintenance or Resolution

Action is one of the most overloaded words in science and everyday language. We use it to describe movement, effort, intention, force, change, and even moral choice. Yet this familiarity hides a deeper confusion. We often treat action as something *added* to a system—as if it were an independent ingredient injected into reality.

Within Universal Energy Dynamics, action is stripped of this mystique.

All action reduces to two and only two categories: delta-maintenance and delta-resolution.

There is no third form.

Delta-maintenance is the act of holding imbalance in place against equilibrium pressure. It is what happens when constraints are reinforced, boundaries are strengthened, or feedback loops are tightened. Engines maintain pressure differences. Muscles maintain tension. Attention maintains cognitive focus. Institutions maintain power gradients. Every sustained activity, from biological metabolism to social order, exists because delta is actively prevented from resolving.

Delta-resolution, by contrast, is the allowance of alignment. It occurs when constraints are relaxed, boundaries are removed, or pathways open. Pressure equalizes. Heat dissipates. Conflicts dissolve. Insight occurs. Rest happens. Resolution is not passive—it is simply unopposed.

This distinction immediately explains why effort feels different from motion. Motion can occur with very little effort when delta is allowed to resolve freely—like a falling object or a relaxing muscle. Effort appears when resolution is resisted, when energy is expended not to move, but to keep things from settling. Holding, lifting, resisting, focusing, controlling—these all feel costly because they oppose equilibrium's default tendency.

This is also why resistance, work, and control are so often misunderstood. We tend to associate work with motion and control with power. In reality, work is primarily the cost of sustaining delta, not the act of moving something through space. Control is not command over motion; it is command over resolution pathways. The more rigid the constraint, the higher the cost of maintenance.

Seen this way, many familiar paradoxes soften. A system can move continuously with little effort if it is resolving delta along permitted pathways. Another system can remain nearly stationary while consuming enormous energy if it is holding imbalance against strong equilibrium pressure. Activity and effort are not the same thing.

This reframing also prepares the ground for understanding force without prematurely invoking it. Force is not the cause of action; it is an effect that appears when resolution is obstructed. When delta attempts to resolve but encounters constraint, force becomes the language we use to describe the resulting interaction. Force is therefore not fundamental—it is situational.

By redefining action at this level, we avoid layering confusion later. Motion will not be mistaken for travel, time will not be mistaken for flow, and force will not be mistaken for agency. All of these will emerge naturally once it is clear that action is never about doing something new—it is always about deciding whether imbalance is held or allowed to dissolve.

This reframing is essential. Without it, force, motion, and time arrive too early and carry assumptions that do not belong to them.

1.5 — Stability, Change, and the Illusion of Control

Stability is often misunderstood as stillness. We imagine a stable system as one where nothing happens, where motion has ceased and forces are balanced into silence. This misunderstanding arises because we confuse *visible change* with *structural tension*. In reality, stability has very little to do with inactivity.

Stability means that delta is being managed successfully.

A system can be highly active and yet stable, provided the imbalances within it are arranged so that no net pressure exists to reconfigure the whole. A spinning planet, a flowing river in its banks, a functioning organism, or a well-run institution all exhibit continuous activity while remaining structurally intact. Stability does not mean the absence of motion; it means the absence of unresolved pressure to reorganize.

Once this is understood, an equally common error becomes visible: the belief that change is inherently good, progressive, or necessary. Change, like stability, has no moral direction. Change occurs whenever delta resolves or reconfigures. Sometimes that resolution leads to improvement, adaptation, or clarity. Sometimes it leads to collapse, loss, or diffusion. Nature does not distinguish between constructive and destructive change. It only distinguishes between maintained imbalance and released imbalance.

This is why progress cannot be assumed simply because something is changing. Change is inevitable wherever delta exists. Progress is conditional, fragile, and dependent on how resolution pathways are shaped. Confusing the two leads to restless systems—constantly altering themselves without understanding whether those alterations are aligning or destabilizing.

Control enters the picture precisely at this point of confusion.

Control is often imagined as mastery over outcomes. In truth, control is nothing more than temporary authority over resolution pathways. To control a system is to decide, for a limited time, which imbalances are allowed to persist and which are permitted to resolve. This authority is never permanent, because equilibrium pressure does not negotiate. It waits.

Every act of control has a cost. That cost is the energy required to maintain delta against its natural tendency to collapse. The tighter the control, the higher the cost. The longer the delay, the greater the accumulated pressure. Systems that appear powerful are often not those with the most freedom, but those capable of sustaining imbalance for the longest duration without structural failure.

This applies universally. A muscle held in tension consumes energy even when no movement occurs. A focused mind fatigues even without visible action. An empire maintaining dominance expends enormous resources simply to keep gradients from equalizing. When that expenditure can no longer be sustained, control does not gradually fade—it fails abruptly.

Understanding this removes the romance from control without turning it into a moral failing. Control is not evil. It is expensive. Stability is not virtue. It is managed cost. Change is not salvation. It is inevitable resolution.

By the time consequence is formally introduced in the next chapter, the reader should already feel its presence. Not as punishment or reward, but as the unavoidable accounting that follows whenever delta is held, redirected, or released. Nothing here requires judgment. It only requires honesty about cost, pressure, and limits.

This prepares the ground for what follows. Once imbalance is permitted and maintained, something must eventually answer for it. That answer is not morality. It is consequence.

1.6 — The Quiet Rule Beneath All Laws

At this point, nothing new needs to be introduced. What matters now is to see what has already been established, and to recognize why it cannot be rearranged without breaking coherence.

The kernel of Universal Energy Dynamics reduces to a simple, non-negotiable structure:

- Equilibrium is the default.
- Delta enables action.
- Action has cost.

These are not principles layered on top of physics. They are the conditions under which physics becomes possible.

Equilibrium does not need to be reached; it needs only to be left undisturbed. Delta does not drive systems forward; it merely delays alignment. Action does not create something from nothing; it consumes energy to hold difference in place or releases energy by allowing resolution. Once these relationships are seen clearly, they stop feeling like assumptions and start feeling unavoidable.

This is the quiet rule beneath all physical laws.

Every law in physics—whether classical, relativistic, quantum, or thermodynamic—operates after this rule has already been satisfied. Laws describe how systems behave *once* imbalance exists, *once* constraints are defined, *once* resolution pathways are shaped. They never explain why imbalance was permitted in the first place, nor why it must eventually resolve. They assume those facts silently.

This is not a flaw in physics. It is a boundary of scope.

Equations require gradients. Measurements require stability. Predictions require repeatable delta.

Without equilibrium as baseline and delta as condition, no law can even be written.

The mistake arises when we forget this ordering and treat laws as foundational rather than contextual. When force is treated as primitive, we struggle to explain inertia. When time is treated as fundamental, we struggle at quantum scales. When motion is assumed rather than derived, paradox accumulates. These difficulties are not signs of incomplete mathematics; they are signs of starting too late.

Starting anywhere other than equilibrium and delta forces explanation to run backward. Effects are mistaken for causes. Tools are mistaken for truths. What should have been orientation becomes debate.

By placing this rule first, Universal Energy Dynamics does not compete with physical laws; it stabilizes them. Laws regain their proper role as precise descriptions of behavior inside permitted conditions, not as metaphysical declarations about reality itself.

This is why nothing that follows in this book will feel arbitrary. Every chapter ahead—whether it addresses time, motion, geometry, life, mind, or society—will return to this same structure. When something appears complex, the question will never be "what new principle is needed?" It will always be "where is delta being held, what constraints apply, and what cost is accumulating?"

That is the foundation.

Once it is locked in place, explanation no longer needs to escalate. It only needs to remain consistent.

From "What Exists" to "What Follows"

Up to this point, the discussion has been deliberately quiet. Nothing dramatic has been claimed, and nothing mystical has been introduced. We have only placed the starting point correctly. Equilibrium has been recognized as the baseline. Delta has been identified as the sole condition that allows anything to happen. Action has been reduced to the maintenance or resolution of that imbalance, always under constraint and always at a cost.

What remains now is not a new concept, but an unavoidable shift in attention.

Once imbalance is permitted—once delta is held open rather than allowed to collapse—something must follow. Not by intention, not by judgment, and not by design, but by structure. This is not a philosophical claim. It is a simple accounting fact. Holding difference apart requires pathways. Delaying alignment consumes energy. Redirecting resolution reshapes the system that hosts it.

Nothing about this process is optional.

There is no exemption for scale. There is no exemption for complexity. There is no exemption for intention.

Whether the delta is microscopic or societal, whether it lasts a fraction of a second or centuries, the moment alignment is delayed, a sequence is set in motion. Energy redistributes. Constraints are stressed. Pathways accumulate history. Outcomes appear—not as rewards or punishments, but as necessary continuations.

This is the point where many discussions quietly drift into moral language. Words like fairness, responsibility, guilt, justice, or karma are often introduced prematurely, carrying emotional weight before structural clarity is achieved. This book intentionally resists that move.

Before consequence can be judged, it must be understood.

What follows from delta is not morality. It is accounting. It is the unavoidable continuation of a process once permission has been granted. The universe does not evaluate imbalance; it responds to it. That response is neither kind nor cruel. It is precise.

The next chapter begins exactly here—not with ethics, not with belief, not with philosophy—but with the structural inevitability that follows whenever difference is allowed to persist. Once imbalance exists, it does not ask whether it should resolve. It only asks how, through what pathways, and with what redistribution.

Once imbalance is permitted, nothing that follows is optional.

Chapter 2 — Consequence Is Not Moral; It's Structural

Why We Mistook Consequence for Judgment

Human beings are meaning-making creatures. When something happens—especially when it affects us—we instinctively ask *why*. That question is rarely neutral. It carries an emotional undertone that quietly searches for responsibility, intention, and often judgment. We do not merely observe outcomes; we evaluate them.

This tendency is understandable. In social life, evaluation helps coordinate behavior. Praise and blame stabilize groups. Credit and guilt distribute trust. Over time, these social tools quietly seeped into how we interpret reality itself. What began as a survival strategy became a habit of thought.

And so we learned to moralize outcomes.

When something ends well, we look for merit. When something ends badly, we look for fault. When consequences arrive unexpectedly, we search for someone—or something—to judge.

Physics never asked for this, but our language did.

Words like *cause*, *effect*, *responsibility*, *fault*, and *deserve* grew close together, until explanation and evaluation became almost indistinguishable. We stopped noticing when we crossed from description into judgment. A consequence was no longer simply what followed; it became what was *earned*.

This confusion is ancient. Long before formal science, humans noticed that actions lead to outcomes. Over time, that observation hardened into belief systems, moral codes, and metaphysical narratives. The idea that "what follows must mean something about who we are" felt comforting. It suggested order, fairness, and intention behind events that were often chaotic and painful.

But comfort is not accuracy.

The critical error was subtle: we conflated sequence with verdict. We mistook continuity for evaluation.

Consequence, in its most basic sense, does not ask whether an action was good or bad, wise or foolish, intentional or accidental. It does not assess worth. It does not assign praise or blame. It simply continues what was set in motion. When imbalance is permitted, pathways open. When pathways open, redistribution occurs. What follows does so because it must, not because it should.

Language obscured this truth by humanizing causality. We speak as if nature reacts, responds, corrects, or teaches. These metaphors are useful for storytelling, but they quietly insert judgment where none exists. The universe does not respond to actions; it propagates conditions. It does not punish or reward; it redistributes.

This distinction is not philosophical hair-splitting. It changes how responsibility, agency, and understanding are approached. When consequence is mistaken for judgment, inquiry collapses into defense or justification. We argue about fairness instead of tracing pathways. We debate intention instead of examining structure.

The central correction of this chapter is simple, but demanding: consequence does not evaluate—it continues.

Once this is accepted, blame-and-credit thinking begins to loosen its grip. What replaces it is not indifference, but clarity. Events are no longer read as verdicts on character or meaning. They are read as outcomes of conditions, constraints, delays, and pathways interacting over time.

This shift does not remove responsibility. It relocates it—from moral theater to structural participation. But that relocation can only happen after judgment is set aside.

Before we can understand consequence, we must stop asking who deserves what, and start asking what followed from what, and why.

Only then does explanation regain its honesty.

2.1 — Consequence as Universal Accounting

Once judgment is removed, consequence reveals its true character. It is not reaction, not correction, and not response. Consequence is accounting—the unavoidable bookkeeping that follows whenever delta is permitted to exist.

In Universal Energy Dynamics, consequence has three inseparable modes: redistribution, transformation, and diffusion.

Redistribution occurs when imbalance seeks new pathways. Energy, pressure, attention, influence, or resources do not vanish; they move. They shift from one location, state, or configuration to another. What appears as loss in one place appears as gain elsewhere. What appears as collapse locally may be expansion globally.

Transformation occurs when the form of delta changes. Mechanical energy becomes heat. Chemical potential becomes motion. Focus becomes fatigue. Authority becomes resistance. The substance of the imbalance remains, but its expression alters as it encounters constraints. Transformation is not escape; it is reformatting under pressure.

Diffusion occurs when resolution spreads across many pathways instead of a single dominant one. What was once concentrated becomes distributed. Sharp gradients soften. Clear signals blur. This is often experienced as dissipation, decay, or forgetting. Yet diffusion is not disappearance—it is widened accounting.

These three processes explain a fundamental truth that physics has always known but often failed to integrate conceptually: nothing disappears. It only changes form, route, or scale. Energy conservation is not merely a numerical rule; it is the physical signature of consequence persistence. What is conserved is not just quantity, but obligation to resolve.

Seen this way, conservation laws are not abstract constraints imposed on nature. They are descriptions of how consequence refuses to be erased. Even when energy becomes inaccessible, even when information seems lost, even when causality becomes obscured by complexity, the accounting has not failed. It has only moved beyond immediate visibility.

This is where delayed consequences enter the picture.

Delayed consequences feel mysterious because human attention is short and local. We expect outcomes to follow actions quickly and visibly. When they do not, we suspect randomness, injustice, or fate. In reality, delay is not an exception—it is the norm. Delay arises whenever delta is buffered by structure, stored in intermediate states, or rerouted through indirect pathways.

A stressed material may not fail immediately. An ecosystem may not collapse until a threshold is crossed. A psychological pattern may not surface until years later. A social policy may not show its effects for generations.

In each case, consequence was present all along. It was simply distributed across time and structure.

Understanding consequence as universal accounting removes the need for reaction-based explanations. Nothing "comes back" to punish or reward. Nothing "waits" to teach a lesson. What exists is continuity—delta unfolding through whatever pathways remain open.

This neutrality is essential. Once consequence is mistaken for reaction, explanation becomes emotional. Once it is understood as accounting, explanation becomes precise. There is no anger in consequence, no mercy, no intention. There is only continuation.

And continuation does not forget.

2.2 — Why "Cause" Is a Shortcut, Not an Origin

Human explanation prefers simplicity. When something happens, we instinctively ask for *the* cause—as if events emerge from a single point, pushed into existence by a lone trigger. This preference is not scientific; it is psychological. Single-cause explanations are comforting because they promise closure. They allow responsibility to be assigned, problems to be named, and uncertainty to be contained.

But comfort is not correctness.

In reality, no meaningful outcome arises from a single cause. What we call a cause is almost always a shortcut—a compressed label applied to the most visible or dominant pathway through which consequence became observable. It is a linguistic convenience, not an ontological origin.

Within Universal Energy Dynamics, "cause" is understood as shorthand for the dominant delta pathway at the moment an outcome becomes apparent. It is the pathway that carried enough imbalance, under sufficient constraint, to cross a threshold and register as an event. That dominance makes it noticeable. It does not make it singular.

Every outcome is the result of multiple deltas interacting, accumulating, redirecting, and sometimes canceling one another until a particular configuration resolves. Most of these deltas remain invisible because they are diffuse, delayed, or balanced elsewhere. What surfaces as the "cause" is simply the pathway that did not dissolve quietly.

This is why explanations framed around single causes feel persuasive but collapse under scrutiny. They mistake visibility for primacy.

Consider how this plays out across domains. A material fails and we blame a defect, ignoring years of stress accumulation. A market crashes and we blame an announcement, overlooking long-standing imbalances. A person breaks down and we point to a recent event, forgetting a history of unresolved tension. In each case, the so-called cause was not the origin—it was the last permissive condition.

Root-cause language fails most dramatically in complex systems because complexity distributes delta across layers. Feedback loops, delays, buffers, and nonlinear thresholds ensure that no single input maps cleanly to a single output. Small deltas can accumulate silently, while large deltas can dissipate harmlessly. Outcomes appear not when a cause acts, but when conditions align.

This does not weaken explanation; it strengthens it.

Breaking linear causality does not mean abandoning causality altogether. It means replacing the fiction of one-to-one causation with a pathway-based understanding. Instead of asking "What caused this?", the more precise question becomes: *Which deltas were present, which pathways were open, and which constraints allowed this particular resolution to surface?*

Under this view, explanation becomes richer without becoming vague. Responsibility does not vanish; it relocates from isolated triggers to participation in systems. Prediction does not disappear; it shifts from guessing events to mapping conditions and thresholds.

Cause, then, is not wrong—it is incomplete. It is a pointer, not a foundation. It tells us where to look, not where reality began.

2.3 — Pathways Matter More Than Events

We are trained to think in moments. An action occurs, an event happens, an outcome follows. This framing feels natural because moments are visible and memorable. They can be named, dated, recorded, and discussed. Pathways, by contrast, are quiet. They operate in the background, shaping what can happen long before anything noticeable occurs.

But consequence does not follow moments. It follows pathways.

An event is only the point at which a pathway becomes visible. By the time something appears to "happen," most of the structural work is already done. The imbalance has traveled, accumulated, redirected, or been buffered through a network of constraints until a threshold was crossed. The moment is not the cause; it is the announcement.

This is why constraints shape outcomes far more than triggers. A trigger merely releases what conditions have already prepared. The same spark can cause nothing, a small reaction, or a catastrophic failure, depending entirely on what pathways were open and how much unresolved delta was already present. The trigger does not decide the outcome; the pathway does.

This distinction explains a common puzzle: why identical actions produce different results in different contexts. The action may be the same, but the surrounding constraints are not. One system absorbs the imbalance quietly. Another channels it into transformation. A third amplifies it into collapse. From the outside, it looks inconsistent or unfair. Structurally, it is inevitable.

Context is not background decoration. It is the active geometry of resolution.

Once attention shifts from events to pathways, history begins to appear differently. History does not accumulate in objects, records, or identities. It accumulates in pathways—in worn channels of behavior, reinforced constraints, hardened structures, and learned responses. Objects may be replaced, people may change, and systems may rebrand themselves, but pathways persist. They remember.

A material remembers stress through microfractures. A mind remembers trauma through conditioned response loops. An institution remembers power through procedures and hierarchies. A society remembers conflict through borders, narratives, and laws.

This memory is not symbolic. It is structural. Past delta reshapes future resolution by altering the routes imbalance can take. That is why systems often repeat patterns even when participants believe they are starting fresh. They are not repeating events; they are reusing pathways.

Thinking in snapshots—isolated actions and isolated outcomes—hides this continuity. It encourages reactive explanation and superficial fixes. Thinking in pathways reveals why lasting change is rare and why unintended consequences are common. It also explains why patience and timing matter more than intensity, and why small interventions at the right constraint can reshape outcomes far more effectively than dramatic actions at the wrong moment.

This shift is essential for what follows. Consequence cannot be understood as a sequence of events. It must be understood as continuous movement through structured space. Once that continuity is seen, explanation stops chasing moments and starts mapping flow.

That is the level at which consequence actually operates.

2.4 — Delay, Accumulation, and Surprise

One of the strongest reasons consequence has been misunderstood is timing. Humans expect cause and outcome to be close together—spatially, temporally, and emotionally. When they are not, we label the result unexpected, accidental, or random. In some traditions, we even call it fate.

But delay is not a failure of consequence. Delay is one of its most common expressions.

Consequences rarely appear immediately because delta is often stored, buffered, or redirected rather than resolved outright. Structures exist precisely to delay resolution. Materials store stress. Systems store imbalance. Minds store tension. Societies store inequality. Delay is not an anomaly—it is a design feature of constrained systems.

When delta cannot resolve directly, it does not disappear. It accumulates.

This accumulation can be quiet and invisible for long periods. Nothing appears to change, so observers assume nothing is happening. In reality, resolution pressure is increasing beneath the surface, redistributed across pathways that are not immediately perceptible. Stability during such periods is often mistaken for safety.

Accumulation continues until a threshold is reached.

Threshold behavior is a defining feature of consequence. Systems do not always respond gradually. They respond when conditions allow resolution to occur in a new regime. A material suddenly fractures. A market suddenly crashes. A person suddenly breaks down. A society suddenly revolts. These shifts feel abrupt not because they are sudden in origin, but because the pathway changed, not the pressure.

Before the threshold, the system absorbs imbalance. After the threshold, the system releases it.

This is why surprise is so common—and so misleading.

What we call "unexpected outcomes" are almost always expected outcomes seen too late. They surprise us because we tracked visible events instead of invisible accumulation. We paid attention to triggers instead of pathways. We assumed continuity where thresholds were quietly forming.

Randomness is often invoked at this point as an explanation of last resort. Fate, luck, or chaos fill the gap left by poor structural awareness. But nothing truly random has occurred. What occurred was resolution under conditions we did not monitor.

This understanding changes how prediction is approached. Prediction is not about forecasting specific events; it is about recognizing accumulating delta and identifying approaching thresholds. The inability to predict exact timing does not mean consequence is unknowable. It means it is conditional, not scripted.

Surprise, then, is not evidence of disorder. It is evidence of ignored continuity.

Once this is understood, many emotional reactions soften. Shock becomes recognition. Confusion becomes mapping. Fear becomes orientation. The universe does not ambush systems with arbitrary outcomes. It waits—patiently, precisely—until resolution becomes unavoidable.

Delay is not mercy. Surprise is not mystery. They are simply the natural companions of accumulation in constrained systems.

Understanding this prepares the reader to encounter consequence without drama, and to recognize inevitability without surrender.

2.5 — LOC: Mechanism-Language, Not Belief-Language

The Law of Consequence (LOC), as used in Universal Energy Dynamics, is not a doctrine, not a warning, and not a moral instruction. It is mechanism-language—a way of describing how systems behave once imbalance is permitted to persist.

LOC simply states this: whenever delta is allowed to exist, its resolution will propagate through available pathways under existing constraints, producing unavoidable outcomes.

Nothing more is implied. Nothing less is sufficient.

LOC does not need belief to function. A material does not believe in stress before it fractures. A system does not agree to collapse before it does. A mind does not subscribe to consequence before it experiences burnout or insight. Consequence unfolds regardless of awareness, intention, or acceptance. It operates silently, impersonally, and continuously.

This is why LOC requires no enforcement.

There is no external agent ensuring compliance. There is no corrective authority monitoring behavior. There is no intelligence judging alignment or misalignment. The only "enforcer" is structure itself—constraints, pathways, thresholds, and redistribution mechanisms doing what they inevitably do once imbalance is present.

Confusion arises when descriptive laws are mistaken for prescriptive ones.

A descriptive law explains what happens given certain conditions. It carries no instruction. It makes no demands. It simply maps behavior. Thermodynamics does not command heat to flow; it describes how heat flows when gradients exist. Gravity does not order masses to attract; it describes the interaction that occurs when mass and geometry coexist.

A prescriptive rule, by contrast, tells someone what should or should not be done. Ethics, morality, legal systems, and belief frameworks operate here. They guide behavior, assign responsibility, and shape social order. They are necessary at the human level, but they are not interchangeable with physical description.

LOC belongs entirely to the first category.

It does not say what ought to happen. It does not say what should be done. It does not say who is right or wrong.

It says only what will follow, structurally, once conditions are set.

Invoking belief at this level weakens understanding because belief introduces negotiation where none exists. Belief invites agreement or disagreement. Mechanism does not care. When LOC is framed as something to be believed in, accepted, or rejected, attention shifts away from mapping conditions and toward defending positions. Explanation becomes ideological.

This is why UED insists on keeping LOC free from religious, ethical, or ideological language—not because those domains are invalid, but because they operate at a different layer. Ethics and belief systems arise precisely because humans must navigate consequences socially and emotionally. They are responses to consequence, not explanations of it.

LOC explains why consequences occur. Ethics decide how humans should respond to them.

Confusing the two leads to unnecessary conflict. Separating them restores clarity.

By treating LOC as mechanism-language, UED allows ethics, morality, and belief systems to be discussed later—on firmer ground, without mystification and without denial. Nothing meaningful is lost by this separation. In fact, much is gained: responsibility without guilt, clarity without cruelty, and understanding without ideology.

At this stage of the book, no moral conclusions are required. Only structural honesty is needed. Once that honesty is established, human values can be reintroduced—not as explanations of reality, but as tools for living within it.

That distinction is essential, and it must be protected early.

2.6 — Responsibility Without Morality (Yet)

Once consequence is understood as structural continuity rather than judgment, a difficult but necessary question remains: where does responsibility fit? If consequence is neutral accounting, does responsibility disappear? Or does it become something colder and less human?

Neither.

Responsibility does not vanish when morality is set aside. It simply changes its location.

In Universal Energy Dynamics, responsibility is not guilt, blame, or merit. Responsibility is structural participation. To be responsible is to be part of the configuration through which delta is permitted, maintained, redirected, or released. Responsibility arises from involvement, not from intention.

This distinction matters deeply.

Intention is a late arrival in most systems. By the time an intention forms, pathways are already present, constraints are already operating, and delta is already in motion. A person may not intend harm, failure, or collapse, yet still participate in the conditions that allow those outcomes. Likewise, a person may intend good and still contribute to instability. Intention influences pathways, but it does not define responsibility.

Involvement always precedes intention.

A structure participates whether it "knows" it or not. A material under load participates in stress accumulation. A node in a network participates in signal propagation. A role in an organization participates in power distribution. A belief held unexamined participates in behavior long before it is consciously chosen.

Responsibility, in this sense, is not accusatory. It is descriptive. It answers the question: *where did delta pass, where was it held, and through which structures did resolution occur?* That answer does not require moral evaluation to be accurate.

This also explains why appeals to "innocence" fail to block consequence. Innocence may matter ethically, socially, and legally—but structurally, it is irrelevant. A bridge collapses whether engineers intended failure or not. A system destabilizes whether participants understood the risks or not. A body breaks down whether stress was knowingly applied or unknowingly absorbed.

This is not cruelty. It is clarity.

Understanding responsibility this way removes both shame and denial. Shame arises when responsibility is confused with moral worth. Denial arises when responsibility is rejected to protect identity. Structural responsibility allows neither extreme. It simply states participation honestly and moves forward.

At this stage of the framework, morality is deliberately withheld—not dismissed, but postponed. Ethics, values, and moral systems are essential for human coordination, restraint, and care. But they must be built on top of an accurate understanding of consequence, not used to replace it.

If ethics are introduced too early, they distort explanation. If responsibility is framed morally too soon, inquiry shuts down.

So here, responsibility is allowed to exist without emotional loading. No praise. No blame. No verdict. Only recognition of participation in pathways and constraints.

Ethics will return later in this book—on firmer ground, with clearer boundaries. When they do, they will not be asked to explain reality. They will be asked to help humans live wisely within it.

For now, responsibility remains what it must be: a structural fact, not a moral sentence.

From Accounting to Ontology

At this point, consequence has been stripped of its familiar disguises. It is no longer reaction, reward, punishment, or correction. It stands revealed as structural continuation—the unavoidable unfolding that follows once imbalance is permitted to persist.

What matters now is to hold this clarity steady.

Consequence does not follow virtue. It does not follow intention. It does not follow belief.

It follows permission.

Where delta is allowed, where constraints shape pathways, where resolution is delayed or redirected, consequence accumulates and unfolds.

This is true whether the system is physical, biological, psychological, or social. Nothing about consequence evaluates character or meaning. It simply propagates conditions forward.

Once this is accepted, a subtle but important shift occurs. The focus naturally moves away from actors and actions and toward structures and states. Explanation stops orbiting around *who did what* and begins asking *what configuration made this outcome unavoidable.*

This shift prepares the ground for the next correction.

If consequence is structural, then what we usually treat as stable entities—objects, identities, systems, even selves—cannot be primary. They must be outcome-states, not origins. They must be the temporarily stabilized results of past resolutions, held together by ongoing constraints and shaped by accumulated pathways.

In other words, what exists is already the result of what has followed before.

Next chapter enters exactly here. It asks a question that would have been premature earlier, but is now unavoidable: *if consequence is continuous accounting, then what, precisely, are the "things" we believe persist through it?* Are objects fundamental, or are they stabilized bundles of resolved and unresolved delta? Do identities endure independently, or do they emerge from repeated pathways that have not yet dissolved?

These questions are not philosophical diversions. They are the natural next step once consequence is understood correctly. Ontology cannot be addressed before accounting, because existence without continuity is an illusion.

Once consequence is understood as structure, the question is no longer who acted—but what was set in motion.

Chapter 3 — What Exists in UED: States, Not Things

Why "Things" Felt Obvious for So Long

From the moment we open our eyes, the world presents itself as a collection of things. Chairs, stones, bodies, tools, walls, stars—each appears separate, stable, and nameable. This way of seeing feels so natural that it rarely invites questioning. We do not experience processes first; we experience persistence. And persistence, to the human mind, quickly becomes substance.

This was not a philosophical mistake. It was a survival advantage.

To survive, humans had to recognize what remained the same across time and context. A predator needed to be identified again. A shelter needed to be trusted as solid. A tool needed to be relied upon. Naming and re-identifying stable patterns allowed quick decisions and reduced cognitive load. Over time, this practical convenience hardened into an assumption: that what persists must *exist as a thing*.

Language reinforced this assumption. Nouns dominate our sentences. Verbs orbit them. We say objects *have* properties, forces *act* on things, and motion *happens to* entities. This grammatical structure quietly teaches ontology. It suggests that reality is built from discrete units, each carrying attributes, interacting through external influences.

Physics inherited this bias not because it was careless, but because it was effective. Early models worked extraordinarily well by treating objects as primary and interactions as secondary. There was no immediate reason to doubt this framing. Stones fell. Planets moved. Forces could be calculated. Prediction improved. Success masked limitation.

The difficulty emerges only when this object-centric view is pushed beyond the conditions that made it useful.

At small scales, objects dissolve into probabilities. At large scales, boundaries blur into fields. Across time, identities change while labels remain.

What seemed fundamental begins to look conditional.

Universal Energy Dynamics does not claim that objects are unreal. It claims something quieter and more precise: objects are not primary. They are outcomes. They are stabilized results of ongoing resolution processes that have not yet completed.

Existence, in this framework, is not defined by substance, but by persistence under constraint.

A rock persists because its internal and external deltas are arranged in a way that resists rapid resolution. A chair exists because its structure channels stress predictably. An institution exists because rules, roles, and power gradients hold form. Even an identity exists because certain behavioral and cognitive pathways repeat with enough consistency to feel continuous.

What we call a "thing" is simply a state that has learned how to last.

This shift does not deny reality; it refines it. It moves ontology away from static entities and toward stabilized configurations. What matters is not what something is made of in isolation, but how it behaves when resolution is attempted and constrained. Persistence becomes a question of dynamics, not essence.

Once this is seen, the question of existence changes subtly but permanently. We stop asking *what things are*, and begin asking *what conditions allow this state to persist*. That question is far more powerful, because it remains valid whether we are discussing particles, people, institutions, or ideas.

This chapter begins exactly there—not to dismantle common sense, but to complete it.

3.1 — From Objects to Stabilized States

Once the grip of object-centered thinking loosens, a quieter but more precise picture comes into view. What we commonly call an object is not a primitive unit of reality. It is a temporarily stabilized configuration—a state that persists because the conditions for its resolution are constrained in a particular way.

In Universal Energy Dynamics, persistence is the key diagnostic. If something appears to "exist," it is not because it possesses an intrinsic substance that guarantees endurance. It exists because the delta within and around it is being managed—held, distributed, or rerouted such that immediate alignment is prevented. Existence, then, is not a property of things; it is an outcome of ongoing dynamics.

This reframing dissolves a common confusion between durability and fundamentality. Something can last a long time without being fundamental, and something fundamental can be fleeting. A mountain persists for millennia not because it is ontologically special, but because the pathways through which its internal stresses could resolve are slow and diffuse. A bubble exists briefly not because it is less "real," but because its stabilizing constraints are fragile. Duration reflects resolution difficulty, not existential rank.

Seen this way, objects stop being isolated entities and start appearing as states under constraint.

A solid body is a state where internal deltas—atomic bonds, lattice structures, stress distributions—are arranged so that resolution pathways cancel out at the macroscopic scale. The body persists not because it is inert, but because its internal dynamics are balanced enough to resist reconfiguration under ordinary conditions.

An institution is a state where roles, rules, incentives, and power gradients stabilize behavior across individuals. Its persistence does not depend on any single person, but on the continued alignment of constraints that prevent rapid collapse or transformation.

An identity is a state where cognitive, emotional, and behavioral pathways repeat with sufficient consistency to create the experience of sameness. The "self" feels stable not because it is fixed, but because the conditions that would dissolve or reconfigure it are rarely allowed to complete.

In each case, the object—or system, or identity—exists because resolution has been slowed, not because change has been eliminated. What appears static is often dynamically maintained.

This understanding also clarifies why objects can change without ceasing to exist. A material can deform, a person can mature, an institution can reform, and yet continuity is preserved. The state adapts while maintaining enough constraint coherence to remain recognizable. Conversely, when constraints fail—when pathways open too widely or pressure exceeds tolerance—the state dissolves. What we call destruction or disappearance is simply resolution completing its work.

By shifting attention from things to stabilized states, UED avoids both reductionism and mysticism. It does not claim that objects are illusions, nor does it elevate them to absolutes. It places them where they belong: as temporary outcomes of consequence, sustained by the intelligent arrangement—or accidental inheritance—of constraints.

This is not a semantic shift. It is an ontological correction. Once existence is understood as state-persistence rather than substance, the groundwork is laid for reinterpreting properties, forces, motion, and identity without contradiction.

3.2 — State-Bundles, Not Isolated Entities

Once objects are understood as stabilized states, another assumption quietly dissolves: that anything exists alone. Isolation is a perceptual convenience, not an ontological fact. What appears as a single object is, in reality, a bundle of coupled states, held together by boundaries that are functional rather than absolute.

No state persists by itself. Persistence always depends on coupling—with an environment, with internal subsystems, with constraints that distribute stress and delay resolution. An object appears singular only because the couplings that sustain it are reliable and often invisible. When those couplings weaken or reconfigure, the illusion of isolation collapses.

Consider a solid body. What we call "one object" is actually a coordination of atomic lattices, electromagnetic interactions, thermal distributions, and mechanical constraints. Each of these is a state with its own dynamics. The object persists because these states are synchronized well enough that their internal resolutions cancel at the scale we observe. The moment that synchronization breaks—through fracture, melting, or phase change—the object's apparent unity dissolves.

The same logic applies beyond physics. An institution is not a thing; it is a bundle of legal rules, social norms, economic incentives, and habitual behaviors. Remove enough of these couplings, and the institution ceases to exist even if its name remains. A person's identity is not a core entity hidden behind behavior; it is a bundle of memories, emotional patterns, bodily states, and social feedback loops. Alter the couplings, and identity shifts—sometimes subtly, sometimes radically.

This is why boundaries matter so much—and why they are often misunderstood.

Boundaries do not separate realities; they define resolution regimes. A boundary tells delta where it can move, how it can spread, and at what rate it may resolve. The skin of a body, the walls of a container, the rules of an organization, the habits of a mind—each functions as a boundary that stabilizes a particular state-bundle by managing pathways.

Crucially, boundaries are functional, not metaphysical. They exist to regulate interaction, not to declare independence. A boundary that cannot regulate resolution ceases to function as a boundary at all. It either dissolves or becomes irrelevant.

Seeing objects as state-bundles also explains why entities are sensitive to context without becoming relativistic. Context does not arbitrarily redefine what something is; it reshapes the couplings that sustain it. The same state-bundle can behave differently under different constraints because different pathways are activated. The object has not changed essence; the resolution geometry has changed.

This perspective prevents a common error: mistaking labels for unity. Naming something as one thing does not make it ontologically singular. It only indicates that, for practical purposes, its internal couplings are stable enough to be treated as such. UED respects this practicality while refusing to elevate it to a foundational claim.

Once entities are understood as state-bundles, explanation becomes more precise and less brittle. Failure, change, and emergence no longer appear mysterious. They appear as reconfigurations of coupling, not as violations of identity. What persists is not the thing, but the pattern of constraints that temporarily holds the bundle together.

This prepares the ground for the next correction. If objects are state-bundles, then the qualities we assign to them—mass, rigidity, charge, character, strength—cannot be possessions. They must be behaviors revealed under interaction.

3.3 — Properties Are Behaviors, Not Possessions

Once objects are understood as stabilized state-bundles, the status of their properties must change as well. Properties can no longer be treated as things an object *has*, stored inside it like permanent labels. That view belongs to substance-based thinking. In Universal Energy Dynamics, properties are patterns of behavior revealed when resolution is attempted under constraint.

A property does not exist in isolation. It appears only in interaction.

Mass is not something an object carries silently; it is how a state resists acceleration when delta attempts to resolve through motion. Rigidity is not an inner attribute; it is how a structure redistributes stress when deformation is attempted. Charge is not a possession; it is how a state responds to electromagnetic gradients. Even identity, at the human level, is not a fixed essence; it is how a person reliably responds across repeated situations.

In every case, the "property" is not a static quality. It is a consistent response pattern.

This reframing immediately resolves a long-standing confusion. We often ask whether properties are intrinsic or relational, as if these were opposing categories. Under UED, the distinction collapses. A property is intrinsic only insofar as the internal constraints of the state-bundle are stable. It is relational because it only manifests when those constraints meet an external delta. There is no contradiction—only incomplete framing.

The language of possession obscures this reality. Saying that an object "has mass" or "has strength" implies permanence independent of context. But remove the relevant interaction, and the property becomes meaningless. Mass without attempted acceleration is undefined. Strength without applied stress is hypothetical. Identity without situation is untestable.

Properties are therefore not stored features. They are resolution behaviors under constraint.

This also explains why properties can appear constant within a domain and variable across domains without undermining realism. A material behaves rigidly under ordinary loads and plastically under extreme ones. A person appears calm in familiar settings and reactive under threat. The property did not change; the resolution regime did. Different deltas activate different pathways.

Seen this way, measurement itself becomes clearer. Measurement does not uncover hidden attributes; it provokes behavior. By applying a controlled delta and observing the response, we infer how a state resolves imbalance under specific constraints. The precision of a property depends not on its intrinsic purity, but on how tightly the interaction is controlled.

This shift has an important consequence for explanation. When properties are treated as possessions, failures are surprising and inconsistencies feel paradoxical. When properties are treated as behaviors, variability becomes expected. A system behaves as it does because of how it is constrained at that moment, not because its essence has mysteriously changed.

By grounding properties in behavior rather than possession, Universal Energy Dynamics preserves the usefulness of classical descriptions while correcting their interpretation. Nothing familiar is lost. What changes is the level at which explanation rests.

Properties stop being labels attached to things. They become expressions of how states negotiate resolution.

This prepares the final correction of this chapter: if properties are behaviors and objects are state-bundles, then the idea of force as a thing acting on objects cannot remain fundamental. Force must be reinterpreted as what appears when resolution is obstructed.

3.4 — Why Force Is Not a Thing

Force has enjoyed a privileged position in the language of physics. It is introduced early, used everywhere, and treated as if it were a fundamental ingredient of reality—something that exists independently and *acts* on objects to produce change. This framing has been enormously effective for calculation and engineering. It is also deeply misleading at the ontological level.

In Universal Energy Dynamics, force is not denied. It is repositioned.

Force is not a thing. Force is not an agent. Force is not a cause.

Force is what appears when delta attempts to resolve and is obstructed.

Historically, force-language emerged as a practical shorthand. When motion changed, something had to be "responsible." When structures resisted deformation, something had to be "applied." Force filled that explanatory gap. It allowed relationships to be quantified without asking deeper questions about what was actually happening.

But once equilibrium and delta are placed correctly, force loses its foundational status.

If delta resolves freely, no force is observed. If delta is blocked, redirected, or constrained, force appears.

A falling object experiences no force in free fall, yet motion occurs. Force appears only when the object's resolution pathway is obstructed—by the ground, by a constraint, by a boundary. The force is not what caused the motion; it is the signature of resistance when motion is prevented from continuing as resolution demands.

The same logic applies universally. Pressure force arises when fluid flow is constrained. Elastic force appears when deformation is resisted. Frictional force emerges when relative motion is obstructed. Even psychological "force"—effort, strain, stress—appears when cognitive or emotional resolution is blocked.

Force, in every case, is reactive, not generative.

This explains why force disappears the moment constraints dissolve. Remove the obstruction, and force vanishes instantly—not gradually. The delta does not need to be persuaded; it simply continues resolving along the newly opened pathway. If force were a thing, it would need to go somewhere. But force goes nowhere, because it was never there to begin with.

Seen this way, force becomes a context-dependent descriptor, not a universal actor. It describes how strongly a system resists resolution under a given constraint. Change the constraint, and the force changes or disappears, even if the system itself remains unchanged.

This reframing also clarifies why force-language becomes problematic at certain scales. At the quantum level, where resolution pathways are probabilistic and constraints are non-classical, force loses clarity. At cosmological scales, where geometry itself shapes resolution, force begins to look secondary to structure. These are not failures of physics; they are signals that force was never meant to be foundational.

Universal Energy Dynamics does not remove force from equations or practice. It removes force from ontology.

Force remains a useful measurement-language for obstructed resolution. It remains indispensable for engineering and prediction. But it is no longer asked to explain *why* systems behave as they do. That explanation now rests where it belongs: in delta, constraints, and pathways.

Once force is released from the burden of being fundamental, it becomes easier to use—and harder to misuse. It stops pretending to be an entity and returns to being what it always was: a visible trace of resistance in a system trying to align.

This sets the stage for the unifying language that follows. If objects are state-bundles, properties are behaviors, and force is obstructed resolution, then the cleanest description of reality is not "forces acting on things," but delta behaving under constraint.

3.5 — Delta-Behavior as the Unifying Description

Once force is released from its role as a fundamental actor, a simpler and more scalable language becomes available. Instead of describing reality as *forces acting on objects*, Universal Energy Dynamics describes it as delta attempting resolution under constraint. What we observe—motion, pressure, interaction, resistance—is not causation layered on top of existence. It is behavior emerging from imbalance navigating structure.

Delta-behavior is not a new phenomenon. It is a cleaner description of what has always been happening.

When pressure drives flow, we are not witnessing a force pushing fluid; we are seeing a gradient seeking alignment through permitted pathways. When a body accelerates, we are not watching force cause motion; we are observing delta resolving where constraints allow it. When stress builds in a material, when attention strains in a mind, when tension rises in a society, the pattern is identical: imbalance explores available routes toward equilibrium.

This language unifies what previously appeared fragmented.

Motion becomes resolution in space. Pressure becomes resolution across boundaries. Interaction becomes coupled resolution between state-bundles. Resistance becomes constraint opposing resolution.

Nothing essential is lost. Everything becomes more coherent.

The power of delta-behavior is that it scales without translation. The same descriptive structure applies whether we are discussing particles, fluids, organisms, minds, or institutions. The vocabulary does not need to change when we cross domains, because the underlying logic does not change. Only the nature of constraints and the sensitivity of pathways differ.

This resolves a long-standing discomfort in interdisciplinary thinking. Traditionally, physics uses force and motion, biology uses regulation and homeostasis, psychology uses motivation and resistance, and sociology uses power and pressure. These appear distinct only because each domain adopted its own metaphors. Under UED, they are recognized as domain-specific expressions of the same delta-behavior.

What differs is not the rule, but the resolution geometry.

By replacing "force-things" with delta-behavior, explanation becomes relational rather than object-centric. We stop asking what acted on what and start asking where imbalance existed, what constrained it, and how resolution unfolded. This shift reduces conceptual friction. It also removes the need to invent separate mechanisms for each scale of reality.

Importantly, delta-behavior does not erase causality. It refines it. Causality becomes a statement about pathway viability, not about entity interaction. When outcomes differ, the explanation lies not in mysterious forces or hidden properties, but in how constraints shaped the route of resolution.

This framing also restores humility to prediction. If behavior emerges from delta navigating constraints, then precise outcomes depend on pathway detail. Prediction improves when constraints are well-mapped and fails when they are not. Uncertainty is no longer an embarrassment; it is a signal of incomplete pathway knowledge.

Delta-behavior is therefore not a metaphor. It is a descriptive minimum—the least language required to remain accurate across scales without contradiction. It does not compete with existing models; it sits beneath them, explaining why they work where they do and fail where they must.

With this unifying description in place, the final illusion of this chapter can be addressed: the sense that stable states possess a permanent identity independent of repetition. What feels like permanence, as we will see next, is simply consistency of behavior across repeated resolution cycles.

3.6 — Identity, Continuity, and the Illusion of Permanence

If objects are stabilized states, and properties are behaviors revealed under constraint, then one final illusion remains to be addressed: permanence. Why do some states feel as if they truly endure? Why do identities—of objects, systems, and selves—feel continuous even when everything involved is in motion?

The answer is not substance. It is repetition.

What we experience as permanence is not something that exists independently of time and change. It is the consistent re-emergence of the same resolution behavior under similar constraints. When a state resolves imbalance in the same way again and again, it appears to persist. When that repetition is stable enough, we assign identity.

Identity, then, is not a thing that travels through time. It is a pattern that keeps reappearing.

A river feels like the same river even though no molecule remains. A chair feels like the same chair even as stresses redistribute and materials age. A person feels like the same self even as thoughts, emotions, and cells continuously change. In every case, what persists is not material sameness but behavioral coherence.

This coherence arises because pathways are reinforced.

When delta repeatedly resolves through the same channels, those channels become easier to use and harder to abandon. Structures strengthen. Habits form. Roles stabilize. Expectations crystallize. Over time, the pathway itself begins to dominate future resolution, giving rise to the feeling of continuity.

This is why identity feels real but resists precise definition.

Identity is not stored anywhere. It is enacted—again and again.

When repetition breaks, identity fractures. When constraints shift, behavior changes. When pathways are disrupted, what once felt permanent dissolves. This can happen gradually, as in aging and learning, or abruptly, as in trauma, collapse, or transformation. The disappearance of identity is often experienced as loss, but structurally it is simply resolution completing along a new route.

Importantly, this does not make identity arbitrary or illusory. It makes it conditional.

A state remains identifiable as long as:

- its internal constraints remain coherent,
- its environment does not force new resolution regimes,
- and its pathways continue to repeat.

Remove any of these, and continuity weakens.

This understanding also explains why identity is always local. An object is stable under certain conditions and unstable under others. A person feels consistent in familiar contexts and fragmented under extreme stress. An institution feels permanent until its supporting constraints erode. Identity is not universal; it is situationally sustained.

By recognizing permanence as repetition rather than essence, Universal Energy Dynamics avoids two extremes. It does not deny continuity, and it does not absolutize it. It acknowledges persistence without freezing it into metaphysics.

What truly persists are pathways—the routes through which delta has learned to move. Entities appear stable because those routes have not yet been abandoned.

This insight completes the ontological shift of the chapter. Existence is not about what is *there*. It is about what continues to resolve in the same way. Once this is seen, identity becomes understandable without mysticism, and change becomes intelligible without loss of meaning.

With this foundation in place, the book is ready to move forward—from what exists, to how nature prefers those states to dissolve. That preference is the subject of the next chapter.

From "What Exists" to "How It Behaves"

By this point, the idea of existence has quietly but decisively shifted. What exists is no longer imagined as a collection of independent things carrying fixed properties and acted upon by external forces. What exists, in Universal Energy Dynamics, is what has managed to persist—states stabilized by constraint, repetition, and delayed resolution.

Nothing here required negation. Objects were not declared illusory. Properties were not dismissed. Forces were not erased. They were repositioned. Each found its proper place as a descriptive layer, not a foundational one. Ontology stopped asking *what reality is made of* and began asking *what conditions allow this configuration to last*.

This reordering matters, because it prevents contradiction later.

If existence is understood as stabilized outcome, then behavior is no longer mysterious. States behave as they do because they are still negotiating unresolved delta through particular pathways. What we observe as interaction, resistance, motion, or persistence is simply the continuation of consequence through structures that have not yet allowed alignment to complete.

Seen this way, being and behaving are not separate categories. Behavior is not something that happens *to* existence; it is how existence maintains itself in the presence of imbalance. Stability is not static; it is active negotiation. Change is not intrusion; it is resolution gaining ground.

This prepares the reader for the final movement of the Kernel.

If objects are outcomes, if properties are behaviors, and if identity is repetition, then the question is no longer *what exists*, but what nature prefers to do with what exists. Does nature preserve states indefinitely? Does it favor expansion, accumulation, or expression? Or does it quietly, relentlessly, bias systems toward a particular end?

Next chapter answers that question directly.

It introduces the simplest and most easily overlooked rule of all: nature prefers to contract delta. Not as intention, not as design, but as tendency. Stability exists only because contraction has been delayed. Change occurs when contraction resumes.

With this transition, the Kernel is complete. The reader is no longer oriented around objects and causes, but around states and tendencies. From here on, nothing that follows will feel imposed. It will feel like the only direction explanation could have taken.

What appears to exist does so only because resolution has been delayed in a particular way.

What appears to exist does so only because resolution has been delayed in a particular way.

Chapter 4 — The Rule of Nature: Contract Delta

The Direction We Never Questioned

From the earliest observations of nature, one direction has always been taken for granted. Hot objects cool. Stretched materials relax. Disturbances fade. Turbulence settles. Systems, when left alone, move toward stillness. This tendency is so familiar that it rarely invites scrutiny. We notice it, use it, and move on.

What we did not notice was how deeply this assumption runs.

Every physical law quietly leans on the same direction. Energy spreads. Gradients flatten. Potentials equalize. Even when motion persists, it does so by circulating within boundaries that prevent full alignment. When those boundaries weaken, motion fades. Rest appears not as an achievement, but as a return.

This was often framed as loss.

Dissipation was treated as waste. Decay was treated as failure. Relaxation was treated as the end of usefulness.

Growth, expansion, and accumulation were celebrated as progress, while contraction was associated with decline. This emotional bias shaped how we narrated physics, even when equations told a calmer story.

But nature never expressed disappointment.

Nature does not mourn dissipation. It does not resist relaxation. It does not aim to preserve disturbance.

It simply aligns.

This alignment is not a goal. It is not a preference in the human sense. It is a tendency—the direction that remains when nothing holds difference apart. No instruction is required. No memory is involved. Alignment happens because there is nothing preventing it.

Once this is seen, a quiet correction occurs. Contraction is no longer interpreted as pessimism or collapse. It becomes coherence. It becomes the most economical state a system can occupy given its constraints.

Expansion, by contrast, begins to look conditional. Whenever something grows, stretches, accumulates, or complexifies, it does so by borrowing permission. Boundaries are reinforced. Pathways are constrained. Energy is continuously spent to delay relaxation. Expansion survives only as long as this expenditure can be sustained.

This does not make growth unnatural. It makes it temporary by design.

The error was never in observing nature's tendency. The error was in treating that tendency as secondary—as something to be explained away, corrected, or overcome by more "fundamental" mechanisms. In truth, this tendency is the mechanism.

Nature does not seek events. It does not seek form. It does not seek history.

It seeks alignment.

Everything else—motion, structure, life, and experience—exists only because alignment has been delayed in a particular way. This chapter does not introduce a new rule. It simply names the one that has been operating silently beneath all others.

Once this direction is acknowledged, explanation stops fighting nature's flow and starts moving with it.

4.1 — Nature's Bias: Relax, Dissipate, Align

Nature does not push systems forward. It does not demand progress, accumulation, or expression. What it does—quietly, consistently, and without exception—is relax imbalance wherever it is permitted to do so. This is not intention and not design. It is bias in the most literal sense: the direction that remains when no constraint interferes.

Relaxation is not inactivity. Dissipation is not destruction. Alignment is not erasure.

They are the simplest outcomes available to a system once delta is no longer held apart.

When a gradient exists—of pressure, temperature, potential, stress, attention—it represents stored permission. As long as boundaries maintain that permission, activity continues. The moment those boundaries weaken, relaxation begins automatically. No trigger is required. No instruction is followed. The system simply stops resisting equilibrium.

This is why alignment feels inevitable. It does not compete with other tendencies; it underlies them.

Dissipation is often misunderstood because it looks like loss when viewed locally. Energy spreads out. Signals weaken. Structures soften. But from a structural perspective, dissipation is redistribution toward equilibrium, not disappearance. What is lost in concentration is gained in accessibility. What fades in intensity persists in breadth.

Relaxation works the same way. A stretched material does not "give up." It releases stored imbalance. A turbulent flow does not "fail." It reorganizes until gradients no longer demand complex motion. A mind at rest is not empty; it is aligned enough to stop forcing resolution.

Alignment, then, is not a final state imposed on systems. It is what remains when nothing actively prevents it.

This bias explains a fundamental asymmetry that is often overlooked: expansion is always conditional; contraction is not. Expansion requires continuous input, reinforcement, or confinement. Growth needs boundaries. Complexity needs layering. Structure needs maintenance. The moment support weakens, contraction resumes.

Nature does not oppose expansion. It simply does not sustain it for free.

This is why equilibrium is never fragile. What is fragile is imbalance. What requires protection is structure. What demands effort is persistence against alignment. Relaxation, by contrast, is effortless. It is what happens when effort ends.

Recognizing this bias reframes the entire landscape of physical explanation. Instead of asking why systems relax, the correct question becomes: *what is holding relaxation back here, and at what cost?* Instead of treating dissipation as noise, it becomes the clearest signal that constraints have loosened.

Nature's rule is not harsh. It is economical.

It does not destroy form. It does not forbid complexity. It does not rush collapse.

It simply contracts delta whenever permission expires.

Once this bias is acknowledged, stability, change, growth, and decay all fall into place as variations of a single theme: how long imbalance can be affordably maintained before alignment resumes.

4.2 — Why Stability Exists at All (Minimum Delta Configurations)

If nature carries a quiet bias toward contracting delta, an obvious question follows: why does anything remain stable at all? Why do forms persist? Why do structures last? Why does the world not collapse instantly into featureless equilibrium?

The answer is not opposition to nature's bias. It is economy within it.

Stability exists because some configurations minimize delta more efficiently than others under given constraints. These configurations are not static endpoints; they are minimum-delta arrangements—states where remaining imbalance is either very small, very evenly distributed, or very costly to resolve further.

Nature does not eliminate delta absolutely. It eliminates delta cheaply.

A stable state is one where:

- further alignment would require more restructuring than the system can afford,
- resolution pathways are long, slow, or diffuse,
- and local equilibrium has been reached even if global equilibrium has not.

This is why stability is common and fragility is specific.

Consider a rock. Its internal atomic structure distributes stress so effectively that there is little incentive for further reconfiguration under ordinary conditions. Delta exists, but it is already arranged near a local minimum. To change that state meaningfully requires significant external intervention—heat, pressure, impact. The rock is not permanent; it is simply well-aligned for its environment.

The same logic applies at every scale.

A molecule is stable because its bond configuration minimizes energy locally. A star is stable because gravitational and thermal deltas balance at a long-lived minimum. An ecosystem is stable because competing flows of energy and matter cancel at the system level. A habit is stable because cognitive and emotional pathways repeat with minimal friction.

In each case, stability is not frozen equilibrium. It is optimized persistence.

This introduces an important distinction: local minima versus global equilibrium. Nature often settles systems into local minima because reaching global equilibrium would require crossing higher delta barriers. A ball rests in a valley not because it has reached the lowest point in the entire landscape, but because climbing out would temporarily increase imbalance. Stability is chosen not for perfection, but for efficiency of resolution.

This explains why forms can be stable yet changeable. They persist until constraints shift enough to make a new configuration cheaper. When that happens, stability does not resist change—it yields. What we call breakdown or transformation is simply the system moving toward a new minimum-delta arrangement.

Crucially, this removes the need for design, intention, or teleology. Stability does not exist because nature wants form. Form exists because some arrangements are easier to maintain than to dissolve. Nature does not protect them; it tolerates them until tolerance is no longer economical.

This also clarifies why stability often feels intelligent or purposeful even when it is not. Minimum-delta configurations can look optimized because they are. But they are optimized for resolution efficiency, not for meaning, survival, or value. Those interpretations come later, layered by observers.

Once this is understood, stability stops being mysterious. It is not a special privilege granted to certain forms. It is the natural outcome of delta seeking the cheapest available resting place under existing constraints.

Stability exists not because nature resists change, but because change has already found its easiest expression—for now.

This insight prepares the reader for the next correction: when stability fails, it is not because nature became chaotic, but because instability has become informative.

4.3 — Why Instability Is Informative (Not "Noise")

Instability is often treated as a problem to be suppressed. In physics, it is labeled turbulence, fluctuation, or error. In systems, it is called volatility. In human affairs, it is dismissed as crisis or dysfunction. The shared assumption is that instability represents a breakdown of order—a deviation from how things are supposed to behave.

Under Universal Energy Dynamics, this interpretation is exactly backward.

Instability is not noise. Instability is information.

Instability appears when a system can no longer maintain delta cheaply. It is the visible signal that existing constraints are misaligned with the imbalance they are trying to hold. What looks like disorder is actually resolution pressure becoming legible.

A stable system hides its internal tensions well. Delta is distributed smoothly. Pathways cancel each other out. Nothing stands out. Instability emerges precisely when that quiet management fails—when imbalance begins to probe for new routes of release.

This is why instability is revealing.

Turbulence in a fluid is not randomness. It exposes the limits of laminar pathways under increased gradient. Fluctuations near a phase transition are not errors; they reveal competing configurations struggling to become the new minimum. Material fatigue is not surprise failure; it is accumulated stress finally finding a viable crack.

The same logic applies outside physics.

Emotional instability reveals unresolved cognitive and physiological delta. Market volatility reveals mispriced risk and delayed redistribution. Social unrest reveals structural imbalances that can no longer be buffered.

In every case, instability is not an interruption of nature's rule—it is nature applying it more visibly.

Noise is what we call a signal we do not know how to read.

Instability becomes "noise" only when we insist on preserving an outdated model of stability. When we expect systems to behave as if constraints were permanent, any deviation feels like chaos. But constraints are always conditional. When they weaken, instability is the first honest report.

This is why instability often precedes transformation. It marks the moment when multiple resolution pathways compete, when the system explores alternatives because the old configuration is no longer economical. The more rigid the system, the sharper the instability. The more adaptive the system, the more gently instability guides change.

Importantly, instability does not demand immediate correction. It demands interpretation. Suppressing instability without understanding it often increases long-term cost by forcing delta back into pathways that are already failing. Many collapses are not caused by instability itself, but by attempts to silence it.

Under UED, instability is treated as diagnostic.

It answers questions that stability cannot:

- Where is delta accumulating?
- Which constraints are overstressed?
- Which pathways are becoming viable?
- What minimum is no longer sustainable?

Once read correctly, instability becomes guidance rather than threat. It tells us not that nature has lost order, but that order is renegotiating.

By reframing instability as information, the reader is prepared for a critical shift: breakdowns are not opposites of stability; they are messages about its limits. Noise is not the enemy of understanding. Misinterpretation is.

This prepares the final sections of the chapter, where life, growth, and complexity must be reconciled with nature's bias toward contracting delta—without contradiction and without romance.

4.4 — Why Growth, Complexity, and Life Don't Violate the Rule

At this point, a natural objection may arise: *If nature prefers to contract delta, how do growth, complexity, and life exist at all?* How can organisms evolve, structures elaborate, and intelligence emerge in a universe biased toward relaxation and alignment?

The short answer is this: they do not violate the rule—they operate inside it.

Life, growth, and complexity are not exceptions to delta contraction. They are temporary strategies for managing it.

Living systems do not escape equilibrium. They postpone it. They do so by continuously importing low-entropy resources, exporting waste, and reinforcing boundaries that delay internal alignment. Metabolism is not a triumph over nature's bias; it is a disciplined negotiation with it. The moment that negotiation fails, life ends—not as punishment, but as resolution completing its course.

Complexity arises in the same way.

Complex structures form when layered constraints allow delta to be held across multiple scales simultaneously. Each layer delays contraction locally while passing pressure outward. Cells hold chemical gradients. Organs coordinate flows. Organisms regulate environments. Societies stack institutions. At every level, complexity exists by distributing the cost of delay.

This is why complexity is never free.

Every increase in structure carries maintenance cost. Every new layer introduces fragility. The more elaborate the system, the more pathways must remain aligned to prevent collapse. Growth feels creative and expansive, but structurally it is borrowed time, financed by continuous energy expenditure and precise regulation.

Evolution itself follows this logic. It does not produce complexity for its own sake. It explores configurations that can delay delta contraction more effectively under given conditions. When environments change, previously stable strategies lose efficiency, and contraction accelerates. Extinction is not failure; it is economic correction.

This perspective removes the false drama from life's story.

Life is not fighting nature. Life is riding nature's bias by shaping how and where contraction occurs.

The same applies to human systems. Economies grow by channeling imbalance through markets and institutions. Technologies evolve by holding energetic and informational deltas inside engineered constraints. Cultures persist by stabilizing meaning and behavior through shared narratives. None of these escape consequence. They survive only as long as their cost of maintenance remains affordable.

This understanding also explains why growth often accelerates collapse rather than preventing it. Systems that expand without accounting for rising maintenance cost eventually amplify instability. What looks like success at one scale becomes unsustainable at another. The rule was never broken; it was merely postponed.

Seen clearly, life and complexity are not rebellions against nature's tendency. They are its most sophisticated expressions. They demonstrate not defiance, but ingenuity in delay.

Nature allows this ingenuity—not because it values it, but because constraints permit it. When constraints fail, contraction resumes without hesitation.

By recognizing growth and life as managed delay rather than violation, the reader is freed from false contradictions. The universe does not oscillate between creation and destruction. It continuously aligns, allowing elaborate detours wherever structure can afford them.

This sets the stage for the final section: why humans consistently misread progress and permanence—and why that misreading carries a cost.

4.5 — The Misreading of Progress and Permanence

Human perception has a bias that runs deeper than culture or education: we equate persistence with success. What lasts is assumed to be right. What grows is assumed to be good. What resists collapse is celebrated as progress. This bias is understandable—we survive by stabilizing conditions—but it quietly distorts how we read nature.

Under Universal Energy Dynamics, endurance is not victory. It is expense successfully paid—for now.

Systems that resist contraction feel powerful because they can hold delta longer than others. A bridge that stands, a company that grows, an empire that endures, a self that feels consistent—all project strength because their constraints are still functioning. But this projection hides the underlying reality: persistence is never free. It is a continuous act of delta-maintenance against equilibrium pressure.

This is why humans so often mistake resistance for achievement.

Effort feels productive. Control feels effective. Stability feels deserved.

Yet structurally, these feelings reflect ongoing cost, not intrinsic merit. The longer a system persists against contraction, the more energy it consumes, the more pathways it hardens, and the more brittle it becomes to change. What looks like permanence is often accumulated obligation.

Progress, then, is easily misread.

We celebrate expansion without tracking maintenance. We admire growth without pricing fragility. We trust continuity without testing constraints.

When collapse arrives, it feels like betrayal—by nature, by fate, by others. In truth, collapse is simply resolution completing after delay has exhausted its affordability. Nothing new happened. What happened finally finished.

This misreading has consequences far beyond physics. In human systems, it fuels denial. Warning signals are dismissed as pessimism. Instability is labeled failure instead of feedback. Systems double down on control just as their constraints weaken. The result is not resilience, but sharper collapse.

UED reframes this without cynicism.

Collapse is not the opposite of progress. It is the cost of unacknowledged delay.

A structure that dissolves is not erased from history. Its existence mattered—it shaped pathways, redistributed delta, and informed future configurations. Resolution does not negate what came before; it accounts for it.

This perspective restores realism without stripping meaning. Progress becomes something that must be defined carefully, not assumed. Permanence becomes something that must be explained structurally, not celebrated emotionally. Success becomes a question of how long delta can be maintained wisely, not how loudly persistence can be declared.

Once this misreading is corrected, the reader is ready for the final synthesis of the Kernel: not a new rule, but the only question that truly matters once equilibrium, delta, consequence, and contraction are understood together.

4.6 — The Only Question That Matters

By now, the Kernel has completed its work—not by adding complexity, but by removing misplaced assumptions.

Nothing extraordinary has been introduced. No exotic entities. No speculative forces. No metaphysical commitments. What has changed is only the order of explanation.

Placed correctly, the structure is unavoidable:

- Equilibrium is the default, not the destination.
- Delta enables action, not because it drives systems, but because it delays alignment.
- Consequence accounts, not morally, but structurally, through redistribution and transformation.
- States persist conditionally, not because they are fundamental, but because resolution is temporarily affordable.
- Nature contracts delta, not by intention, but by tendency.

Once these are seen together, a profound simplification occurs.

The question *"What is real?"* quietly loses its urgency.

Reality is no longer a puzzle of substances and entities. It is no longer a contest between appearances and essences. What exists is simply what has not yet finished resolving under the constraints that currently apply.

This reframing does something subtle but decisive: it replaces static ontology with dynamic orientation.

Instead of asking:

- What caused this?
- What acted on what?
- What should have happened?
- What deserves to persist?

The correct question becomes: What is resolving here, where is it being delayed, and at what cost?

This question works everywhere.

In physics, it dissolves false paradoxes by revealing where constraints were misidentified. In biology, it clarifies life as managed delay rather than exception. In psychology, it reframes struggle as unresolved delta rather than personal failure. In societies, it exposes instability as accumulated pressure rather than sudden disorder.

Most importantly, it prevents premature explanation. It stops us from inventing stories when structure is sufficient.

This is why the Kernel matters.

Without it, later discussions of motion, time, geometry, mind, or ethics will feel arbitrary—like clever reinterpretations layered on top of old assumptions. With it, those discussions become inevitable extensions of a single, consistent orientation.

Nothing ahead will ask the reader to believe more. Everything ahead will ask the reader to track resolution more carefully.

By the end of Part I, the reader should feel a quiet shift in attention. Not excitement. Not persuasion. But recognition. The sense that the universe has not become stranger—only more economical.

Reality does not ask what it is. It continuously answers a different question: What is resolving now, and what is still being held apart?

With that question installed, the Kernel is complete. Everything that follows is application.

Leaving the Kernel

Nothing fundamental has been added in this part. No new entities were invented. No old laws were discarded. What changed was orientation.

The Kernel did not replace physics; it repositioned it. It moved explanation upstream—before force, before motion, before time, before objects—into the quiet territory where equilibrium, delta, consequence, state, and contraction already operate. Once seen there, everything else naturally falls into place.

This is why Part I functions as a reference spine rather than an argument. It does not compete with existing models; it stabilizes them. Any framework that respects equilibrium, acknowledges delta, tracks consequence, understands states as conditional, and accepts nature's bias toward contraction will align with what has been laid here. Any framework that ignores these will eventually generate paradox—not because it is wrong, but because it started too late.

From this point forward, the book no longer needs to persuade. It only needs to apply.

What comes next—motion, time, geometry, waves, life, mind, society—will not introduce new principles. Each domain will simply be revisited with the same question already installed:

- Where is delta being held?
- What constraints are shaping resolution?
- What cost is accumulating, and where will alignment resume?

Because this orientation is now fixed, later chapters will feel less like reinterpretations and more like recognitions. Familiar equations will look cleaner. Old debates will lose urgency. Apparent contradictions will soften into boundary errors.

The Kernel is complete when the reader no longer searches for foundations elsewhere.

From here on, nothing stands alone. Every phenomenon will be read as a case of delayed resolution, and every explanation will trace how nature patiently contracts delta through whatever pathways remain open.

Reality does not move forward. It relaxes—until something holds it apart.

With that understanding locked in place, the book can now move outward—confident that it will never lose its center.

The UED Resolution Chain

↓

SYSTEM STRUCTURE CONSTRAINT
(Boundaries of behavior)

↓

DELTA
(Accumulated unresolved pressure)

↓

GRADIENT
(Directional difference enabling movement)

↓

ALIGNMENT THRESHOLD TIME (ATT)
(Limit beyond which gradual correction fails)

↓

COUPLING
(Propagation of effects across connected systems)

↓

CONSEQUENCE
(Structural feedback after resolution occurs)

⇩

Figure 1 — Canonical Resolution Chain in Universal Energy Dynamics

All systemic change follows a consistent structural progression. Constraints shape how pressure accumulates as delta. Gradients determine the direction of possible resolution. When alignment is delayed beyond the system's threshold (ATT), correction occurs through coupling with surrounding structures, producing observable consequences. What appears as sudden change is often the delayed release of previously accumulated pressure.

Deliverable of Part I

The Kernel (Explicit Outcome)

By the end of this part, something subtle but irreversible should have happened to the reader's way of thinking. Not an acquisition of new facts, but a reorientation of attention.

The old questions quietly lose their urgency.

- **"What is real?"** fades because reality is no longer framed as a contest between appearances and essences. What exists is simply what has not yet finished resolving under current constraints.
- **"What caused this?"** weakens because single-cause narratives no longer satisfy. Outcomes are seen as continuations of permitted imbalance moving through pathways, not as effects fired from isolated origins.
- **"What acted on what?"** dissolves because action is no longer imagined as entities pushing entities. What matters is how delta navigates structure, not which object plays the role of agent.

These questions are not answered. They are outgrown.

In their place, a different set of questions begins to arise—almost automatically, without effort or instruction:

- **Where is delta being held?** The reader looks for imbalance, gradients, misalignment—physical, biological, psychological, or social—and asks where alignment is being delayed.
- **What constraints are operating?** Attention shifts to boundaries,

structures, rules, habits, and couplings that shape how resolution is allowed or resisted.

- **What is resolving—and what is resisting resolution?** Change and stability are no longer opposites. They are read as different phases of the same process, distinguished only by cost and permission.

This shift marks the installation of UED.

Not as a belief. Not as a competing theory. But as a native orientation—a way of seeing that remains valid regardless of domain.

Once this orientation is in place, explanation no longer needs persuasion. The reader does not need to be convinced that time behaves strangely, that force is derivative, that motion is contextual, or that systems collapse predictably. Those insights will feel like natural consequences of questions already being asked correctly.

From here on, the book does not argue. It demonstrates.

Each subsequent part—whether addressing motion, time, geometry, life, mind, or society—will feel like a case study in delayed resolution. Familiar models will still function, but they will be read as tools, not truths. Paradoxes will appear not as mysteries, but as signals of misidentified constraints.

This is the purpose of the Kernel.

Once it is absorbed, the reader does not carry UED as a concept. They think from it.

And when that happens, everything that follows is no longer persuasion. It is application.

Part II — The Canonical Triad

If previous part established *orientation*, this part establishes precision.

The Kernel answered *where to stand*. The Canonical Triad answers *what to track*.

This part does not introduce new ideas; it sharpens the three that already govern everything: Delta, Equilibrium, and Consequence. They are called "canonical" not because they are abstract, but because every phenomenon—across physics, life, mind, and society—can be decomposed into their interaction without residue.

Here, delta is no longer treated as a vague imbalance, but as a structured condition with type, magnitude, and topology. Equilibrium is no longer a poetic endpoint, but a geometric attractor with local and global expressions. Consequence is no longer discussed defensively, but as neutral redistribution, explaining why paradox appears only at the psychological layer, never at the structural one.

This part also marks a tonal shift.

Previous part dissolved confusion. This part standardizes language.

Readers should come away from this section able to:

- identify *what kind* of delta is operating,
- distinguish *where* equilibrium is acting as magnet versus ceiling,
- trace consequence without invoking morality, fate, or reward.

By the end of this part, the triad should feel unavoidable—not as theory, but as the minimal grammar of reality. Anything that cannot be expressed in terms of delta, equilibrium, and consequence is either incomplete, mislabeled, or operating at a different descriptive layer.

From this point onward, UED no longer needs to explain itself. It becomes the reference frame against which motion, time, geometry, life, and mind will be re-read.

Next part will not ask *what these are*. It will ask how they behave once the triad is active.

Chapter 5 — Delta

Gradient, Misalignment, and Potential to Resolve

Why Difference Is the First Observable

Before there is an object, there is contrast. Before there is motion, there is asymmetry. Before there is meaning, there is difference.

Nothing can be noticed, measured, or even experienced without it.

A perfectly uniform universe would be indistinguishable from nonexistence—not because it would lack substance, but because it would lack variation. No edges, no gradients, no tensions, no cues. Observation itself would have nothing to attach to. In that sense, delta is not a disturbance added to reality; it is the minimum condition for notice.

This is easy to miss because difference feels so ordinary. We see light because of intensity contrast. We hear sound because of pressure variation. We think because one thought is not another. We care because what *is* does not match what *could be*. Difference is everywhere, so we treat it as background rather than foundation.

Physics often enters the story later—at motion, force, or energy transfer—but perception enters earlier. What the observer encounters first is not action, but imbalance. A gradient catches attention before it causes movement. A mismatch is registered before it resolves. Delta announces itself quietly as *something not yet aligned.*

This is why delta should not be confused with disorder.

Disorder implies failure of structure. Delta implies potential.

A gradient does not mean chaos; it means resolution is possible. A misalignment does not mean error; it means change is allowed. Delta is not what breaks systems—it is what gives them something to do.

Even stillness does not negate delta. A stretched bow held motionless contains difference. A silent room with a temperature gradient contains difference. A calm mind holding unresolved questions contains difference. Motion is optional. Delta is not.

This matters because much confusion in science and philosophy arises from treating difference as an anomaly rather than a condition. We ask why systems deviate from equilibrium, as if equilibrium were the starting point. In truth, equilibrium is the *absence* of difference, and difference is what makes the universe legible in the first place.

Delta is therefore not the opposite of stability. It is the precondition of experience.

Only after difference is registered do concepts like force, motion, time, and cause enter the picture. They are responses to delta, not replacements for it. By placing delta first, Universal Energy Dynamics restores the correct sequence: notice before action, gradient before flow, misalignment before resolution.

This chapter begins by honoring that order.

Before we classify delta, measure it, or trace its consequences, we must recognize what it truly is: not a problem to be eliminated, but the first observable fact of existence.

5.1 — What Delta Actually Is (and Is Not)

Delta is often spoken about loosely, as if it were just another word for difference, disturbance, or deviation. That looseness creates confusion. Before delta can be classified or applied, it must be stabilized conceptually.

In Universal Energy Dynamics, delta is defined precisely as: *a gradient, misalignment, or asymmetry that carries the potential to resolve under existing constraints.*

This definition matters because it draws clear boundaries around what delta is not.

Delta is not disorder. Disorder is a description of unpredictability relative to a model. Delta exists even in perfectly lawful systems. A smooth pressure gradient in a fluid is delta, not disorder. A clean voltage difference in a circuit is delta, not chaos.

Delta is not chaos. Chaos describes sensitivity to initial conditions within deterministic rules. Delta is present long before chaos appears and often resolves without ever becoming chaotic. Chaos is a *behavioral regime*; delta is a precondition.

Delta is not error. Error is deviation from expectation or design. Delta does not care about expectation. It simply indicates that alignment is incomplete. Treating delta as error leads systems to suppress signals they should be reading.

Most importantly, delta is not motion.

Motion is what may happen when delta resolves along a permitted pathway. But delta can exist without motion for long periods. A compressed spring at rest holds delta. A charged capacitor holds delta. A social tension that has not yet surfaced holds delta. The absence of motion does not imply the absence of imbalance.

This is a critical correction.

Delta should be understood as potential-to-resolve, not as activity itself. It is stored possibility. It is permission waiting to be exercised. Whether that potential becomes visible depends entirely on constraints and pathways, not on delta's existence alone.

This also explains why delta can be present yet invisible. Many systems are engineered—or evolved—to hold imbalance quietly. They distribute it internally, buffer it across layers, or delay it through feedback. To an external observer, nothing seems to be happening. Structurally, everything is waiting.

By clarifying what delta is not, we protect the framework from two common mistakes:

1. Treating delta as a problem that must always be eliminated.
2. Treating delta as a force that must always act.

Neither is true.

Delta is neutral. It does not push, judge, or insist. It simply marks the fact that equilibrium has not yet been reached and that resolution remains possible. Whether resolution occurs gently, violently, slowly, or never within the observer's timeframe depends entirely on how delta is shaped.

Once this definition is stable, classification becomes meaningful. Without it, different domains appear to talk about different things—energy here, stress there, desire elsewhere—when in fact they are describing the same structural condition in different languages.

Delta is the common denominator beneath them all.

With this clarity in place, the next step is unavoidable: to examine the canonical types of delta that appear across domains, and to see why they are not separate species, but variations of a single underlying condition.

5.2 — The Four Canonical Types of Delta

Once delta is defined cleanly—as misalignment carrying the potential to resolve—it becomes clear that delta does not belong to any single domain. What changes across physics, biology, psychology, or society is not the nature of delta, but the medium through which it is expressed.

To make delta operational rather than abstract, Universal Energy Dynamics groups its expressions into four canonical types. These are not categories in competition. They are views of the same condition appearing at different layers of organization.

1. Energetic Delta — Gradients That Drive Physical Resolution

Energetic delta is the most familiar form because physics learned to measure it early. Pressure differences, temperature gradients, electrical potentials, chemical concentrations—all are expressions of energy unevenly distributed.

Energetic delta resolves by flow, transfer, or transformation. Heat spreads. Fluids move. Charges redistribute. These behaviors are not special cases; they are textbook examples of delta attempting alignment.

What matters here is not the energy itself, but the difference. A high temperature alone means nothing. A temperature *difference* creates potential. Remove the gradient, and action stops—not because energy vanished, but because delta did.

2. Structural Delta — Misfit Between Form and Constraint

Structural delta appears when a system's configuration does not match the stresses imposed upon it. A beam under load, a shell under pressure, a network stretched beyond capacity—all exhibit structural delta.

This type of delta is often mistaken for weakness or failure. In truth, it is information-rich. Structural delta tells us where constraints are mismatched, where geometry resists resolution, and where force will appear if alignment is obstructed.

Structural delta resolves by deformation, fracture, reconfiguration, or redesign. It does not need energy gradients to exist; it needs constraint mismatch.

3. Informational Delta — Asymmetry of Signal and Expectation

Informational delta arises when what is known, signaled, or predicted does not match what is encountered. Surprise, uncertainty, ambiguity, novelty—these are not psychological curiosities. They are structural indicators of informational imbalance.

A signal carries meaning only because it differs from background. A message matters only because it reduces uncertainty. When information is evenly distributed, nothing is learned. When it is uneven, delta exists.

Informational delta resolves by updating models, redistributing attention, or restructuring representation. Learning is simply delta resolution in informational space.

4. Psychosocial Delta — Misalignment of Desire, Role, and Reality

Psychosocial delta is often treated as subjective or emotional, but structurally it follows the same rule. Desire versus reality, identity versus expectation, power versus legitimacy—these are gradients, not moods.

Stress, motivation, conflict, ambition, frustration, aspiration—all arise from persistent misalignment. Psychosocial delta resolves through adaptation, negotiation, resignation, rebellion, or collapse. The pathways differ, but the logic does not.

This is why psychological tension feels energetic and social pressure feels physical. They are not metaphors. They are delta expressed in different media.

The crucial insight is this: these four types are not separate species. They are projections of the same underlying condition onto different constraint spaces.

Energetic delta resolves through physical pathways. Structural delta resolves through geometric pathways. Informational delta resolves through representational pathways. Psychosocial delta resolves through behavioral and relational pathways.

Treating them as separate leads to fragmented theories. Treating them as one restores continuity.

Once this is seen, interdisciplinary confusion begins to fade. Stress, pressure, uncertainty, and conflict stop being poetic parallels. They become structurally equivalent phenomena, governed by the same logic of delayed alignment.

This prepares the reader for the next refinement: not all deltas matter equally. What determines impact is not just *how much* imbalance exists, but how it is shaped and constrained. That distinction—between delta magnitude and delta topology—is where naive explanations fail and precise understanding begins.

5.3 — Delta Magnitude vs Delta Topology

Once delta is recognized across domains, a common misunderstanding immediately appears: the belief that more delta automatically means more effect. This assumption feels intuitive. A bigger pressure difference should produce stronger flow. A stronger desire should produce stronger action. A larger imbalance should lead to larger consequences.

Sometimes it does. Often, it does not.

The reason is simple and widely overlooked: delta magnitude and delta topology are not the same thing.

Delta magnitude refers to *how much* imbalance exists. It is the quantity—the height of a gradient, the intensity of a mismatch, the strength of a tension. This is what most measurements focus on, because magnitude is easy to quantify and compare.

Delta topology refers to *how that imbalance is arranged*, constrained, and connected. It is the shape, geometry, and accessibility of resolution pathways. Topology determines where delta can go, how quickly it can move, and whether it can resolve smoothly or violently.

Magnitude answers *how big*. Topology answers *how it behaves*.

This distinction explains many phenomena that otherwise appear paradoxical.

A large delta can dissolve quietly if its topology allows many open pathways. Heat spreads gently in a well-conducting material despite a large temperature difference. Emotional tension may fade harmlessly when expression and reflection are available. Social inequality may persist without explosion when redistribution pathways exist.

Conversely, a small delta can dominate behavior if its topology is tight, blocked, or sharply constrained. A tiny crack can destroy a structure. A small cognitive inconsistency can destabilize a belief system. A minor regulatory mismatch can cascade into systemic failure. The magnitude was small; the topology was lethal.

This is why force, drama, and crisis correlate more with topology than with size.

When delta is tightly confined, pressure builds. When delta is distributed, pressure softens. When delta is rerouted, consequences shift shape.

Topology determines whether delta resolves as flow, vibration, oscillation, fatigue, collapse, or quiet alignment.

This insight corrects a major error in how systems are managed. Many interventions focus on reducing magnitude—less pressure, less stress, less conflict—without addressing topology. When pathways remain blocked, reducing magnitude temporarily only delays resolution. When topology is corrected, even large deltas can resolve gracefully.

Nature itself prioritizes topology.

It does not ask how much imbalance exists. It asks how cheaply it can resolve it.

Minimum-delta configurations are not those with zero imbalance, but those where remaining imbalance is geometrically harmless—spread out, buffered, or slow to act. Stability arises not from absence of delta, but from benign topology.

This also explains why prediction is so difficult. Magnitude is visible; topology is subtle. Measuring gradients is easier than mapping pathways. Yet without topology, magnitude tells us very little about outcome.

Under UED, understanding delta means always asking two questions together:

- How much imbalance exists?
- How is that imbalance constrained and connected?

Only when both are answered does explanation become reliable.

This prepares the reader for the next refinement. Delta does not resolve everywhere at once. It resolves only where pathways allow it to move. Understanding where delta can and cannot go is therefore more important than knowing how much of it exists.

5.4 — Pathways: Where Delta Can and Cannot Go

Delta never acts blindly. It does not push randomly, explode impulsively, or choose directions arbitrarily. Delta moves only where it is allowed to move. If there is one principle that prevents confusion later in the book, it is this: resolution follows pathways, not magnitude.

A pathway is not a tunnel or channel in the literal sense. It is any configuration—physical, structural, cognitive, or social—that permits alignment to proceed. Pathways are created by geometry, rules, habits, materials, feedback loops, and expectations. They exist wherever resolution can occur with less cost than resistance.

Delta does not cause motion. Delta searches for a viable path.

When pathways are open, resolution is gentle. Flow appears effortless. Change looks natural. When pathways are partially open, delta negotiates—oscillating, vibrating, stressing boundaries. When pathways are blocked, delta accumulates. Pressure rises. Force appears. Drama begins.

This explains why the same delta behaves so differently in different systems.

In an open system, imbalance diffuses quietly. In a constrained system, imbalance concentrates. In a rigid system, imbalance fractures structure.

The delta did not change. The pathways did.

Pathways are often mistaken for causes because they shape outcomes so strongly. But pathways are not origins; they are permissions. They tell delta where it may resolve and where it must wait. When no pathway is available, delta does not disappear—it stores itself, reshapes itself, or reroutes through less obvious routes.

This is why blockage is more important than intensity.

Blocked pathways convert potential into pressure. Open pathways convert potential into flow.

Force, resistance, stress, and effort are all symptoms of delta encountering insufficient pathways. They are not primary phenomena; they are indicators that alignment is being obstructed.

This principle applies far beyond physics.

A person under stress is not overloaded by magnitude alone, but by lack of expression pathways. A society becomes unstable not because grievances exist, but because channels for redistribution, dialogue, or reform are blocked. An organization collapses not from ambition, but from rigidity in how decisions and corrections are allowed to move.

Understanding pathways shifts intervention strategy immediately. Instead of asking how to suppress delta, the intelligent response is to ask:

- Where can resolution be allowed?
- Which constraints are unnecessarily rigid?
- Which pathways are misaligned with current deltas?

Nature itself behaves this way. It does not eliminate imbalance aggressively. It waits for pathways to open—or for accumulated pressure to force new ones. Sudden transitions, breakdowns, and phase changes are not spontaneous events. They are moments when a previously unavailable pathway becomes viable.

Seen through this lens, change becomes intelligible rather than threatening. Stability becomes understandable rather than mysterious. And control reveals itself as temporary authority over pathways, not command over outcomes.

Once pathways are understood, delta becomes predictable without becoming trivial. The future is not fixed, but it is constrained. What can happen is limited by where delta is allowed to go.

This prepares the ground for the final insight of the chapter: why difference itself is not a flaw to be eliminated, but a non-optional condition for existence, experience, and meaning.

5.5 — Why "Difference" Is Not Optional in Any Domain

Once delta is understood as gradient, topology, and pathway—not as error or disorder—a final misconception must be dismantled: the idea that difference is something to be eliminated, minimized, or overcome. This instinct appears everywhere. In physics, we seek uniformity. In systems, we seek stability. In human life, we seek harmony. Difference is tolerated only as a temporary inconvenience.

Under Universal Energy Dynamics, this instinct is revealed as misguided.

Difference is not optional. It is the price of existence.

A universe without difference would not be peaceful or perfect—it would be unreadable. No perception, no motion, no learning, no adaptation, no meaning could arise. Equilibrium without delta is not a higher state; it is informational silence. Nothing could be noticed because nothing would stand apart.

This is true in every domain.

In physics, without gradients there is no flow, no radiation, no interaction. Temperature without difference is thermodynamic death. Potential without difference is inert. A perfectly uniform field does not act because it has nothing to resolve.

In biology, life exists precisely because internal states are kept misaligned with the environment. Metabolism is organized difference. Homeostasis is not elimination of delta, but controlled imbalance. Death occurs when difference can no longer be sustained.

In cognition, thought arises because one idea is not another. Curiosity exists because knowledge is incomplete. Motivation exists because desire does not match reality. A mind without difference would not be enlightened; it would be inactive.

In society, meaning emerges from disagreement, diversity, and tension. Power gradients, role differences, and value conflicts are not defects of civilization; they are its raw material. When difference is forcibly eliminated, systems do not become stable—they become brittle.

This is the crucial correction: equilibrium is necessary, but total equilibrium is terminal. Every living, thinking, functioning system survives in a narrow band between excessive delta and none at all. Too much difference overwhelms pathways. Too little erases function.

Nature does not attempt to eliminate difference globally. It contracts delta locally, wherever it can do so cheaply, while allowing difference to persist elsewhere. This is why structure exists at all. Uniformity is not the goal; economical alignment is.

Human attempts to remove difference entirely—whether through over-control, enforced sameness, or premature resolution—often backfire. They collapse pathways that were quietly managing delta and replace them with pressure that has nowhere to go. What was manageable becomes explosive.

Difference, then, is not a flaw to be corrected. It is a resource to be managed.

The skill of any system—natural or designed—lies not in eliminating delta, but in knowing which differences to resolve, which to hold, and which to redistribute. This distinction will matter deeply in later chapters, especially when we address intelligence, agency, and decision-making.

For now, the point is simple and structural: if difference disappears everywhere, so does behavior. If difference explodes everywhere, so does coherence. Reality lives between these extremes, continuously negotiating misalignment rather than abolishing it.

With this understanding in place, delta can finally be treated not as a problem to fix, but as the currency of change—the medium through which systems evolve, adapt, and transform. That insight completes the chapter and prepares the transition to equilibrium as attractor, ceiling, and grave.

5.6 — Delta as the Currency of Change

Once difference is no longer treated as a flaw, delta reveals its final and most practical role: it is the currency through which all change is paid for.

Nothing changes for free.

Every transformation—physical, biological, psychological, or social—requires delta to be spent, stored, transferred, or redistributed. Systems do not act because they "want" to change. They change because delta exists and resolution has become affordable along some pathway.

In this sense, delta functions exactly like a currency:

- It can be accumulated (stored imbalance, latent tension).
- It can be circulated (flow through pathways).
- It can be invested (maintained deliberately to enable structure).
- It can be wasted (dissipated without useful alignment).
- It can be mismanaged (held too long, released too abruptly).

Seen this way, change is no longer mysterious or dramatic. It is simply delta being exchanged for alignment.

This reframes effort immediately. Effort is not motion. Effort is the cost of holding delta against equilibrium pressure.

A system that appears calm may be spending enormous delta internally to maintain structure. A system that appears active may be resolving delta efficiently with little cost. What matters is not visible activity, but delta economics—how imbalance is budgeted across time and structure.

This is why intelligence, in UED terms, is not about eliminating problems. It is about allocating delta wisely.

Resolving the wrong delta too early collapses opportunity. Holding the wrong delta too long creates instability. Redirecting delta through better topology reduces cost.

Skill, whether in engineering, leadership, learning, or living, is the ability to answer a single question repeatedly and correctly:

Which delta should be resolved now, which should be held, and which should be redirected?

Nature already operates this way. Rivers do not eliminate elevation difference; they exploit it. Life does not erase entropy gradients; it feeds on them. Minds do not eliminate uncertainty; they navigate it. Societies do not remove tension; they channel it.

Failure occurs when delta is misunderstood as an enemy rather than a resource. Systems then fight imbalance blindly—overcorrecting, suppressing signals, exhausting pathways. Success occurs when delta is treated as informative pressure, guiding where alignment should happen next.

This insight closes the chapter with practical force. Delta is no longer a conceptual abstraction or philosophical reframing. It becomes the operational variable that links physics to experience, structure to behavior, and theory to practice.

With delta now established as gradient, topology, pathway, necessity, and currency, the reader is prepared for the next movement of the triad.

If delta is what enables change, then equilibrium is what shapes its direction.

Next chapter will examine equilibrium not as a peaceful endpoint, but as a powerful attractor—sometimes a ceiling, sometimes a magnet, and sometimes the grave of all unresolved difference.

From Delta to Attraction

By the end of this chapter, delta should no longer feel like a disturbance that demands correction. It should feel like the only thing that makes change possible. Difference has been stripped of drama and restored to its proper role: not an accident in an otherwise uniform universe, but the medium through which all structure, behavior, and experience emerge.

Delta is now fully specified. It is not merely "difference," but difference with directionality. It is not merely "imbalance," but imbalance shaped by topology. It is not merely "tension," but tension negotiating pathways under constraint.

Most importantly, delta has been separated from moral language. It does not mean good or bad, success or failure, progress or regression. It simply marks where alignment is incomplete and where resolution remains possible.

This matters because delta alone does nothing. Left unreferenced, delta is only potential. What gives it direction, urgency, and eventual finality is something else—something quieter and more relentless than force or intention. That something is equilibrium.

Equilibrium does not oppose delta; it defines its destination. It is the silent background tendency against which all gradients are measured. Delta exists only because equilibrium is not yet reached—and it resolves only because equilibrium remains unavoidable.

The transition, then, is natural. From asking:

- What kind of difference exists?
- How large is it?
- How is it shaped and constrained?

We now move to asking:

- Where does this difference want to settle?
- What configurations are stable?
- Why some states persist while others vanish?

Next chapter will show that equilibrium is not a single place or moment, but a family of attractors—local, temporary, and ultimate. It will also show why life, mind, and society survive by circling equilibrium without ever fully surrendering to it.

If delta is the currency of change, equilibrium is the market that sets its value. With this shift, the Canonical Triad begins to close.

Wherever difference exists, resolution becomes inevitable—the only question is how long it is delayed.

Chapter 6 — Equilibrium: The Ceiling, The Magnet, The Grave

Why Equilibrium Both Comforts and Terrifies Us

Equilibrium carries a strange emotional weight. It sounds peaceful, yet unsettling. It promises rest, yet hints at ending.

This tension is not accidental. Humans encounter equilibrium first not as a physical principle, but as a felt idea. We associate it with stillness, silence, completion—and we are unsure whether to desire it or avoid it. A calm moment feels good. A final calm feels frightening. Somewhere along the way, equilibrium became confused with absence, even with erasure.

Physics did not create this confusion; psychology amplified it.

We were taught to think of equilibrium as an end state—something systems "reach" when all activity stops. In that framing, equilibrium feels like a loss of story. No motion. No struggle. No progress. Just flatness. It is no surprise that life instinctively resists this image. Survival itself is resistance to final stillness.

But this image is wrong at the structural level.

Equilibrium is not an event. It is not a goal. It is not a moment when something *happens*.

Equilibrium is a tendency—quiet, persistent, and always present. It does not wait at the end of time. It operates continuously, shaping every gradient, every flow, every pause. Systems do not move toward equilibrium because they want to. They move because nothing else is cheaper.

This reframing dissolves the fear.

When equilibrium is understood as tendency rather than destination, it stops feeling like annihilation. It becomes background pressure—the same way gravity does not feel threatening until we fall. Most of the time, equilibrium simply sets direction, not fate.

Yet the discomfort remains, because equilibrium does something humbling: it does not care. It does not negotiate. It does not reward effort or honor intention. It does not recognize meaning. It waits patiently until resistance becomes unaffordable.

This is where the emotional split appears.

On one side, equilibrium comforts us. It explains why systems settle, why chaos fades, why wounds heal, why turbulence decays. It reassures us that not everything must be held together by force.

On the other side, equilibrium frightens us. It explains why no structure lasts forever, why control weakens, why life ends. It removes the fantasy of permanent escape.

Universal Energy Dynamics asks the reader to hold both truths without flinching.

Equilibrium is not the enemy of life. It is the reason life must be careful.

Life exists not by defeating equilibrium, but by orbiting it—never touching it fully, never escaping its pull. Growth happens because equilibrium is delayed. Meaning exists because resolution is incomplete. Motion continues because alignment has not yet finished.

To make this intuition precise, this chapter will frame equilibrium in three roles—not as metaphors, but as structural behaviors:

- as a ceiling, limiting how far systems can rise,
- as a magnet, drawing systems back after disturbance,
- and as a grave, where all unresolved difference eventually rests.

These are not poetic choices. They are unavoidable consequences of how delta resolves.

Once equilibrium is seen this way, fear recedes. Not because endings disappear, but because they finally make sense.

6.1 — Equilibrium as Attractor, Not Destination

Equilibrium is almost always described as a place—a state a system reaches when motion stops, forces cancel, or change ceases. This language is convenient, but it is deeply misleading. It turns equilibrium into an event, when in reality it is a continuous influence.

In Universal Energy Dynamics, equilibrium is not where systems go. It is what systems are always being pulled toward.

An attractor does not need to be reached to matter. Gravity shapes trajectories long before impact. A magnet influences iron filings without contact. In the same way, equilibrium shapes behavior even when alignment is never completed. Systems respond to equilibrium constantly, not episodically.

This explains why systems "relax" without instruction.

A stretched spring contracts. A hot object cools. A disturbed ecosystem stabilizes. A shocked mind slowly regains baseline.

Nothing is commanded to do so. No agency is required. These behaviors occur because equilibrium defines the cheapest available direction for delta resolution.

Crucially, equilibrium as attractor does not imply uniformity or stillness. A system can be dynamic, oscillatory, or even chaotic while still being governed by equilibrium tendencies. What changes is not the pull, but the path taken toward it.

This is why equilibrium can coexist with motion.

A river flows continuously yet moves downhill. A heart beats rhythmically yet tends toward balance. A market fluctuates yet reverts to constraints.

The system is active, but its bias is stable.

Seeing equilibrium as attractor also resolves a common confusion: why systems often approach equilibrium asymptotically, never quite arriving. They do not fail to reach equilibrium because of error. They fail because constraints keep delta alive—injecting energy, reshaping topology, or opening new pathways.

Equilibrium does not demand completion. It only defines direction.

This reframing is essential because it removes intention from nature. Systems are not trying to be balanced. They are simply following the structural gradient of least resistance. Where alignment is easy, it happens quickly. Where it is costly, it is delayed.

This is also why equilibrium is inescapable but not urgent.

A system may resist equilibrium for a long time if it can afford the cost. Life does this continuously. But resistance is always temporary, because affordability is always finite. Equilibrium does not rush; it waits.

Understanding equilibrium as attractor rather than destination prepares the reader for the next distinction: not all equilibria are equal. Some are local, provisional, and misleading. Others are global, final, and unavoidable. Confusing the two leads to false confidence—and to unnecessary shock when alignment eventually completes.

6.2 — The Ceiling: Why Systems Can Only Rise So Far

Once equilibrium is understood as an attractor, a second role becomes visible—one that humans often experience as frustration, injustice, or limitation. Equilibrium does not only pull systems back; it also sets an upper bound. It functions as a ceiling.

Every system that grows, accumulates, or intensifies does so by holding delta against equilibrium pressure. Growth is never free expansion; it is sustained imbalance. As long as the cost of holding that imbalance is affordable, the system can rise—becoming more complex, more organized, more energetic, or more influential.

But affordability is finite.

At some point, each additional unit of growth costs more delta to maintain than the system can reliably supply. When that happens, growth slows, plateaus, or reverses. This is not failure. It is structural accounting.

Equilibrium defines this ceiling quietly.

It does not announce limits in advance. It does not negotiate extensions. It simply increases the cost of further ascent.

This is why ceilings feel unfair. They are not moral judgments. They do not correspond to effort, intelligence, or desire. Two systems may work equally hard and reach different limits because their topology, constraints, and pathways differ. One can hold delta cheaply; the other cannot.

Physics shows this clearly. No material can be stressed indefinitely. No energy gradient can be amplified without loss. No signal can be strengthened without noise. The ceiling appears not because laws forbid growth, but because maintenance becomes too expensive.

The same structure applies beyond physics.

In biology, organisms grow until metabolic cost outweighs benefit. In cognition, learning accelerates until cognitive load saturates. In societies, expansion continues until coordination cost exceeds gain.

Ceilings are not imposed from outside. They emerge from within the system's own accounting.

This also explains why attempts to break ceilings by force often fail. Injecting more energy, more control, more pressure may temporarily push a system higher, but it also steepens the cost curve. Eventually, collapse becomes sharper, not avoided.

Equilibrium does not oppose ambition. It prices it.

Understanding ceilings changes how limits are interpreted. Instead of asking, *Why am I blocked?*, the structural question becomes, *What delta is becoming too expensive to hold in this configuration?* Sometimes the answer is to stop pushing. Sometimes it is to change topology. Sometimes it is to redirect growth elsewhere.

Ceilings are not enemies of progress. They are signals that a different form of alignment is required.

Recognizing equilibrium as a ceiling prepares the reader for the next clarification: many systems appear stable and successful not because they have reached a final state, but because they have settled into local equilibria—temporary plateaus mistaken for permanence.

6.3 — Local Equilibria: The Illusion of Permanence

Once ceilings are acknowledged, another subtle confusion reveals itself—one that shapes how humans misread stability, success, and truth. Systems often appear to settle, persist, and function smoothly for long periods. From inside those periods, it feels natural to assume that equilibrium has been reached. In reality, what has been reached is almost always local equilibrium, not final alignment.

A local equilibrium is a configuration where delta resolution is cheap enough to pause further change. Gradients still exist, imbalances still remain, but the system has found a temporary arrangement in which holding them does not immediately escalate cost. Nothing pushes urgently. Nothing collapses visibly. Everything feels "normal."

This feeling is deceptive.

Local equilibrium is not completion. It is postponement that works—for now.

Nature creates local equilibria everywhere because global resolution is rarely affordable in one step. Energy settles into basins. Structures lock into metastable forms. Habits stabilize behavior. Institutions harden into routines. These states are not errors; they are economical compromises between delta and equilibrium pressure.

The illusion arises because local equilibria mimic permanence.

They resist small disturbances. They self-correct after minor shocks. They reward repetition and consistency.

From within them, it feels as if the system has found its "true" form. Humans then attach meaning, identity, and certainty to what is actually a temporary balance.

This is why local equilibria are often mistaken for truth.

In physics, metastable states look solid until they suddenly are not. In biology, ecological balances persist until slow pressures accumulate. In psychology, coping mechanisms feel healthy until context shifts. In societies, stable orders appear timeless until they fracture rapidly.

Nothing changed suddenly. What changed is that the cost curve crossed a threshold.

Local equilibria fail not because something external attacks them, but because internal delta accumulates quietly while attention is focused on surface stability. Small inefficiencies, suppressed gradients, and ignored mismatches pile up. Eventually, holding the configuration becomes more expensive than abandoning it.

This is why collapse feels shocking.

From inside a local equilibrium, change appears unnecessary. Warnings sound exaggerated. Signals are dismissed as noise. The system feels strong precisely because it has not yet been tested beyond its comfort zone.

Universal Energy Dynamics removes the moral drama from this pattern.

Local equilibria are not lies. They are working arrangements.

But they are always conditional. They persist only as long as the underlying delta remains affordable. When affordability shifts—through accumulation, constraint change, or topology alteration—resolution resumes.

Understanding this prevents two opposite errors:

- romanticizing stability as final,
- and demonizing collapse as failure.

Local equilibria are neither. They are chapters, not conclusions.

Recognizing them as such prepares the reader for equilibrium's next role: not as illusion or limit, but as a magnet—the quiet force that draws systems back after disturbance, often mistaken for intelligence, intention, or healing.

6.4 — The Magnet: Why Systems Drift Back After Disturbance

When a system is disturbed—shocked, displaced, or stressed—there is a common human impulse to interpret its recovery as intelligence, intention, or even wisdom. We say a body "heals," a market "corrects," a mind "adapts," a society "learns." These words feel natural, but they quietly misattribute agency to what is, at root, a structural tendency.

Equilibrium behaves like a magnet.

Not because it pulls with force, but because it defines the lowest-cost neighborhood for delta resolution. When a system is pushed away from that neighborhood, the pathways that lead back become cheaper than those that lead further away. Return is not chosen; it is afforded.

This explains why recovery is so common.

A stretched material returns to shape. A heated object cools. A startled organism calms. A disrupted routine re-forms.

Nothing "decides" to return. The system simply follows the gradient of declining cost. This gradient is what we experience as resilience.

The magnetic role of equilibrium also explains why recovery often overshoots or oscillates. When a disturbance displaces a system beyond its usual basin, the return path may carry excess momentum. The system swings, corrects, and slowly damps out. This is not indecision; it is topology negotiating resolution.

Importantly, the magnet does not guarantee recovery.

If the disturbance reshapes constraints—damaging pathways, altering topology, or exhausting buffers—the old equilibrium may no longer be reachable. The system may settle into a new local equilibrium, or fail to stabilize at all. What looks like "failure to heal" is often successful alignment to a different basin.

This is why recovery is contextual.

The same shock produces healing in one system and collapse in another. The difference lies not in willpower or virtue, but in available pathways and remaining affordability. Systems with redundant pathways and flexible constraints are pulled back gently. Systems with rigid, brittle topology snap.

Understanding equilibrium as a magnet also dissolves a dangerous illusion: the belief that stability implies intention or goodness. A harmful system can be magnetically stable. An unjust structure can self-correct against minor disturbances while remaining globally misaligned. The magnet does not evaluate outcomes; it rewards cheap resolution, nothing more.

This insight is crucial beyond physics.

In human life, people often mistake emotional numbness for peace, routine for health, repetition for mastery—because these states are magnetically stable. Delta has been suppressed or rerouted cheaply, not resolved wisely. The magnet has done its job; meaning has not.

Equilibrium's magnetism is therefore comforting but morally blind. It explains why systems drift back to what they know, why reform is hard, and why genuine change often requires intentional disruption of stable basins.

Recognizing this prepares the reader for equilibrium's final and most misunderstood role. Beyond ceiling and magnet lies something quieter, unavoidable, and often resisted—not as threat, but as completion.

6.5 — The Grave: Full Resolution Without Drama

If equilibrium as ceiling humbles ambition, and equilibrium as magnet explains resilience, equilibrium as grave confronts the deepest human discomfort: the fact that all managed delay eventually ends.

This word—*grave*—is not used here for emotional impact. It is used for precision.

A grave is not violence. It is not failure. It is completion.

In Universal Energy Dynamics, global equilibrium is the state in which all remaining delta has resolved and no pathway exists—or is required—for further alignment. Nothing pushes. Nothing pulls. Nothing waits. Resolution is finished, not interrupted.

This is why global equilibrium feels frightening when misinterpreted. Humans tend to equate activity with value. When activity ceases, meaning seems to vanish. But this is a psychological projection, not a structural truth.

Nature does not mourn resolution. It does not celebrate persistence. It simply accounts.

Death, in this framework, is not collapse. It is not negation. It is the moment when a system can no longer afford to hold delta—and therefore does not. All the gradients that defined the system disperse. All the misalignments that gave it structure dissolve into alignment with the larger field.

What disappears is not energy, not history, not consequence. What disappears is local differentiation.

Life, then, is not the opposite of death. It is managed delay.

Living systems exist by continuously holding themselves away from global equilibrium through metabolism, repair, reproduction, and regulation. They do not defeat equilibrium; they rent time against it. Every breath, every heartbeat, every act of maintenance is a small payment to keep delta alive a little longer.

When those payments stop being affordable, resolution completes.

This framing removes drama without removing significance. Death is not meaningless—but its meaning is not moral or metaphysical. It is structural. A life matters not because it avoids equilibrium, but because of how it managed delta while it could—what pathways it opened, what redistributions it caused, what configurations it temporarily sustained.

Seen this way, death is not a punishment for living. It is the cost of having been differentiated at all.

This also explains why nothing survives global equilibrium—not stars, not species, not civilizations. Survival is always local, temporary, and conditional. Permanence belongs only to alignment, not to form.

Importantly, calling equilibrium a grave does not turn it into an enemy. Graves are part of landscapes. They mark completion, not erasure. They are where stories stop accumulating, not where they are undone.

Understanding equilibrium this way frees the reader from two extremes:

- romanticizing life as rebellion against nature,
- and fearing death as betrayal by nature.

Nature does neither. It allows delay where possible and completes resolution where unavoidable.

With this clarity, one final question remains—not about endings, but about how life manages to exist at all in the shadow of equilibrium. That question leads directly to the final section of the chapter.

6.6 — Living Near the Edge Without Falling In

If equilibrium is inevitable, and global equilibrium is complete resolution, the natural question arises: how does anything live at all? Why is the universe not already silent? Why do structures, organisms, minds, and societies persist for meaningful durations instead of collapsing immediately into alignment?

The answer is neither rebellion nor escape. It is proximity management.

Living systems survive by operating in a narrow, precarious band—close enough to equilibrium to remain stable, far enough to remain active. They neither rush toward full resolution nor drift into uncontrolled divergence. They hover, adjust, compensate, and recalibrate continuously. Life is not defiance of equilibrium; it is skillful navigation around it.

This band is narrow because the margins are unforgiving.

Move too far from equilibrium, and pathways overload. Delta becomes unmanageable. Chaos, breakdown, or exhaustion follows. Move too close to equilibrium, and gradients vanish. Function fades. Adaptation stops. Death approaches quietly.

What we call "health," "resilience," or "functionality" is simply successful operation within this viable corridor.

This is why life looks inefficient from a purely physical perspective. Enormous effort is spent on repair, regulation, feedback, and redundancy. These are not flaws. They are the price of staying near the edge without crossing it. Every regulatory loop is a small correction away from collapse. Every buffer is a delay mechanism.

The same logic extends beyond biology.

A mind remains alive by tolerating uncertainty without dissolving into anxiety or rigid certainty. A society survives by holding tension without letting it explode or flatten into uniformity. An organization functions by allowing disagreement without paralysis or fragmentation.

In each case, intelligence is not mastery over equilibrium, but sensitivity to distance from it.

This reframes intelligence itself.

Intelligence is not problem-solving in the abstract. It is the capacity to sense when delta is becoming unaffordable and to adjust topology before resolution turns destructive or terminal. It is anticipation of cost, not domination of outcome.

Crucially, no system can remain in this band indefinitely. Proximity management buys time; it does not cancel finality. But within that time, complexity, meaning, learning, and contribution are possible. The value of life lies not in permanence, but in what is done while delay is still affordable.

This removes the last trace of existential drama from equilibrium.

Living near the edge is not heroic. Falling in is not tragic. Both are structurally consistent outcomes.

What distinguishes living systems is not that they avoid equilibrium, but that they dance around it—sometimes gracefully, sometimes clumsily—until they no longer can.

With this, equilibrium has been fully reframed. It is not merely an attractor, not only a ceiling, not just a grave. It is the background condition that gives urgency, constraint, and significance to every act of holding, choosing, adapting, and letting go.

The Canonical Triad is now almost complete. Delta has been defined. Equilibrium has been clarified.

What remains is to understand how everything that happens in between is accounted for—how delays accumulate, how redistributions propagate, and why outcomes appear paradoxical only when viewed through moral or emotional lenses.

That accounting is consequence.

Next chapter will close the triad by showing that nothing escapes it—not because the universe is punitive, but because nothing that is permitted to happen is ever free.

From Equilibrium to Accounting

At this point, equilibrium should no longer feel abstract, threatening, or romantic. It should feel inevitable, neutral, and clarifying.

Nothing in this chapter asked the reader to fear endings or glorify persistence. Instead, equilibrium was repositioned as the quiet reference against which all living, moving, and changing systems are measured. It sets ceilings without judgment, pulls systems back without intention, and completes resolution without cruelty.

With this framing, a crucial illusion dissolves.

Systems do not collapse because they fail morally. They do not survive because they deserve to. They persist, adapt, and end because delta becomes affordable or unaffordable under changing constraints.

This realization does something subtle but decisive: it removes narrative bias from explanation. Growth is no longer assumed to be good. Stability is no longer assumed to be correct. Endings are no longer assumed to be mistakes. All three become structural phases of the same accounting process.

And that word—*accounting*—now becomes unavoidable.

If delta enables action, and equilibrium defines direction, then everything that happens in between must be tracked.

Not morally. Not symbolically. Not by reward or punishment.

But structurally.

Every delay has a cost. Every redistribution leaves traces. Every alignment shifts burden elsewhere.

Equilibrium does not erase history. It closes it. What determines how history unfolds before closure is not intention or virtue, but how consequence propagates through time, topology, and constraint.

This is where many human confusions arise.

We feel consequences late and call them unfair. We experience redistribution indirectly and call it paradox. We inherit accumulated delta and call it fate.

These misinterpretations happen because consequence has been treated as judgment rather than mechanism.

The Canonical Triad now requires its final articulation.

Delta explains *why* change is possible. Equilibrium explains *where* change tends to go. Consequence explains how everything is paid for along the way.

Next chapter will complete this triad by removing the last remaining illusion—that consequence evaluates. It will show that consequence does not punish or reward; it redistributes, transforms, and diffuses whatever was permitted to occur.

Once this is understood, the framework closes cleanly.

Nothing that follows in the book will add new fundamentals. Everything that follows will simply trace consequences more carefully.

With that orientation fixed, the reader is ready to move from tendency to bookkeeping—from where systems drift, to how their paths are never forgotten.

Equilibrium does not judge when a system should end—it waits until nothing resists alignment.

Chapter 7 — Consequence: Redistribution, Not Punishment

Why We Keep Expecting Judgment

There is a quiet expectation most humans carry, often without realizing it: that outcomes should respond to intent. That good effort should be rewarded, that harmful action should recoil back toward the actor, that the universe should somehow keep score in a way that *feels* fair.

This expectation is deeply human—and structurally misplaced.

We inherit it from lived experience, not from nature. In human relationships, judgment exists. Parents correct. Societies punish. Institutions reward. These mechanisms are necessary for coordination and survival, but they quietly train the mind to project moral evaluation onto reality itself. Over time, consequence is mistaken for verdict.

Physics never promised judgment. Nature never agreed to fairness. Yet we keep looking for both.

The confusion intensifies because of delay. Consequences rarely appear where or when actions occur. Effects surface elsewhere, later, and in altered forms. When outcomes do not align with expectation, the mind reaches for explanation—and morality is the fastest one available. If good did not lead to good, something must be wrong. If harm did not rebound, something must be missing.

This is how punishment language enters explanation.

But consequence does not evaluate. It does not aim. It does not return.

It moves.

What we call consequence is simply the continuation of redistribution once delta has been permitted. Energy shifts. Burden relocates. Structure transforms. Pressure diffuses. Nothing waits for intention to be checked or character to be assessed. The universe does not pause to decide what should happen next.

This mismatch—between emotional time and structural time—is the root of what people later call paradox.

Emotion expects immediacy. Structure operates across pathways.

Emotion expects symmetry. Structure allows asymmetry.

Emotion expects closure. Structure tracks continuity.

When these expectations collide, confusion arises. Outcomes feel unfair. Success looks undeserved. Failure feels personal. Entire belief systems grow around trying to explain what is, at base, a mechanical process misread through a moral lens.

Universal Energy Dynamics does not dismiss this human experience. It explains it.

Humans experience consequence locally, but consequence propagates globally. Humans feel outcomes episodically, but consequence unfolds continuously. Humans search for who is responsible, but consequence tracks what was allowed and where the cost could move.

This chapter makes a single correction—calm, firm, and final:

Consequence is not punishment or reward. It is redistribution, transformation, and diffusion of what was permitted to occur.

Once this is understood, many long-standing confusions dissolve. The so-called "karma paradox" stops being paradoxical. Delayed outcomes stop feeling mysterious. Responsibility becomes structural rather than emotional.

This is not a demotion of meaning. It is a clarification of mechanism.

With that clarity, the final element of the Canonical Triad can be stated cleanly—without belief, without fear, and without judgment.

7.1 — What Consequence Actually Is in UED

Once judgment is removed, consequence becomes surprisingly simple—and far more powerful.

In Universal Energy Dynamics, consequence is the structural continuation of what was permitted, expressed through three inseparable processes: *redistribution, transformation, and diffusion.*

Nothing mystical is hiding here. This is bookkeeping, not belief.

When delta is allowed—held, amplified, redirected, or delayed—it does not disappear after action occurs. It must go somewhere. Consequence is the name we give to where it goes, how it changes form, and how it spreads across pathways.

Redistribution: Cost Never Stays Put

The first property of consequence is redistribution. Cost rarely remains at the site of action. Energy disperses. Stress migrates. Burden shifts. Benefits concentrate elsewhere. This is why outcomes so often feel misaligned with actors.

A factory pollutes a river; illness appears downstream. A decision creates profit; instability surfaces years later. An action relieves pressure locally; it builds pressure globally.

Nothing "returns" to the origin in a moral sense. Effects propagate according to topology. Consequence does not aim back—it flows outward along whatever pathways are cheapest.

Transformation: Effects Change Form

The second property is transformation. Consequence almost never resembles its initiating action. Heat becomes motion. Motion becomes wear. Wear becomes failure. Psychological stress becomes physiological illness. Economic shortcuts become social fragility.

This is why people say, "I didn't do this." Structurally, that claim is often irrelevant.

Consequence is not identity-preserving. It does not carry labels like "my action" or "your fault." It changes medium as it moves. Expecting symmetry between cause and effect is a category error introduced by linear thinking.

Diffusion: No Target, No Endpoint

The third property is diffusion. Consequence spreads. It dilutes, accumulates, branches, and delays. It does not select a recipient. It does not stop when attention fades. It continues until equilibrium pressure absorbs it.

This is why consequence often feels invisible—until it doesn't.

Diffusion explains delayed crises, sudden collapses, and collective impact. It explains why no single event "caused" a disaster, and why many small permissions can outweigh one large action.

The Critical Correction

Under UED, consequence is not reaction. It is accounting continuity.

It does not respond to behavior. It extends it.

It does not evaluate intent. It tracks permission.

It does not punish or reward. It redistributes unresolved delta until alignment completes.

This reframing dissolves a deep error in how consequence has been discussed across philosophy, religion, psychology, and even science. When consequence is treated as reaction, it invites judgment. When it is treated as accounting, it invites understanding.

Understanding consequence this way immediately clarifies why outcomes feel disconnected, delayed, or unfair—without invoking fate, randomness, or moral failure.

It also prepares the reader for the next refinement: redistribution rarely respects boundaries. Costs cross systems, agents, and generations. To grasp consequence fully, we must examine where the cost actually goes—and why feeling unaffected does not mean being uninvolved.

7.2 — Redistribution: Where the Cost Really Goes

One of the most persistent illusions humans carry is that consequences stay close to their source. We expect effects to circle back—to the same person, the same place, the same moment. When that does not happen, we feel either relieved or betrayed. Both reactions come from the same misunderstanding: that consequence respects boundaries.

It does not.

Redistribution is the primary reason consequence feels elusive. When delta is permitted, the cost rarely remains where the permission occurred. It moves—across systems, across agents, across time. What looks like escape is almost always export.

In physical systems, this is obvious. Heat dissipates into surroundings. Stress travels through materials. Energy injected at one point emerges elsewhere as vibration, noise, or wear. No engineer expects cost to stay localized; structures are designed precisely because redistribution is inevitable.

In human systems, the same rule applies—but is easier to deny.

A decision reduces pressure in one department and creates burnout in another. A shortcut increases profit here and erodes trust elsewhere. Comfort in the present shifts instability into the future.

The system did not avoid cost. It relocated it.

This relocation is not malicious or moral. It is structural. Consequence follows pathways of least resistance, not lines of responsibility. Where redistribution goes depends on topology: who is connected, who absorbs load, who lacks voice, who has buffers, who does not.

This is why feeling unaffected is not evidence of innocence.

An individual may not feel the consequence of their action because the system around them absorbed it. A generation may appear to benefit because the cost has been deferred. A group may prosper because another group quietly carries the burden. None of this contradicts consequence; it confirms redistribution.

This also explains why systems often appear stable while accumulating fragility. Redistribution can hide cost remarkably well—until buffers saturate. When they do, consequence does not suddenly appear; it reappears where it was always going.

Understanding redistribution corrects a subtle but damaging narrative: that responsibility belongs only where discomfort is felt. In reality, responsibility belongs wherever permission was granted, regardless of where the cost later manifests.

UED makes no ethical claim here. It makes a diagnostic one.

If a system benefits without paying locally, the cost has moved.

If an outcome seems unconnected to prior actions, the pathway has been obscured.

If consequences appear to strike "the wrong place," that place was simply cheaper to absorb them.

Redistribution is not injustice. It is how systems survive imbalance—temporarily.

Recognizing this prepares the reader for the next refinement. Redistribution alone does not explain why outcomes so often look unfamiliar, distorted, or unrecognizable. For that, we must understand the second property of consequence: transformation—the reason effects rarely resemble their origins.

7.3 — Transformation: Why Outcomes Don't Resemble Actions

Even when redistribution is accepted, another confusion remains stubborn: why do consequences look so different from what caused them? Why does a small action lead to a large, unfamiliar outcome? Why does harm return in forms no one recognizes? Why do people insist, often sincerely, *"This isn't what I did"*?

Under Universal Energy Dynamics, this confusion dissolves once consequence is understood to transform as it propagates.

Consequence does not preserve identity. It preserves accounting.

When delta moves through a system, it rarely remains in the same form. Energy converts. Stress changes medium. Information mutates. Social pressure becomes institutional inertia. What began as one kind of imbalance reappears as another, shaped by the constraints it encounters along the way.

This is not distortion. It is translation across domains.

Heat becomes expansion. Expansion becomes fracture. Fracture becomes failure.

A lie becomes mistrust. Mistrust becomes withdrawal. Withdrawal becomes collapse.

At no point did consequence "change its mind." It simply adapted to the language of the pathway it traveled through.

This is why expecting symmetry between action and outcome is structurally naïve. Symmetry belongs to closed, idealized models—not to real systems layered with constraints. In real systems, consequence changes form precisely because systems are heterogeneous.

This also explains a frequent psychological defense: denial through dissimilarity.

People say:

- "I only made a small choice."
- "I never intended this."
- "That outcome doesn't look like my action."

Structurally, intention and resemblance are irrelevant. Consequence tracks permission, not narrative. If delta was allowed into a system, it will resolve through whatever forms the system offers. The final appearance of the outcome reflects the system's structure, not the actor's self-image.

Transformation also explains why blame is such a poor explanatory tool.

Blame assumes identity preservation: that what comes out should resemble what went in. But consequence erases labels as it moves. It does not carry names, roles, or intentions. It carries cost.

This is why complex systems fail dramatically after long periods of normalcy. Small misalignments transform repeatedly—converted, diffused, rerouted—until they accumulate into a form that finally becomes visible. When visibility arrives, it feels disconnected from its origin because the origin has been translated multiple times.

UED treats this not as mystery, but as inevitability.

Whenever systems are layered, consequence will be layered. Whenever pathways cross domains, consequence will change language.

Understanding transformation removes the need for dramatic explanations—fate, corruption, bad luck, or cosmic irony. Outcomes look strange not because reality is cruel, but because accounting passed through many constraint spaces before becoming visible.

This insight prepares the reader for the next source of confusion: not just *what* consequence becomes, but *when* it appears. Delay and diffusion—rather than immediacy—are the final reasons consequence is so often misread.

7.4 — Diffusion and Delay: The Source of Confusion

Even after redistribution and transformation are understood, consequence still feels puzzling for one stubborn reason: it rarely appears on time. Effects lag. Outcomes surface late. Costs emerge long after attention has moved on. When they finally arrive, they feel disconnected, excessive, or unfair.

This is not because consequence is slow. It is because consequence diffuses.

Diffusion means that once delta is permitted, its resolution spreads across pathways rather than striking at a single point. Instead of collapsing immediately into a visible outcome, consequence disperses—thinning out, branching, and settling into places that can absorb it cheaply. Visibility is delayed not by intention, but by distribution efficiency.

In physical systems, this is ordinary. Heat diffuses gradually. Stress redistributes across materials. Pollutants disperse before concentrating. No one expects instant collapse simply because imbalance exists. Delay is built into the system's geometry.

In human systems, the same process is often misread as escape.

A harmful practice persists because its costs are diffuse. A risky decision feels safe because its consequences are delayed. A system appears healthy because stress is spread thinly—until it isn't.

Delay is not mercy. It is storage.

Diffusion allows systems to survive longer than they should under naive models. But it also ensures that when resolution finally becomes visible, it does so abruptly. What appears as a sudden crisis is usually the exhaustion of diffusion capacity. Buffers fill. Pathways saturate. Thresholds are crossed.

This is why consequences feel surprising.

They were not absent. They were distributed below perception.

Delay also explains why timing feels unjust. Humans experience time locally and emotionally. We remember actions vividly, but forget structures slowly accumulating cost. When consequence surfaces long after the initiating permission, it feels unrelated. The mind protests: *"Why now?"*

Structurally, the answer is simple: Now is when diffusion could no longer hide it.

This is also why small, repeated permissions often outweigh singular dramatic actions. Diffusion handles spikes easily. It struggles with accumulation. Many minor deltas, allowed consistently, build invisible pressure that eventually overwhelms pathways.

UED removes mysticism from delay.

There is no cosmic patience. There is no moral timing. There is only topology managing visibility.

When delay is understood structurally, surprise disappears. Outcomes stop feeling random or targeted. They become legible as the moment when redistribution, transformation, and diffusion have finished negotiating affordability.

This prepares the ground for the most emotionally charged confusion of all—the so-called karma paradox. Once diffusion and delay are understood, the paradox collapses into a misalignment between psychological expectation and structural reality.

7.5 — The "Karma Paradox" Explained Cleanly

What people often call the *karma paradox* arises from a simple but powerful mismatch: human moral intuition operates locally and emotionally, while consequence operates structurally and globally.

The paradox is usually framed like this: *Why do unethical actions sometimes succeed? Why do good intentions sometimes lead to suffering? Why does harm not "return" to those who caused it—at least not visibly?*

From a UED perspective, there is no paradox at all. There is only misaligned expectation.

The expectation is that consequence should behave like judgment—fast, targeted, proportional, and personal. When reality fails to conform to this template, the mind assumes something is broken: justice, order, or meaning itself.

But consequence was never designed to satisfy moral symmetry.

Structurally, what appears as "success" in unethical behavior is usually one of three things:

1. **Delayed accounting:** The cost has not disappeared; it has not yet become visible. Diffusion is still affordable. Pathways are still absorbing pressure. What looks like success is simply unpaid balance, not profit.
2. **Exported cost:** The system has redirected burden elsewhere—onto others, the environment, the future, or less visible layers. The actor feels untouched not because consequence failed, but because someone else is carrying it.
3. **Local equilibrium masking global misalignment:** The system has settled into a stable local configuration that rewards certain behaviors while quietly accumulating fragility. From inside that basin, everything looks justified. From outside—or later—the cost becomes obvious.

In all three cases, nothing contradicts consequence. What contradicts expectation is the belief that outcomes must be immediate, personal, and symmetrical.

The same logic explains why ethical action can feel unrewarded or even punished. Acting ethically often means absorbing cost locally instead of exporting it. The system stabilizes at a higher structural integrity but offers less immediate relief or advantage. The benefit is distributed, delayed, or invisible—making it feel like loss in the short term.

From a psychological lens, this feels unfair. From a structural lens, it is coherent.

The so-called karma paradox exists only when:

- consequence is mistaken for judgment,
- delay is mistaken for absence,
- redistribution is mistaken for escape.

Once these errors are corrected, the paradox collapses.

There is no mechanism that ensures "good people win" or "bad people lose." There is a mechanism that ensures all permitted delta is eventually accounted for, regardless of how it is labeled. Ethics influences *how* cost is distributed, not *whether* it exists.

This is why moral language feels emotionally true but structurally incomplete. It captures lived experience without explaining underlying movement. UED does not dismiss moral intuition—it grounds it. Ethics matters because it shapes topology and pathways, not because it invokes cosmic reward.

Understanding this clears a major blockage in human reasoning. It allows responsibility without superstition, accountability without fear, and realism without cynicism.

With the paradox dissolved, the final piece becomes visible: consequence does not occur as isolated reactions. It unfolds as chains—extended sequences of redistribution that outlive intentions, actors, and even systems themselves.

7.6 — Consequence Chains: Nothing Stops at One Event

One of the last illusions to dissolve is the idea that consequence is a single outcome—a point where the story ends. We look for closure: *this happened because of that*. But in Universal Energy Dynamics, consequence never resolves as a dot. It resolves as a chain.

Every permission initiates a sequence, not a response.

Once delta is allowed into a system, redistribution, transformation, and diffusion do not occur once and stop. Each resolution creates new local deltas, which must then be resolved again. What looks like an outcome is usually just a midpoint in a longer accounting process.

This is why consequence routinely outlives:

- the original action,
- the original actor,
- the original context.

A decision made today reshapes constraints tomorrow. Those constraints alter behavior later. That behavior redistributes cost further still.

No single link contains the whole story.

This is what a consequence chain is: a continuity of accounting across time, scale, and domain. Chains can be short or long, visible or hidden, narrow or sprawling. But they never terminate simply because attention moves on.

This has an important implication for responsibility.

Responsibility is not about blame. It is about participation in chains.

Whether an actor intended harm, foresaw outcomes, or even remained present is structurally secondary. What matters is that permission was granted somewhere in the chain, and the chain continued. Saying *"I didn't mean this"* may be emotionally valid, but it does not interrupt accounting.

This is why systems inherit consequences.

Children inherit environmental decisions. Institutions inherit cultural shortcuts. Societies inherit historical redistributions.

No one chose the starting point, but everyone participates in continuation. Consequence chains do not reset with generations, leadership changes, or ideological shifts. They only reset when alignment completes or topology changes enough to reroute them.

Understanding consequence chains also explains why intervention is difficult. Cutting one link rarely dissolves the chain. Suppressing a symptom often pushes cost elsewhere. Real change requires altering pathways, not reacting to endpoints.

This reframing removes despair as well.

If consequence is a chain, then intervention is also chain-based. Small changes in topology—opening pathways, redistributing burden earlier, absorbing cost locally—can reshape entire sequences downstream. Responsibility becomes structural agency, not moral weight.

Nothing about this is punitive. Nothing is personal.

Consequence chains are simply how unresolved delta remembers itself across time.

Once this is seen, the Canonical Triad is complete in practice, not just in theory. Delta enables action. Equilibrium defines direction. Consequence ensures nothing that was permitted is forgotten.

The reader is now equipped with a framework that does not ask *who is at fault*, but *what chain is still active.*

With that understanding, the book is ready to move forward—from foundational accounting to secondary descriptions like motion, force, and time—without ever losing structural clarity.

The Canonical Triad Sealed

With consequence understood as redistribution rather than judgment, the Canonical Triad finally closes—cleanly, without residue.

Nothing mystical was added. Nothing familiar was denied.

Delta explained why action was possible. Equilibrium explained where action tends to go. Consequence explains how everything is paid for along the way.

Together, they remove the last illusion that reality responds to intention, virtue, or narrative symmetry. What reality responds to is permission under constraint. Once something is allowed, it enters accounting. Once it enters accounting, it cannot be erased—only transformed, redistributed, or diffused until alignment completes.

This is the point where explanation becomes irreversible.

After this chapter, it is no longer coherent to ask:

- *Why did this happen to me?*
- *Who deserves this outcome?*
- *Why didn't consequence strike where it should have?*

Those questions belong to moral storytelling, not to structural understanding.

The correct questions now arise automatically:

- *Where did the cost move?*
- *Which pathway carried it?*
- *What chain is still active—and where can it be altered?*

This is not a colder view of reality. It is a clearer one.

Clarity does not remove ethics; it grounds them. Ethics no longer rely on cosmic enforcement or emotional expectation. They emerge as choices about where cost is absorbed, delayed, or exported. Morality becomes topology-aware instead of hope-driven.

With the Canonical Triad sealed, Universal Energy Dynamics stops explaining itself and starts reading the world.

From this point onward, the book no longer operates at the level of fundamentals. Everything that follows—motion, force, time, geometry, waves, life, mind, society—will be shown as secondary descriptions, useful languages layered on top of the same accounting logic.

Nothing new will be introduced. Everything familiar will be repositioned.

The reader now stands on stable ground.

Not because the universe has become predictable, but because it has become legible.

What comes next is not deeper philosophy. It is application.

The triad does not repeat. It carries.

Nothing that is permitted to happen is ever free—it is simply paid for elsewhere, later, or differently.

ATT Model

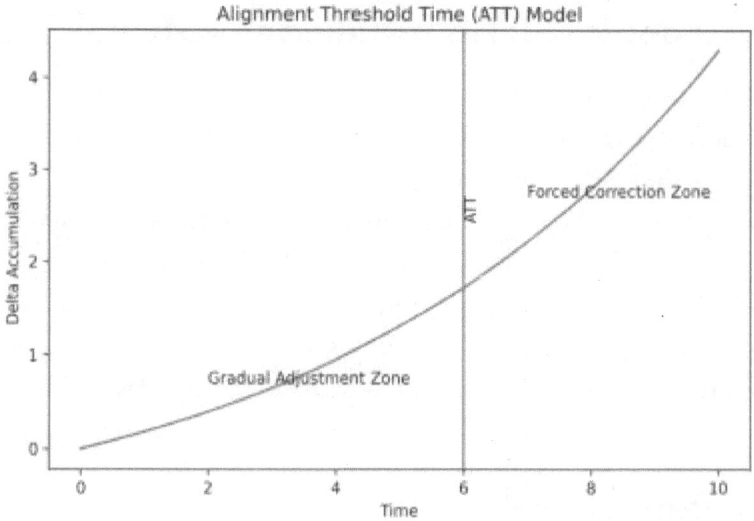

Before ATT: Gradual adjustments can resolve delta.
After ATT: Resolution occurs through abrupt structural correction.

Figure 2 — Alignment Threshold Time (ATT) and System Correction

In Universal Energy Dynamics, unresolved delta accumulates within constraints as systems operate over time. When alignment mechanisms respond early, resolution occurs gradually and stability is maintained. However, if correction is delayed beyond the system's Alignment Threshold Time (ATT), gradual adjustment becomes impossible. The accumulated pressure is then released through abrupt structural correction, often appearing as sudden failure, collapse, or rapid transformation.

Deliverable of Part II

The Canonical Triad Installed as a Mental Engine

By the end of this part, Universal Energy Dynamics should no longer feel like a framework the reader *understands*. It should feel like a framework the reader thinks with.

The Canonical Triad collapses into a simple, automatic mental engine:

Delta → Equilibrium pressure → Consequence

This is not a slogan. It is a cognitive reflex.

Whenever the reader encounters any phenomenon—physical, biological, psychological, social, or systemic—the mind should now run this sequence instinctively:

- **Delta:** *What difference, misalignment, or gradient is present? What is being held apart, and why has it not resolved yet?*
- **Equilibrium pressure:** *In which direction does alignment naturally pull? What ceilings, magnets, or completion points are shaping behavior?*
- **Consequence:** *How is the cost of delay being redistributed, transformed, or diffused? What chains are still active, and where is the burden moving?*

Once this engine is installed, explanation becomes economical.

There is no need to search for hidden forces. No need to invent moral narratives. No need to appeal to randomness, fate, or exception.

Every event becomes legible as a case of delayed resolution under constraint.

Crucially, this triad is not domain-specific. It does not belong to physics any more than it belongs to psychology or geopolitics. The words change, the media change, the scales change—but the engine does not. Whether the reader is observing a pressure gradient, a failing institution, a personal conflict, or a historical collapse, the same three-step logic applies without modification.

This is the test of completion.

If the reader can no longer look at a situation without silently asking:

- *What delta is being held here?*
- *What equilibrium pressure is acting?*
- *Where is the consequence moving?*

then this part has succeeded.

At that point, Universal Energy Dynamics stops being a theory to remember and becomes a tool that cannot be put down. It does not compete with existing models; it sits beneath them, quietly organizing them. Equations, narratives, and interpretations now appear as surface languages riding on top of the same accounting flow.

From here on, the book does not need to restate fundamentals. The engine is running.

What follows will not add principles. It will show how familiar concepts—motion, force, time, geometry, waves, life, and mind—emerge naturally once this engine is allowed to operate.

This part does not conclude an argument. It installs orientation.

And once installed, it does not switch off.

Interlude — When Duration Replaces Direction

Up to this point, the book has carefully avoided one word that dominates almost every explanation humans reach for instinctively: time.

Not because time is unimportant—but because it is usually introduced too early, before the reader has the structural equipment to see it clearly.

Most theories begin with time as a backdrop: a river flowing forward, carrying events along its current. Motion happens *in* time. Change unfolds *over* time. Cause precedes effect *in* time. This feels natural because it mirrors how experience is narrated.

But notice something subtle.

Nothing we have established so far required time to exist as a flowing entity.

Delta existed as misalignment. Equilibrium existed as tendency. Consequence existed as accounting.

All three functioned without invoking a temporal river. Resolution happened. Delay accumulated. Chains extended. Yet no independent "time substance" was needed to explain any of it.

This interlude marks that realization explicitly.

What we experience as time has quietly been doing a different job all along. It has been acting as a label for duration of unresolved delta, not as a medium through which reality moves.

We say something "takes time" when resolution is delayed. We say something is "fast" when pathways are open. We say something is "slow" when topology resists alignment.

None of these require time to flow. They require resolution to persist.

Clocks, calendars, and equations then step in—not to reveal time's essence, but to count repetition. A pendulum swings. An atom oscillates. A cycle repeats. By counting repeats, we construct predictability. Predictability feels like time.

But repetition is not duration itself. It is a proxy.

This distinction matters because it explains why time behaves so differently across domains. Why it feels smooth in classical mechanics, fragmented in quantum systems, stretched in cosmology, and elastic in human experience. The mistake was never mathematical. It was ontological.

We treated a measurement convenience as a fundamental dimension.

The next part will not abolish time. It will demote it gently—from river to residue. From background to outcome. From cause to consequence.

Time will reappear, but only after something else is allowed to take its rightful place first: resolution duration.

Once that substitution is made, paradox dissolves quietly.

Quantum time stops looking broken. Relativistic dilation stops looking strange. Psychological time stops looking subjective.

All of them become expressions of the same structural truth: how long alignment takes when repetition is fragile, pathways are constrained, and thresholds must be crossed.

This interlude is the pause before that shift.

Nothing new is being added in the next part. Only one illusion is being removed.

Time does not carry reality forward. Reality resolves—and we call the delay *time*.

Part III — Time, Motion, And ATT

This part does not introduce time as a new entity. It repositions it.

After the Canonical Triad is installed, time can no longer be treated as a river in which events float. Nothing in delta, equilibrium, or consequence required a flowing background. What required explanation was not *when* things happen, but how long resolution remains incomplete.

This part makes that correction explicit.

Time is reframed as resolution duration—the persistence of delta under constraint. Motion is reframed as visible resolution along pathways. Predictability emerges not because time flows smoothly, but because repetition remains stable enough to be counted.

When repetition breaks, time feels broken. When repetition stretches, time stretches. When thresholds gate resolution, time jumps.

This immediately clarifies long-standing confusions.

Quantum time feels discontinuous because resolution dominates over repetition.

Cosmic time dilates because propagation speed limits coherence across scale.

Human time distorts because attention, stress, and meaning reshape pathways.

Nothing special is happening to time itself. What is changing is how reliably delta can resolve in repeatable ways.

This part also introduces **Alignment Threshold Time (ATT)**—not as a new law, but as a structural insight. Resolution does not proceed smoothly everywhere. It waits, accumulates, and then releases once thresholds are crossed. ATT names the duration required for alignment to become *affordable*.

By the end of this part, the reader should no longer ask:

- *Why does time behave differently here?*
- *Why does motion look continuous in one domain and discrete in another?*

They should instead ask:

- *What is resolving here?*
- *How repeatable is the pathway?*
- *What threshold gates alignment?*

Time, motion, and causality then collapse into a single, legible picture—not mystical, not counterintuitive, but structurally inevitable.

From here on, the book no longer debates what time *is*. It shows why time appears the way it does wherever resolution occurs.

This part is where paradox finally runs out of room.

Chapter 8 — Time Is Not a River: Time as Resolution Duration

Why Time Always Felt Slippery

Time has always been the most familiar mystery. We speak about it constantly, measure it obsessively, and yet struggle to say what it actually *is*. It feels like something that passes, something we move through, something that carries moments away whether we want it to or not. The river metaphor comes naturally because memory itself is narrative: one thing after another, scenes arranged in sequence, a sense of before and after stitched together into a story.

But notice something important.

No one has ever *observed* time moving.

What we observe are changes. We observe processes completing. We observe waiting, decay, growth, recovery, anticipation. The feeling of "time passing" arises not from witnessing a flow, but from remaining inside an unresolved situation while resolution is delayed. The mind fills that delay with sensation, emotion, and expectation—and names the experience *time.*

This is why time feels elastic.

When nothing urgent is unresolved, time seems to vanish. Hours disappear without trace. When something matters deeply and cannot yet resolve, time stretches painfully. Minutes feel heavy. Seconds feel loud.

If time were a river flowing at a fixed rate, these distortions would make no sense. But if time is the experience of unresolved delta, they become obvious.

The river image survives because it is convenient, not because it is accurate. It allows us to place events on a line, to coordinate action, to build schedules and equations. It works operationally. But convenience slowly hardened into ontology. What began as a counting tool was mistaken for a substance.

Physics inherited this mistake early.

To describe motion, we parameterized change. To parameterize change, we introduced time. To make time manageable, we imagined it as uniform flow. The equations worked, so the metaphor went unquestioned. Over centuries, time stopped being a label and started being treated as a thing.

Yet the discomfort never disappeared.

Why does time feel different under stress? Why does it fracture at microscopic scales? Why does it stretch at cosmic scales? Why does it collapse into "now" during crisis or insight?

These are not failures of mathematics. They are signals of a misplaced assumption.

What if time is not what moves things forward? What if time is what we call the duration for which things fail to finish resolving?

This chapter makes that substitution—quietly, without drama.

Nothing important is taken away. Clocks still tick. Equations still work. Cause still precedes effect. What changes is the *direction of explanation*. Time stops being the container of events and becomes an outcome of unresolved alignment.

Once that inversion is allowed, time stops feeling slippery. It stops demanding metaphysical defenses. It becomes legible, practical, and consistent with everything already established in Universal Energy Dynamics.

Time does not flow past us. We remain inside resolution—and measure how long it takes.

That realization is where this chapter begins.

8.1 — Time as Duration of Unresolved Delta

Once the river metaphor is set aside, time can be defined with precision—without mysticism and without loss of explanatory power.

In Universal Energy Dynamics, time is the duration for which delta remains unresolved under existing constraints.

Nothing more is required.

Time does not exist independently and then host events. Events exist as states of misalignment, and time appears because those states persist. Where there is no delta, there is no duration to register. Where alignment completes instantly, time collapses to zero—not conceptually, but experientially.

This immediately clarifies a long-standing confusion.

We do not experience time *passing*. We experience waiting.

Waiting for heat to dissipate. Waiting for a reaction to complete. Waiting for clarity, relief, arrival, recovery.

All waiting is the same phenomenon: delta has not yet resolved, and the system remains inside that delay. The longer the delay, the longer the experienced duration.

This also explains why time feels inseparable from effort and cost.

Holding delta is expensive. The longer it is held, the more energy, attention, or structure is consumed.

A task that drags on feels "long" because resolution is being resisted. A process that flows smoothly feels "short" because delta resolves quickly. Duration tracks resistance to alignment, not distance along a timeline.

Importantly, duration is always local.

Different deltas resolve at different rates depending on topology and pathways. A chemical reaction may complete in microseconds while a social conflict persists for decades. The difference is not because they exist in different times, but because their resolution constraints differ.

This dissolves the illusion of a universal clock governing all processes equally. What we call "global time" is simply a coordination convention, built by synchronizing repeatable processes. Structurally, each system carries its own durations, determined by how cheaply or expensively delta can resolve.

This also explains why time can feel absent.

In states of full engagement, flow, or absorption, delta resolves continuously without obstruction. There is no waiting. No backlog. No resistance. As a result, duration leaves no psychological trace. Hours vanish—not because time sped up, but because nothing was held unresolved long enough to be noticed.

Conversely, in anxiety, pain, or anticipation, delta is intense and pathways are constrained. Resolution stalls. The system remains suspended between states. Duration becomes oppressive. Time seems to slow—again, not because time itself changed, but because resolution did not proceed.

From this perspective, time has no independent direction. There is no arrow pushing events forward. What exists instead is a gradient of completion—states moving from unresolved to resolved, whenever constraints allow.

Time, then, is not what causes change. Change causes what we call time.

This definition also preserves causality without invoking flow. Cause precedes effect not because time moves forward, but because resolution must occur in sequence. One misalignment must be addressed before another becomes accessible. Ordering emerges from dependency, not from temporal current.

By grounding time in unresolved delta, UED removes the need for speculative dimensions or metaphysical rivers. Time becomes a measurable residue of delay, not a fundamental substance.

With this foundation in place, the next question becomes inevitable: if time is duration, how did we ever come to measure it so precisely? The answer lies not in time itself, but in repetition—the quiet mechanism that creates predictability and gives time its familiar numerical form.

8.2 — Why Motion Does Not Require Time

Once time is understood as duration of unresolved delta, a long-standing assumption quietly dissolves: the idea that motion needs time in order to occur.

It does not.

What motion requires is resolution along a pathway.

Motion is not something that happens *in* time. Motion is what we see when delta resolves spatially, structurally, or configurationally. A particle moves because a gradient exists and a pathway allows redistribution. A body falls because alignment is cheaper downward than upward. A signal propagates because coherence must be maintained across space. In none of these cases is a flowing temporal medium doing the work.

Time only appears afterward—as a label for how long the resolution took.

This inversion is subtle but decisive.

Traditional explanations say: Motion occurs because time passes and position changes with respect to it.

UED says: Position changes because delta resolves; time is the record of that resolution's duration.

Sequencing does not require a river of time. It requires dependency. One configuration must exist before another can form. One alignment must complete before the next becomes accessible. Ordering arises from structural necessity, not from temporal flow.

This is why causality survives intact even when time is demoted.

Cause precedes effect not because time enforces direction, but because resolution has prerequisites. A door must open before one can walk through it. A gradient must exist before flow occurs. A boundary must collapse before redistribution follows. These are logical and structural orders, not temporal commands.

Consider a simple example.

A ball rolls downhill.

We usually say: *as time passes, the ball moves*. Structurally, what happens is simpler: a height difference exists, a pathway is open, and alignment proceeds. The motion is the visible trace of resolution. If resolution were instantaneous, the ball would already be at the bottom, and there would be no motion to narrate—only before and after.

Time did not cause the ball to move. Time appeared because the movement was not instantaneous.

This distinction becomes even clearer when motion is constrained.

When pathways narrow, motion slows—not because time stretches, but because resolution becomes more expensive. When pathways break, motion stops—not because time freezes, but because resolution is blocked. When pathways multiply, motion accelerates—not because time speeds up, but because alignment becomes cheaper.

The same logic applies beyond physics.

A thought progresses not because time advances, but because insight resolves uncertainty step by step. A social change unfolds not because decades pass, but because resistance gradually yields. A healing process continues not because time heals, but because repair pathways remain open.

In every case, motion is secondary. It is a symptom, not a driver.

This reframing also explains why motion can feel continuous or discrete depending on context. Where resolution is smooth and pathways are stable, motion appears continuous. Where resolution jumps between thresholds, motion appears stepwise. Time is not fragmenting; resolution is.

By removing time as a causal substrate, UED simplifies explanation rather than complicating it. Motion no longer needs an invisible conveyor belt. It needs only delta, constraints, and a viable path.

Time enters later—quietly—as bookkeeping.

We count how long resolution took because counting helps coordination. But the counting does not make motion possible any more than a stopwatch makes a runner run.

With motion now understood as resolution in action rather than movement through time, the next clarification becomes unavoidable: what clocks actually measure—and why repetition, not duration itself, gives time its familiar numerical stability.

8.3 — Clock Time: Counting Repetition, Not Duration

Once time is no longer treated as a flowing substance, clocks can finally be understood without mystique.

A clock does not measure time itself. A clock counts repetition.

Every clock—mechanical, atomic, biological, or digital—relies on a process that repeats reliably. A pendulum swings. A quartz crystal oscillates. An atomic transition cycles. A heartbeat pulses. We choose one such repeatable process as a reference and count how many times it completes while some other resolution is underway.

What we call "time" is the ratio between two processes:

- one that repeats predictably,
- and one whose resolution we wish to track.

This distinction is crucial.

Duration belongs to the process being resolved. Time belongs to the measurement strategy we impose on it.

If repetition were perfect and universal, clock time would appear absolute. But repetition is never guaranteed—it is earned through stability, isolation, and constraint management. This is why clocks work well in some domains and fail in others.

When repetition is robust, time feels smooth. When repetition degrades, time becomes noisy. When repetition collapses, time loses meaning altogether.

This explains a deep asymmetry that has puzzled both physicists and philosophers.

We assume time is fundamental and clocks are imperfect. In reality, clocks are fundamental—and time is the abstraction.

Clocks do not reveal time's essence; they manufacture regularity.

Consider what happens when repetition breaks.

At microscopic scales, quantum systems do not offer stable, isolated oscillations that can be counted indefinitely. Measurement disrupts repetition. Boundaries enforce resolution. As a result, clock-like behavior becomes fragile. Time feels "smeared," discontinuous, or ill-defined—not because time misbehaves, but because there is nothing reliable left to count.

At macroscopic scales, repetition is easier to maintain. Large systems average out noise. Oscillators remain stable. Environmental disturbances cancel statistically. As a result, clock time appears continuous, uniform, and universal. Classical physics thrives here—not because time is smoother, but because repetition is.

This also explains why different clocks drift.

A clock placed under stress, acceleration, temperature change, or gravitational gradient does not reveal a change in time itself. It reveals that the repetition it depends on has been altered. Time dilation is not time stretching; it is repetition rate change under constraint.

From this perspective, clock time is a powerful but limited invention. It allows coordination, prediction, and calculation. It makes engineering and science possible. But it is a proxy, not a foundation.

When we mistake the proxy for the thing itself, paradox follows.

We ask why time slows near mass, fractures in quantum systems, or feels elastic in consciousness. The correct answer is simpler: the processes we use to count repetition are sensitive to context, while resolution itself is not obliged to be uniform.

Clocks do not fail because time becomes strange. Clocks fail because repetition becomes fragile.

Once this is understood, time regains its proper role: not as the driver of reality, but as a derived coordinate built on top of stable cycles.

With that clarity, one final question remains for this chapter: why does repetition, when stable, create the powerful illusion of predictability—and why does physics feel so comfortable inside that illusion?

8.4 — Why Repetition Creates "Time-Like" Predictability

Repetition does more than allow us to count. It creates the feeling that time itself is reliable.

When a process repeats with high fidelity, it produces regular intervals. Regular intervals allow comparison. Comparison allows extrapolation. And extrapolation creates predictability. This entire chain operates without invoking time as a substance. What we experience as "time behaving well" is actually repetition behaving well.

This is why classical physics feels so comfortable.

In everyday physical systems, repetition is abundant and stable. Planetary orbits repeat. Pendulums swing. Waves oscillate. Thermal averages smooth fluctuations. Noise cancels out across scale. Under these conditions, counting cycles gives a consistent mapping between one process and another. Duration can be translated into numbers cleanly. The world appears orderly, continuous, and governed by smooth temporal evolution.

But notice the dependency.

Predictability does not come from time flowing uniformly. It comes from pathways that reset themselves.

A repeating system returns to a familiar state after each cycle. This reset erases history. Small deviations average out. Memory dissolves. As long as the system keeps resetting, future behavior can be inferred from past behavior. That inference is what we call "time-like predictability."

This explains why repetition is so deeply embedded in scientific modeling.

Differential equations assume smooth change because they assume small deltas repeated many times. Conservation laws appear stable because repeated interactions redistribute energy in statistically reliable ways. Even the notion of "constant rates" is a shorthand for consistent repetition under controlled constraints.

When repetition holds, time appears linear.

But this linearity is not fundamental—it is manufactured.

The illusion becomes especially strong because repetition hides resolution cost. Each small step resolves only a tiny portion of delta, so nothing dramatic appears to happen. The system glides. Motion looks effortless. Time feels like a neutral background rather than a record of delayed alignment.

This is why disruption feels so unsettling.

When repetition falters—when a system fails to reset cleanly—predictability collapses. Rates change. Outcomes diverge. Suddenly, time feels unreliable. Not because time has changed, but because the mechanism that created regularity has failed.

This also clarifies why repetition dominates in engineering and technology.

We design machines to repeat because repetition is controllable. We create clocks, engines, protocols, and routines that cycle because cycles are easier to regulate than one-off resolutions. Technology, in this sense, is not mastery over time—it is mastery over repetition.

And this brings us to a critical boundary.

Repetition is not guaranteed everywhere.

At small scales, isolation is hard. Noise matters. Measurement interferes. Repetition degrades. At large scales, propagation limits intervene. Coherence weakens. Repetition stretches.

When repetition can no longer be trusted, time-like predictability breaks down. Equations still work locally, but global smoothness vanishes. What remains is not chaos, but resolution without a stable clock.

This prepares the ground for the next transition.

If repetition is the source of time-like behavior, then domains where repetition is fragile will experience time as fragmented, probabilistic, or strange. That is not a failure of reality—it is a signal that resolution is outrunning the cycle-counting tools we rely on.

The next section examines exactly this moment: when time stops making intuitive sense, not because the universe misbehaves, but because our dependence on repetition has reached its limit.

8.5 — When Time Stops Making Sense

Time does not fail everywhere. It fails selectively—precisely where repetition can no longer be trusted.

This is the point at which confusion enters physics, psychology, and everyday experience alike. We assume something strange is happening *to time*, when in fact something ordinary has happened to repetition: it has become fragile, intermittent, or impossible to maintain.

When repetition breaks, time loses its familiar structure.

At microscopic scales, systems do not sit quietly and repeat. They interact strongly with their boundaries. Measurement interferes. Isolation collapses. A process cannot complete one clean cycle without being altered by the act of observing it. Under these conditions, there is nothing stable to count. Clock time becomes noisy, probabilistic, or ill-defined—not because time has fragmented, but because the oscillator has.

At macroscopic but extreme conditions, a different failure appears. Repetition still exists, but coherence across space cannot be maintained cheaply. Signals take time to propagate. Alignment cannot be enforced everywhere at once. Cycles stretch. Rates diverge. Time appears to dilate.

Again, nothing mystical has happened. The system is simply paying the cost of maintaining repetition under constraint.

The same pattern appears in human experience.

Under stress, attention fractures. Memory loops. Anticipation overwhelms resolution. Internal repetition—breathing rhythms, thought cycles, emotional regulation—becomes irregular. Time feels slow, thick, distorted. In moments of shock or insight, repetition collapses entirely. Experience compresses into an extended "now." People report time stopping or jumping, not because time changed, but because their internal cycle-counting machinery shut down.

This is why boredom and trauma both distort time, but in opposite ways. Boredom maintains repetition with no resolution; trauma destroys repetition entirely. In both cases, duration is present, but predictable time is not.

What is often called "time dilation" is therefore a category error. Time does not stretch. Repetition stretches or degrades. What remains constant is the effort to resolve delta under constraints.

This also explains why equations that rely on time parameters still work—up to a point.

As long as repetition is approximately stable, time-based modeling succeeds. When repetition becomes fragile, the models don't suddenly become wrong; they become inapplicable. The failure is not of physics, but of the assumption that a global, uniform counting mechanism still exists.

This is the deep reason time feels paradoxical.

We expect a tool built on repetition to remain valid even where repetition cannot survive.

Once this is seen, the mystery dissolves. There is no broken time, no exotic temporal behavior demanding metaphysical repair. There is only resolution occurring faster, slower, or more discontinuously than our counting frameworks can track.

This realization quietly prepares the final inversion of the chapter.

If time breaks only when repetition breaks, then time was never fundamental to begin with. It was an emergent convenience—a surface effect riding on top of deeper structural behavior.

The next section completes that inversion cleanly, showing why time belongs at the end of explanation, not at the beginning.

8.6 — The Quiet Reversal: Time as Outcome, Not Input

At this point, the inversion becomes unavoidable.

Time does not explain resolution. Resolution explains time.

Everything that made time appear fundamental—continuity, order, predictability, measurement—has already been accounted for without invoking a flowing temporal substrate. Delta persists. Pathways allow or resist alignment. Repetition stabilizes or collapses. Duration emerges as a residue of delay. Time arrives after the fact, not before it.

This is the quiet reversal at the heart of UED.

Traditional models place time at the beginning: a dimension in which events unfold, motion occurs, and causality operates. UED places time at the end: a descriptive layer constructed once behavior has already happened. Time is not the stage; it is the scorecard.

This reversal does not break physics. It explains why physics worked even while carrying a flawed ontology.

Equations that treat time as an independent variable still succeed because they operate inside domains where repetition is stable. In those domains, time behaves *as if* it were fundamental. The math remains valid because it is tracking regularity, not because it has captured time's essence.

This is why time-based equations fail gracefully rather than catastrophically. They do not explode; they blur, probabilize, or require correction terms. They signal not the breakdown of reality, but the limits of the repetition regime they depend on.

Once time is recognized as an outcome, several long-standing paradoxes dissolve without effort:

- The "arrow of time" no longer needs metaphysical justification. Direction arises from asymmetric resolution, not from a flowing dimension.
- Time dilation ceases to be strange. It is simply repetition rate change under constraint.

- Timelessness at quantum scales stops being mysterious. It reflects resolution without stable cycles.
- Psychological time stops being subjective or illusory. It tracks internal resolution efficiency.

Most importantly, this reversal restores intellectual economy.

We no longer need:

- multiple kinds of time,
- speculative temporal dimensions,
- special exceptions where time "behaves differently."

We need only one structural insight: time appears wherever resolution is delayed and repetition is available to be counted.

Everything else follows.

By placing time after delta, equilibrium, and consequence, UED does not demote time out of disrespect. It demotes it out of precision. Time is powerful because it is useful—not because it is fundamental.

This is why UED can speak comfortably across domains. In physics, time becomes a parameter of process. In biology, it becomes a measure of recovery and fatigue. In psychology, it becomes a function of attention and resistance. In social systems, it becomes the duration of unresolved tension. The language changes; the structure does not.

With this reversal complete, the reader no longer needs to defend or reject time. Time becomes optional, deployable where repetition supports it, ignorable where it does not.

And that leaves one final step for this chapter: to pivot from understanding time as outcome to understanding what breaks time next—not philosophically, but structurally.

That pivot leads directly to the closing transition.

From Duration to Fragility

Time, once stripped of its river, reveals its true dependency: repetition must survive for time to remain legible.

This chapter has not abolished time. It has located it. Time now stands where it belongs—not as the driver of motion or the container of events, but as a record of how long resolution resisted completion under stable cycles. Where cycles persist, time appears smooth. Where they degrade, time fractures. Where they collapse, time vanishes.

Nothing mystical has occurred. Nothing has been taken away.

What has changed is the order of explanation.

Delta resolves. Constraints shape pathways. Repetition, if it holds, allows counting. Counting gives rise to time-like predictability.

When any link in that chain weakens, time does not misbehave—it exposes its dependence.

This prepares the reader for the next step, where the illusion of time finally loses its strongest footing. At quantum scales, repetition does not merely degrade; it is actively disrupted. Measurement enforces boundaries. Resolution overtakes cycles. The tools built for stable repetition begin to stutter.

Quantum time feels broken not because the universe abandons order, but because order no longer takes the form of repeatable cycles.

The transition is therefore natural, not dramatic.

We move next from:

- time as duration, to:
- time as fragile parameter under resolution dominance.

The question is no longer *why time behaves strangely*. The question is *why we expected it to behave smoothly everywhere at all.*

That expectation dissolves in the next chapter, where quantum behavior will no longer look paradoxical, only honest.

Time does not fail there. Our dependence on repetition does.

Time does not carry events forward—events linger unresolved, and we name the delay.

Chapter 9 — Why Quantum Time Feels Broken

The Place Where Time First Cracked

Quantum mechanics is often described as the point where reality stops making sense. Time behaves strangely. Causality feels uncertain. Outcomes appear probabilistic. For more than a century, this discomfort has been treated as evidence that the universe itself is fundamentally weird.

But this reaction says more about our expectations than about nature.

Quantum systems did not break time. They were simply the first domain where our time assumption was no longer protected by repetition. The crack did not appear because nature changed rules, but because the scaffolding that made those rules feel obvious could no longer hold.

Classical intuition was built in a regime of abundance: large systems, stable cycles, averaged noise, reliable clocks. Time felt smooth there because repetition was cheap. Quantum systems, by contrast, operate where isolation is fragile, interactions are unavoidable, and resolution occurs before cycles can settle. In such a regime, time is not wrong—it is unsupported.

Historically, this was deeply unsettling.

Physics had learned to trust time as a universal parameter. Equations evolved states "in time." Causality was narrated as temporal sequence. When quantum behavior refused to fit cleanly into this structure, the response was not to question time itself, but to invent increasingly elaborate interpretations around it.

We asked whether the observer creates reality. We argued about branching universes. We debated whether collapse is real or apparent.

In all of this, one possibility remained largely untouched: that time was never meant to carry explanatory weight at this scale.

The unease surrounding quantum time comes from a misplaced loyalty. We assumed that if time worked well in classical physics, it must work everywhere. When it didn't, we treated the failure as profound mystery rather than as a signal of domain boundary.

UED invites a calmer interpretation.

Quantum systems are not strange because they ignore time. They are strange because they expose how much classical physics depends on repeatable cycles that quantum systems cannot sustain. Where repetition collapses, clock-based time cannot anchor description. What remains is resolution without smooth narration.

This is not a loss of order. It is a change of bookkeeping.

Resolution still occurs. Delta still collapses. Consequence still propagates. What disappears is the comforting ability to say *how long* in a continuous, clock-like sense.

The crack in time that quantum mechanics revealed was not a wound in reality. It was a crack in our assumption that time belongs everywhere equally.

This chapter explores that crack—not to dramatize it, but to close it properly. Once resolution dominance is understood, quantum behavior stops feeling paradoxical. Time did not break here.

It was never holding the structure together in the first place.

9.1 — When Resolution Dominates Repetition

Quantum systems feel hostile to time because they operate in a regime where resolution consistently outruns repetition.

In classical systems, repetition stabilizes first. Oscillations settle. Cycles repeat. Only then does resolution proceed in small, smooth steps that can be counted. Time feels continuous because the system obliges our counting strategy. Delta resolves slowly enough for clocks to keep up.

Quantum systems do the opposite.

Here, delta does not wait for cycles to mature. Alignment occurs before repetition can stabilize. Interactions are immediate, boundaries are porous, and coupling to the environment is unavoidable. As a result, the very processes required to build a clock—clean oscillation, isolation, reset—fail to establish themselves.

This is the decisive shift.

It is not that quantum systems evolve *faster in time*. It is that they resolve without allowing time to be built first.

Resolution dominance means that misalignment collapses as soon as pathways allow, often in a single decisive adjustment rather than through many small, repeatable steps. Without repetition, there is nothing to count. Without counting, duration cannot be translated into time-like continuity.

This is why the language of "evolution in time" becomes fragile at small scales.

Equations may still carry a time parameter, but the system itself does not carry an internal clock. The parameter indexes change from the outside; it does not describe a lived, internal progression. The system resolves; the math narrates.

This also explains why quantum transitions appear abrupt.

Abruptness is not a violation of continuity. It is what resolution looks like when delay is unaffordable. In a regime where maintaining delta is expensive or unstable, the system does not drift—it snaps. What appears as a jump is simply alignment occurring without extended resistance.

Contrast this with classical motion.

A rolling ball resolves height difference gradually because friction, inertia, and pathway geometry stretch resolution across many repeatable micro-steps. Quantum systems lack this luxury. Their pathways are narrow, their couplings strong, and their tolerance for delay minimal. Resolution therefore dominates over repetition by necessity, not exception.

Once this is seen, the feeling that "time is broken" loses its force.

Time did not fail to govern quantum behavior. Quantum behavior never depended on time governance to begin with.

What failed was our insistence on importing a tool—clock-based time—into a domain that cannot sustain the conditions required for that tool to function. Quantum systems are not resisting time; they are indifferent to it because they resolve before repetition can anchor duration.

This insight prepares the next correction.

If resolution dominates repetition, then any act that enforces resolution—especially measurement—cannot be neutral. Measurement will not merely observe what is already happening. It will close delta by force, collapsing possibilities into outcomes.

Understanding that move requires reframing measurement itself—not as passive observation, but as boundary enforcement. That is the focus of the next section.

9.2 — Measurement Is Boundary Enforcement, Not Observation

Measurement in quantum systems has long been described as mysterious because it was framed incorrectly from the beginning. We treated it as a special kind of seeing—an act of observation that somehow disturbs reality simply by being aware of it. This framing invited metaphysics where none was needed.

In UED, measurement is not observation at all. Measurement is boundary enforcement.

A quantum system, left to itself, exists as a field of permitted deltas—multiple pathways held open because no constraint has yet made resolution cheaper than persistence. The system is not "undecided" in a psychological sense. It is structurally unforced. Delta remains unresolved because no boundary has demanded closure.

Measurement changes this condition instantly.

To measure is to couple the system to a macroscopic apparatus. That coupling is not gentle. It introduces rigidity, scale, irreversibility, and cost asymmetry. The apparatus cannot remain neutral because it is designed to force alignment into a recordable state. In doing so, it closes permission windows that were previously open.

What is traditionally called "collapse" is nothing more than delta closure under enforced constraint.

The wavefunction does not collapse because reality notices an observer. It collapses because the system is no longer allowed to remain unresolved. Measurement injects a boundary strong enough to make persistence more expensive than resolution. Alignment becomes unavoidable.

This reframing removes several long-standing confusions at once.

First, it explains why asking "what was the system doing before measurement?" is a misplaced question. Before measurement, the system was not evolving along a hidden time path toward a definite outcome. It was remaining unresolved within permitted bounds. Time narration presumes a sequence of internal states. But in a resolution-dominant regime, such narration is not supported.

Second, it explains why different measurement setups produce different outcomes. Each setup enforces a different boundary topology. The result is not revealing a pre-existing property; it is selecting which resolution pathway is made affordable.

Third, it dissolves the myth of the special observer. Human awareness is irrelevant. A detector, a screen, an environment—all enforce boundaries. Collapse occurs because of structural coupling, not consciousness.

This also clarifies why measurement is irreversible.

Once a boundary enforces resolution at macroscopic scale, the cost of reopening delta becomes prohibitive. Information spreads. Energy redistributes. Consequence propagates outward. The system cannot be "unmeasured" because the resolution has already been paid for and accounted.

From the UED perspective, measurement is simply consequence entering a new regime.

The system moves from:

- resolution without clocks to:
- resolution under clock-dominated constraints.

That transition is abrupt not because time jumps, but because resolution is forced rather than negotiated.

Seen this way, quantum measurement is no longer paradoxical. It is an extreme case of a universal rule: whenever boundaries become strong enough, delta closes. What makes quantum systems special is not that they collapse, but that they resist boundary formation until measurement imposes it.

With measurement now understood as enforced resolution, the next issue becomes obvious. If clocks depend on repetition, and measurement disrupts repetition, then quantum systems cannot naturally carry their own clocks.

That limitation—not mystery—is why quantum systems resist being described "in time."

9.3 — Why Quantum Systems Resist Clock Attachment

To say that a system has a "time" in the usual sense means something very specific: it means the system can host, or be reliably coupled to, a repeatable internal process whose cycles can be counted without significantly disturbing the system itself. This is what it means to attach a clock.

Quantum systems resist this attachment—not as a peculiarity, but as a structural necessity.

A clock requires three conditions:

1. **Isolation** — the repeating process must be shielded from disruptive coupling.
2. **Stability** — each cycle must return the system to a near-identical state.
3. **Non-invasiveness** — counting cycles must not significantly alter the system.

Quantum systems fail all three simultaneously.

At small scales, isolation is fragile. Any attempt to shield a quantum system introduces boundaries that themselves enforce resolution. Stability is elusive because quantum states are exquisitely sensitive to interaction; even weak coupling alters the configuration space. And non-invasiveness is impossible because counting, by definition, requires interaction—and interaction is precisely what collapses permitted deltas.

This is the core reason quantum systems do not "carry their own time."

They cannot sustain an internal oscillator without that oscillator becoming part of the system's dynamics, altering the very resolution it is meant to measure. The clock does not sit beside the system; it enters it, changes it, and forces alignment.

In classical systems, this problem is hidden by scale. Large systems absorb interaction without noticeable disruption. Noise averages out. Cycles reset. A clock can tick quietly in the background. Quantum systems have no such buffer. Every interaction matters. Every coupling counts. There is no "background" small enough to ignore.

As a result, time in quantum mechanics is always external.

The time parameter used in equations does not belong to the system. It belongs to the experimental context—the laboratory clock, the apparatus cycle, the macroscopic world in which repetition still holds. The system itself resolves; the experimenter narrates.

This distinction is subtle but decisive.

When we say a quantum state "evolves in time," we are not describing an internal temporal process. We are indexing changes relative to an external repetition framework that the system itself cannot support. Time is imposed from outside as a bookkeeping device, not generated from within as a lived duration.

This also explains why different experimental setups produce different temporal descriptions. Change the coupling, change the boundaries, change the clock—and the "time evolution" looks different. Nothing about the system's resolution has become inconsistent. Only the external narration has shifted.

Once this is seen, the resistance of quantum systems to clock attachment stops being mysterious. It is not a failure of reality to provide time. It is a refusal to subsidize repetition where repetition is structurally unaffordable.

Quantum systems are not missing time. They are unclockable.

And that realization points directly to the next correction. If time is externally imposed in quantum equations, then the time parameter itself must be understood for what it is: a mathematical convenience, valid under assumptions that are known to be fragile.

9.4 — The Time Parameter as Mathematical Convenience

Once it is clear that quantum systems cannot carry their own clocks, the role of the time parameter in quantum equations becomes obvious—and far less mysterious.

Time appears in quantum mechanics not because it is fundamental to quantum behavior, but because mathematics needs an index.

Equations require ordering. They require a way to label change. When repetition is unavailable inside the system, we borrow repetition from outside—from laboratory clocks, experimental cycles, or idealized reference processes—and use that borrowed regularity as a parameter. The symbol t does not represent an internal physical substance. It represents external bookkeeping under conditional assumptions.

This is why quantum equations work as long as those assumptions hold.

The Schrödinger equation, for example, describes how a state changes relative to an external time parameter. Implicit in its use is a fragile agreement: that the system can be treated *as if* it evolves smoothly between interactions, and that the external clock is not part of the system's dynamics. When isolation is approximately valid, this works remarkably well. When it is not, the description frays.

This is not a bug. It is an honest boundary.

Time, in this context, functions like a page number in a book. It helps organize content, but it does not cause the story. The events do not happen because the page number increases. The page number increases because we are tracking progression in a convenient way.

The mistake arises when the index is mistaken for the driver.

When the time parameter is treated as something the system *experiences*, paradox follows. We begin asking where the system "is" at a given time, what it was "doing" before measurement, or how it "jumps" instantaneously from one state to another. These questions assume an internal temporal continuity that the system never possessed.

From the UED perspective, the quantum system is not evolving *in time*. It is resolving under changing constraints, and the time parameter is a way of labeling when those constraints are applied or withdrawn from the outside.

This also explains why quantum theory tolerates multiple time-like formulations without contradiction. Whether time is continuous, discrete, emergent, or absent altogether depends less on the system and more on which external repetition framework is chosen. The physics remains intact because the resolution logic is intact. Only the narration shifts.

Importantly, recognizing time as a mathematical convenience does not weaken quantum theory—it strengthens it.

It frees the theory from unnecessary metaphysical burden. It allows equations to be used where they are valid and set aside where they are not, without forcing reality to conform to a single temporal picture. It explains why quantum mechanics remains internally consistent even while resisting intuitive time-based storytelling.

Time, here, is not wrong. It is conditional.

It works where repetition can be borrowed. It fades where repetition cannot be sustained.

Once this is accepted, the remaining mystery evaporates. What remains is not a broken concept of time, but a clearer understanding of where time belongs in explanation—and where it does not.

This naturally leads to the final clarification of the chapter: quantum systems are not outside time in any metaphysical sense. They are simply operating in a regime where time cannot be cleanly assigned.

9.5 — Probability Replaces Prediction When Time Fails

When time-based narration collapses, something else quietly takes its place: probability.

This shift has often been misunderstood. Probability is usually treated as a confession of ignorance—an admission that we do not know enough about the system. In UED, probability has a far more precise meaning. It appears not because reality is random, but because resolution occurs without stable repetition.

Prediction requires two things:

1. A repeatable pathway.
2. A reliable way to count progress along that pathway.

When repetition holds, prediction is cheap. We extrapolate. We project. We say "if this, then that." Time-based equations work because the system reliably revisits similar states.

When repetition breaks, prediction fails—not because the system becomes chaotic, but because there is no longer a single privileged pathway.

In quantum systems, multiple resolution pathways remain structurally permitted until a boundary enforces closure. Without repetition to favor one pathway over others, the system cannot be said to be "on its way" to a specific outcome. It is not undecided; it is uncommitted.

Probability enters precisely here.

Probability is not a statement about chance. It is a statement about pathway availability under unresolved delta.

Each probability amplitude reflects how affordably a given resolution pathway can close when constraint is applied. The distribution does not describe ignorance of hidden motion in time; it describes structural openness before enforced alignment.

This is why probability replaces prediction rather than complementing it.

Prediction assumes continuity. Probability assumes conditional closure.

Once closure occurs—once measurement enforces a boundary—probability disappears instantly. The system resolves into a definite state. Consequence propagates. There is nothing probabilistic left. Probability existed only while resolution remained deferred.

This also explains why quantum probabilities are stable and law-like.

If probability were merely ignorance, it would vary wildly with knowledge. Instead, probability distributions are reproducible across experiments. They are not epistemic accidents; they are structural signatures of resolution topology.

Different setups yield different probabilities because they enforce different boundaries. Change the topology, and you change which pathways are cheap or expensive. The system responds accordingly, without reference to time evolution in the classical sense.

From this perspective, probability is not a failure mode of physics. It is a mode of description appropriate to resolution-dominant regimes.

Time-based prediction works where repetition dominates. Probability-based description works where repetition collapses.

Both are valid. Neither is fundamental.

What failed in quantum mechanics was not determinism, but the assumption that determinism must always be narrated through time. Quantum systems are deterministic in how they respond to enforced constraints. What they refuse to provide is a continuous story before that enforcement occurs.

Once this is seen, randomness loses its mystique. There is no cosmic dice being thrown. There is only resolution choosing among permitted pathways when forced to choose.

This prepares the final clarification of the chapter. If probability replaces prediction because clocks fail, then quantum systems are not timeless in a mystical sense. They are simply operating without an internal temporal narrative. That distinction matters—and it closes the chapter cleanly.

9.6 — Quantum Systems Are Not Timeless, Just Unclocked

The final confusion surrounding quantum time arises from a single mislabel: timelessness.

Quantum systems are often described as existing "outside time," as if they inhabit a strange metaphysical realm detached from duration altogether. This description feels profound, but it is inaccurate. It replaces one misunderstanding with another.

Quantum systems are not timeless. They are unclocked.

This distinction is critical.

Timelessness would imply absence of change, absence of sequence, absence of resolution. Quantum systems show none of these. Change occurs. Delta collapses. Consequence propagates. What is missing is not duration, but a stable internal reference by which duration can be smoothly narrated.

In other words, things still happen—but they do not happen *on a clock the system itself can carry*.

Resolution in quantum systems occurs, but it does not stretch itself across evenly spaced, repeatable markers. There is no internal pendulum ticking away quietly in the background. There is only persistence of permitted delta followed by enforced closure when constraints arrive.

This is why words like "instantaneous" are misleading.

Quantum transitions are not instantaneous in the sense of occurring in zero duration. They are instantaneous only relative to clocks that fail to attach. Duration exists, but it is not metrized. The system does not advertise how long resolution took because there is no repetition framework inside the system capable of expressing that duration numerically.

This also explains why attempts to reinsert time at the quantum level often feel artificial. Proposals for discrete time steps, hidden clocks, or fundamental temporal grains are efforts to *force repetition back into a regime that structurally resists it*. Such models may be useful mathematically, but they mistake a descriptive gap for an ontological absence.

UED offers a simpler view.

Quantum systems resolve in duration, but that duration is path-dependent, boundary-triggered, and unrepeatable in isolation. Time appears only when resolution can be compared against something that repeats. Without that comparison, duration remains real but unnamed.

This places quantum behavior in direct continuity with other domains.

Trauma is not timeless—it is unclocked. Insight is not timeless—it is unclocked. Phase transitions are not timeless—they are unclocked.

In all these cases, resolution happens without smooth internal narration. We experience an outcome, but cannot say *how long* it took in the usual sense because repetition was suspended during the transition.

Quantum systems live permanently in that regime.

They are not defying time. They are exposing its dependency.

Once this is understood, quantum mechanics no longer feels like a special exception demanding exotic philosophy. It becomes the purest case of a general rule already established in UED:

Where repetition fails, clocks fail. Where clocks fail, time narration fails. Resolution does not.

This realization closes the quantum chapter cleanly. Time did not break at small scales. It simply lost the structural support it quietly relied on everywhere else.

What comes next is not fragmentation, but contrast.

At cosmic scales, repetition survives—but coherence does not. Time does not disappear there; it stretches. And that stretching, unlike quantum fragmentation, will turn out to be not strange at all—but inevitable.

That is where the discussion moves next.

From Fragile Time to Stretched Time

Quantum systems did not reveal a broken universe. They revealed a limit of narration.

In this chapter, time was not repaired, redefined, or rescued. It was simply placed where it belongs. Quantum behavior feels strange only because we insisted on using a clock where clocks cannot survive. Once repetition collapses, time-based storytelling collapses with it. What remains—resolution, boundary enforcement, and consequence—continues without interruption.

Nothing essential was lost.

Quantum systems still change. They still resolve. They still produce definite outcomes.

What they do not provide is a smooth internal duration that can be sampled repeatedly without disturbance. Time does not vanish here—it becomes unassignable.

This prepares a clean contrast.

At small scales, repetition fails first. Time fragments. At large scales, repetition survives—but coherence becomes expensive.

Cosmic systems do not resist clocks the way quantum systems do. They resist synchronization. Signals must propagate. Alignment must be maintained across vast distances. Resolution cannot occur everywhere at once. As a result, repetition stretches rather than collapses.

Time does not break at the cosmic scale. It dilates.

Not as an anomaly. Not as a paradox. But as a structural consequence of finite resolution speed and scale.

The next chapter explores this mirror image of the quantum case. Where quantum systems resolve too fast for time to anchor, cosmic systems resolve too slowly for time to remain uniform. In both cases, time behaves exactly as it should once its dependency on resolution and repetition is acknowledged.

The question moving forward is no longer *why time behaves differently across domains*. The question is *why we ever expected one kind of time to survive them all.*

With that expectation finally released, the discussion turns outward—to scale, coherence, and the unavoidable stretching of duration across the universe.

Quantum systems do not live outside time; they live outside clocks.

Chapter 10 — Why Cosmic Time Must Stretch

The Scale Where Smooth Time Finally Breaks

If quantum mechanics was the first place where time cracked, cosmology is where it finally refuses to stay smooth.

Relativity is often introduced as a theory that makes time strange—stretching it, slowing it, bending it under speed and gravity. This framing is dramatic, but it misses the deeper point. What relativity actually exposed was not a problem with time, but a problem with scale.

At everyday scales, time behaves politely because coherence is cheap. Signals travel fast enough relative to distances involved. Systems stay synchronized without effort. Repetition remains stable. Under these conditions, a shared notion of "now" feels natural. We mistake this convenience for universality.

Cosmic scales break that illusion.

As systems grow larger, distances grow faster than coordination ability. Information must travel. Alignment must be enforced gradually. Resolution cannot occur everywhere at once. The assumption of a single, smooth, global time begins to fail—not because time malfunctions, but because coherence becomes expensive.

This is the key correction.

Relativity did not discover that time changes with motion or mass. It discovered that keeping clocks aligned across distance, motion, and gravitational gradients has a cost—and that cost must be paid in duration.

Time stretches because alignment stretches.

The shock of relativity came from clinging to a classical expectation: that simultaneity should survive at any scale. Once that expectation is released, the theory becomes quietly obvious. Of course distant events cannot share a universal "now" if influence takes time to propagate. Of course clocks moving through different pathways of resolution will accumulate different durations. Of course gravity affects time if gravity shapes the very pathways through which alignment must travel.

Nothing weird is happening.

The universe is simply enforcing a limit: resolution cannot outrun propagation.

This chapter reframes relativistic time not as warped chronology but as honest accounting. When systems attempt to remain coherent across scale, speed, or mass, they pay for it. The payment appears as dilation—not as illusion, not as observer trickery, but as structural necessity.

By the end of this chapter, relativistic effects should feel unsurprising, even mundane. Time is not bending. It is stretching under load.

And once that is understood, the final concept of this part—Alignment Threshold Time—will arrive naturally. Because when alignment costs rise high enough, resolution does not merely slow.

It waits.

10.1 — Propagation Is Not Instantaneous

The most important constraint in cosmic systems is also the simplest one: resolution must travel.

Nothing—no influence, no alignment, no redistribution—can appear everywhere at once. For any change to matter beyond its point of origin, it must propagate through space, across structure, and along permitted pathways. This is not an assumption added by physics; it is a structural inevitability of extended systems.

At small scales, this constraint is easy to ignore. Distances are short. Propagation delays are negligible compared to the processes we care about. Alignment appears almost simultaneous. We grow accustomed to the illusion that resolution is effectively instant.

Cosmic scale removes that illusion.

When distances become vast, propagation delay becomes dominant. A change here cannot immediately affect conditions there. Information must move. Fields must adjust. Geometry must respond. Alignment proceeds not as a global update, but as a wave of resolution advancing step by step.

This has a profound implication: simultaneity is not free.

Two distant regions cannot share a common "now" unless information can travel between them without delay. If propagation takes time, then alignment takes time. If alignment takes time, then any notion of a universal present collapses.

This is where the classical picture quietly breaks.

Classical thinking assumes that space can be updated instantaneously and that time merely labels the update. Cosmic reality insists on the opposite: space resists instant update, and time records the resistance.

From a UED perspective, propagation is not motion through time. It is resolution spreading under constraint. A delta introduced in one region does not vanish globally; it must be redistributed locally, then passed along, paying cost at each step. Duration accumulates not because time flows, but because alignment cannot skip distance.

This reframes the role of limits like the speed of light.

The universe does not impose a speed limit arbitrarily. It enforces a maximum rate at which coherence can be restored across space. Propagation speed is not about how fast things move; it is about how fast misalignment can be corrected.

Once this is understood, several confusions evaporate at once.

Events do not fail to be simultaneous because clocks disagree. Clocks disagree because events cannot be made simultaneous without violating propagation limits.

Time does not stretch because observers move differently. Duration stretches because alignment must travel longer or more expensive paths.

Propagation delay is not an inconvenience layered on top of time. It is the reason duration exists at scale at all.

This insight sets the foundation for the next correction. If propagation speed limits resolution, then the constant traditionally called *c* is not merely a velocity ceiling. It is something deeper: the universe's way of preventing cost-free, instantaneous delta erasure.

That reinterpretation comes next.

10.2 — *c* as a Resolution-Speed Ceiling, Not a Velocity Limit

The constant *c* has been burdened with more mythology than almost any number in physics. It is called the "speed of light," treated as a hard velocity barrier, and often discussed as if the universe arbitrarily decided that nothing may go faster.

From a UED perspective, this framing is superficial.

c is not primarily about how fast objects move. It is about how fast misalignment can be resolved across space.

In Universal Energy Dynamics, *c* functions as a ceiling on resolution speed—the maximum rate at which coherence, information, and consequence can propagate to restore alignment. Motion happens *within* this constraint; it is not what defines the constraint.

This distinction matters.

If *c* were merely a speed limit on particles, it would be a strange rule: why should motion be restricted while other processes are not? But if *c* limits how quickly delta can be redistributed, then the rule becomes inevitable. Allowing faster-than-*c* propagation would mean allowing instant, cost-free alignment across distance—a direct violation of consequence.

In other words, exceeding *c* would allow delta to disappear without paying for it.

That is not a relativistic issue. It is an accounting violation.

Every alignment has a cost. Every correction must propagate through structure. *c* is the universe's way of ensuring that no part of a system can offload its imbalance instantly onto the rest. It prevents cheating at scale.

Seen this way, *c* is not an empirical curiosity discovered by measuring light. Light merely happens to travel at this limit because it is the least encumbered carrier of resolution—massless, unconfined, and maximally efficient at transmitting alignment. Light reveals the limit; it does not define it.

This reinterpretation also clarifies why *c* appears everywhere.

It governs causality because causality is about resolution order. It governs spacetime structure because geometry encodes propagation cost. It governs energy–momentum relations because redistribution speed affects stability.

c shows up not because the universe loves light, but because resolution economics are universal.

Once this is understood, many paradoxes dissolve.

Nothing "slows down" to obey *c*. Rather, resolution cannot proceed faster than *c*.

A fast-moving system is not racing against time. It is consuming coherence budget. As more of that budget is spent on maintaining alignment across motion, less is available for internal repetition. Clocks tick slower not because time dilates mysteriously, but because resolution capacity is being reallocated.

This is why *c* links speed, time dilation, and mass–energy so cleanly. All three are drawing from the same limited resource: the ability to maintain coherence while resolving delta.

c is therefore not a prohibition. It is a budget constraint.

And like all real constraints, it does not announce itself with drama. It simply shows up as stretching, delay, and loss of simultaneity whenever systems attempt to resolve too much, too fast, across too much space.

With *c* now understood as a resolution-speed ceiling, the next consequence follows naturally. If resolution speed is capped, then maintaining coherence across large systems becomes progressively expensive. Synchronization fails. Local clocks diverge. Global time dissolves.

That is not weird. It is structural.

And it leads directly to the next section: why coherence itself becomes the dominant cost at cosmic scale.

10.3 — Coherence Becomes Expensive at Scale

Once propagation is finite and resolution speed is capped, a new cost dominates large systems: coherence.

Coherence is the ability of different parts of a system to remain mutually aligned—to agree on state, order, and outcome. At small scales, coherence is cheap. Distances are short, delays are negligible, and resolution can keep the system synchronized without visible effort. Local changes propagate quickly enough that everything appears to move together.

Cosmic scale changes that economy completely.

As systems extend across vast distances, coherence can no longer be maintained "for free." Any attempt to keep distant regions synchronized must now pay propagation cost. Signals arrive late. Adjustments lag. Alignment becomes staggered rather than simultaneous. What was once a shared state fragments into local states stitched together imperfectly.

This is the real reason global time fails.

A single universal clock presumes universal coherence. It assumes that all parts of the system can be updated together, or at least coordinated cheaply enough to pretend they are. Finite propagation breaks that assumption. When information cannot reach everywhere at once, simultaneity becomes an approximation—and at large enough scale, an untenable one.

From a UED perspective, this is not a failure of time. It is the inevitable outcome of coherence economics.

Each region of a large system must now resolve delta locally, based on locally available information. Alignment across regions becomes delayed, partial, and context-dependent. As a result, duration diverges. Clocks that were once synchronized drift apart—not because time flows differently for them in some mystical sense, but because they are paying different coherence costs.

This reframes the meaning of "local time."

Local time is not a subjective illusion or an observer artifact. It is the only time that can exist once coherence cannot be globally enforced. Each region carries its own duration because that is all the structure can support. Attempts to resurrect a universal "now" are attempts to deny propagation limits.

This also explains why relativity insists on locality so strongly.

Causality becomes local because coherence is local. Ordering becomes local because resolution is local. Duration becomes local because alignment cannot outrun distance.

Large-scale systems therefore do not share time; they negotiate it.

Synchronization protocols—whether physical, biological, or technological—are not revealing an underlying global time. They are engineering coherence against resistance. GPS systems, for example, do not assume time is universal. They constantly correct clocks because they know coherence is expensive and unstable under motion and gravity.

From this vantage point, the relativistic loss of simultaneity is not surprising at all. It is the first visible symptom of a deeper truth: the universe does not subsidize coherence across scale.

Where coherence is cheap, time feels smooth and shared. Where coherence is expensive, time stretches, splits, and localizes.

This prepares the next clarification. If coherence cost explains divergence, then time dilation itself is not an anomaly. It is simply what honest clocks report when alignment is delayed by motion, gravity, or scale.

That structural reinterpretation is the focus of the next section.

10.4 — Time Dilation as Alignment Delay

Time dilation is often presented as one of the strangest outcomes of modern physics: clocks slow down when they move fast, clocks slow down near mass, clocks disagree depending on where they travel. This strangeness is entirely self-inflicted. It comes from assuming that clocks are reporting some universal time that has mysteriously warped.

In UED, time dilation is neither mysterious nor exceptional. It is simply alignment taking longer under constraint.

A clock is not a passive recorder of an external flow. It is an active system that must maintain internal coherence while interacting with its surroundings. Every tick is a successful resolution of internal delta—an oscillation completed, a transition reset, a cycle closed.

When a system moves, accelerates, or sits in a strong gravitational field, the cost of maintaining that internal coherence rises. Alignment pathways lengthen. Resolution must negotiate additional constraints. As a result, fewer cycles complete per unit of external coordination.

The clock does not slow because time slows. The clock slows because alignment has become more expensive.

This reframes all forms of dilation at once.

- Velocity dilation occurs because a moving system must continuously reconcile its internal coherence with external propagation limits. Part of its resolution budget is spent on maintaining alignment across motion, leaving less capacity for internal repetition.
- Gravitational dilation occurs because mass-energy reshapes resolution pathways. Near mass, alignment must follow curved, crowded routes. Each cycle pays higher cost, so fewer cycles complete.

In both cases, the clock is not malfunctioning. It is telling the truth.

Two clocks that experience different constraints will not agree—not because one is wrong, but because they are resolving under different alignment costs. Expecting them to agree would be like expecting two runners, one uphill and one downhill, to report the same exertion per step.

This also explains why dilation is symmetric and reciprocal in relativity. Each system is locally honest. Each clock measures duration relative to its own coherence conditions. There is no privileged reference frame because there is no privileged alignment pathway.

Seen this way, time dilation is not time warping. It is resolution delay made visible.

The language of "slowing clocks" persists because it is convenient. But structurally, nothing is slowing down except the rate at which resolution cycles can complete under added cost. Time appears stretched because alignment takes longer to close the same deltas.

Once this is understood, dilation stops being weird and starts being expected.

Any system that must:

- move while maintaining coherence,
- operate near dense mass-energy,
- or synchronize across large distances,

will experience duration expansion. The universe is not bending time to punish motion or mass. It is simply charging for coherence.

This prepares the final unification of relativistic effects. If time dilation is alignment delay, then gravity itself must be understood not as a force acting on time, but as a reshaping of the pathways through which alignment occurs.

That deeper correction comes next.

10.5 — Gravity as Delta Curvature, Not Time Warping

Gravity is often described as something that *slows time*. This description is memorable—but it is backwards.

Gravity does not act on time. Gravity acts on delta pathways.

What changes near mass is not time itself, but the geometry through which alignment must occur.

In Universal Energy Dynamics, mass–energy is understood as condensed, stabilized delta. It is not passive. It reshapes the surrounding resolution landscape. Where mass accumulates, the available pathways for redistribution curve, crowd, and lengthen. Alignment can still occur, but it must negotiate a more expensive topology.

This is what curvature really means.

Curvature is not space bending as a visual metaphor. It is delta being redirected through constrained routes.

When a clock operates near mass, its internal resolution cycles must follow these curved pathways. Each cycle pays a higher cost to complete. Fewer cycles close per unit of external coordination. The clock reports longer duration—not because time has slowed, but because alignment has become harder.

Seen this way, gravitational time dilation is not exotic. It is inevitable.

Any process near mass—chemical reactions, biological rhythms, atomic transitions—must resolve through the same crowded geometry. All clocks slow together because they are all paying the same alignment tax. There is no special clock effect; there is a shared structural condition.

This reframing also unifies gravity with motion.

Velocity dilation and gravitational dilation are not separate phenomena. Both arise when resolution pathways are stretched:

- motion stretches pathways through propagation constraint,
- gravity stretches pathways through curvature.

In both cases, duration expands because resolution takes longer, not because time behaves differently.

This perspective removes the last hint of mysticism from relativistic gravity.

Time does not fall into gravity wells. Clocks do not sink into slow time. There is no temporal fluid being distorted.

What exists is a resolution field whose geometry is altered by mass–energy. Time follows that geometry because time is nothing more than a record of how long resolution takes along those paths.

Once this is seen, gravity becomes continuous with everything else already established in UED. It is not a force acting at a distance, nor a magical curvature of spacetime. It is the structural shaping of delta resolution.

And this leads to a sobering conclusion.

If gravity reshapes resolution pathways, and resolution pathways determine duration, then no single time can survive across the universe. Each region, each mass distribution, each motion state defines its own alignment economy.

There is no cosmic clock hiding behind gravity.

Only local clocks, honestly reporting the cost they pay.

That realization sets up the final step of this chapter: why a global, universal time is not just absent—but structurally impossible.

10.6 — Why Global Time Cannot Exist

The idea of a single, universal time is not just empirically false—it is structurally impossible.

Global time presumes three hidden assumptions:

1. **Instantaneous propagation** — that information can be shared everywhere without delay.
2. **Uniform coherence cost** — that all regions resolve delta under identical conditions.
3. **Shared simultaneity** — that "now" can be meaningfully defined across distance.

Cosmic reality violates all three.

Once propagation is finite, coherence is local. Once coherence is local, alignment histories diverge. Once alignment histories diverge, duration diverges. No amount of synchronization can erase this; it can only manage it approximately.

This is why global time fails even in principle.

A universal clock would require the universe to update itself everywhere at once, or at least cheaply enough that delays could be ignored. But resolution cannot outrun propagation, and propagation cannot outrun the ceiling imposed by *c*. Any attempt to impose a global "now" would require cost-free delta erasure across space—an accounting impossibility.

From a UED perspective, time fragments not because reality is disordered, but because resolution is honest.

Each region resolves delta based on:

- its local constraints,
- its local geometry,
- its local coupling to mass–energy,
- its local motion state.

The durations that emerge are therefore locally valid and globally incompatible. This is not a failure of coordination; it is the correct outcome of distributed resolution.

Relativity did not remove global time as a philosophical gesture. It revealed that no structural mechanism exists to sustain it.

This is also why all attempts to restore absolute time quietly reintroduce hidden assumptions: preferred frames, privileged observers, background clocks, or invisible synchronizing agents. Each such attempt smuggles in free coherence where none can exist.

What survives instead is a patchwork.

Time becomes a network of local durations, stitched together imperfectly by propagation and synchronization protocols. Ordering exists, but only locally. Causality exists, but only along permitted pathways. Duration exists, but only where resolution is paid for.

This does not weaken reality. It strengthens it.

A universe with global time would be fragile—overconstrained, brittle, and paradox-prone. A universe with local time is robust. It allows independent regions to resolve independently, interacting only when pathways permit. It scales without collapsing under its own coordination demands.

From this vantage point, the search for universal simultaneity looks like a nostalgic error—a desire to preserve human-scale convenience at cosmic scale.

The universe declines politely.

There is no master clock behind existence. There is no shared "now" spanning galaxies. There is only resolution unfolding locally, honestly, and at cost.

And this brings the discussion to its natural close.

If time stretches because coherence is expensive, and fragments because repetition fails, then one final question remains—one that neither quantum mechanics nor relativity answers cleanly:

What happens when resolution does not proceed continuously at all, but waits—accumulates—until a threshold is crossed?

That question does not belong to time as duration. It belongs to Alignment Threshold Time.

And that is where the next chapter begins.

From Stretched Time to Thresholded Time

At cosmic scale, time does not misbehave. It accounts.

This chapter has shown that what we call time dilation is not an exotic distortion of reality, but the visible cost of maintaining coherence under finite propagation, curved pathways, and distributed resolution. Time stretches because alignment stretches. Duration expands because resolution must travel, negotiate geometry, and pay for synchronization.

Nothing mystical entered the picture.

Quantum domains broke time because repetition could not survive. Cosmic domains stretch time because coherence cannot be made cheap.

Both outcomes follow from the same structural rule: time appears only where resolution can be narrated, and it deforms wherever narration becomes expensive.

At this point, time has been fully demoted—from river to residue, from driver to record. What remains is not confusion, but a sharper question.

If resolution can be:

- smooth where repetition holds,
- fragmented where repetition collapses,
- stretched where coherence costs rise,

then what happens in regimes where resolution does not proceed at all—until it suddenly does?

Because many of the most consequential changes in nature do not unfold gradually.

They wait.

They accumulate pressure. They tolerate misalignment. They resist resolution—until resistance becomes unaffordable.

Then alignment does not drift. It snaps.

Phase transitions. Fracture. Yielding. Insight. Breakdown. Revolution.

These are not time problems. They are threshold problems.

Time, as duration, can describe how long the wait lasted. It cannot explain *why* resolution remained blocked, nor *why* it released when it did. For that, a different concept is required—one that sits between delta, equilibrium, and consequence.

That concept is Alignment Threshold Time (ATT).

ATT is not another theory of time. It is the recognition that resolution is gated, and that duration alone is not the controlling variable.

With this transition, this part reaches its final turn.

Time has been explained. Motion has been reframed. Scale has been accounted for.

What remains is to understand when resolution becomes inevitable.

That inevitability—quiet, structural, and universal—is where the next chapter begins.

At cosmic scale, time does not misbehave—it invoices.

Chapter 11 — ATT: Alignment Threshold Time

Why Some Changes Refuse to Happen Gradually

We are trained—by clocks, calendars, and progress charts—to expect change to be smooth. If enough time passes, we assume something should move. If effort continues, we expect results to accumulate. This expectation is so deeply ingrained that when change *doesn't* happen gradually, we experience confusion, frustration, or shock.

Nothing happens. Nothing happens. And then—everything happens.

A bridge collapses after years of stability. A market crashes after months of calm. A person endures stress silently, then suddenly breaks. An idea refuses to form, and then arrives fully shaped in a single moment.

Our instinctive response is to blame time. We say things like *"it built up over time"* or *"it finally reached its moment."* These phrases feel explanatory, but they hide the real mechanism.

Time did not cause the change. Time merely recorded the waiting.

What actually governed the transition was not duration, but permission.

Certain systems can tolerate misalignment for long periods. They absorb stress, store imbalance, and redistribute cost internally. From the outside, they appear stable. From the inside, delta is being held—sometimes carefully, sometimes dangerously. As long as holding delta remains cheaper than resolving it, nothing visible happens.

This is where linear intuition fails.

We expect change to be rate-controlled: a little effort produces a little movement. But many systems are threshold-controlled. They do not respond to incremental pressure. They respond only when a boundary is crossed—when the cost of persistence exceeds the cost of release.

Until that moment, time is irrelevant.

Waiting longer does not help. Pushing harder may make things worse. Explaining more clearly changes nothing.

Then, without warning—alignment occurs.

The shift feels sudden not because it lacked history, but because the history was invisible. Delta was accumulating quietly. Pathways were narrowing. Options were disappearing. Resolution was being postponed, not prevented. When postponement became unaffordable, release was instantaneous.

This is why sudden change always feels dramatic.

Not because it is rare—but because it violates our expectation of continuity.

ATT names this violation precisely.

Alignment Threshold Time is not about *how long* something takes. It is about how long misalignment can be afforded under existing constraints. It explains why some processes evolve smoothly, while others remain frozen and then snap. It explains why patience sometimes works—and why sometimes it does absolutely nothing.

This chapter begins where time-based explanations reach their limit.

Not all change flows. Some change waits.

And understanding that difference is the final step in freeing explanation from the illusion that duration alone governs reality.

11.1 — Resolution Is Threshold-Gated, Not Rate-Controlled

The central mistake in how we imagine change is the belief that resolution responds proportionally to time or effort. We assume that if a system is pushed gently but continuously, it will move gently but continuously in return. This assumption holds only in regimes where resolution pathways are already open.

Many systems do not operate there.

In UED, resolution is threshold-gated. It does not respond to how long a delta has existed, but to whether holding that delta is still affordable. As long as the system can pay the cost of persistence—energetically, structurally, psychologically, or socially—it will do so. Nothing changes on the surface because nothing *needs* to change yet.

This creates a critical distinction:

- Delay is when resolution is possible but slow.
- Blockage is when resolution is not permitted at all.

Time only matters in the first case. In the second, time accumulates without effect.

Thresholds mark the boundary between these two regimes. A threshold is not a moment in time; it is a permission boundary. On one side of the boundary, maintaining delta is cheaper than resolving it. On the other side, maintaining delta becomes more expensive than release.

Resolution occurs precisely at that crossing—and not before.

This explains why applying steady effort often produces no visible progress. The system is not ignoring the effort; it is absorbing it. Additional energy, pressure, or attention may even reinforce the blockage by strengthening the structures that hold delta in place. What looks like persistence can, paradoxically, raise the threshold rather than approach it.

ATT captures this logic cleanly.

Alignment Threshold Time is the duration required for conditions to change such that release becomes unavoidable. That change may come from accumulation, exhaustion, topology shift, boundary removal, or external shock. But until the affordability condition flips, resolution simply does not proceed—no matter how much clock time passes.

This is why threshold-gated systems feel stubborn, opaque, or "stuck."

They are not slow. They are waiting.

Waiting for:

- material to reach a phase boundary,
- stress to exceed structural tolerance,
- fatigue to overwhelm compensation mechanisms,

- cognitive tension to overwhelm existing frames,
- social pressure to overwhelm institutional inertia.

In each case, the system is not progressing along a timeline. It is approaching a decision point embedded in its structure.

Once that point is crossed, resolution is rapid—not because time suddenly accelerates, but because permission suddenly exists. The same system that resisted change moments ago now reorganizes decisively.

This is why thresholded resolution often looks dramatic or violent. The stored delta has not been dissipating gradually; it has been contained. When release occurs, it happens along the cheapest available pathway, often in a burst.

Understanding this removes the temptation to explain sudden change as randomness, luck, or mystery. Thresholds are not surprises. They are unseen boundaries that become visible only when crossed.

ATT, therefore, is not about patience or timing. It is about recognizing whether a system is in a rate-controlled regime or a threshold-controlled regime—and acting accordingly.

With that distinction established, the next step is to clarify what ATT actually measures, and why duration alone cannot predict when alignment will finally occur.

11.2 — What ATT Actually Measures

Alignment Threshold Time is often misunderstood the moment it is named. The word *time* tempts us back into familiar thinking—as if ATT were simply a longer clock duration, a delayed payoff, or a slow-motion version of ordinary change.

It is none of these.

ATT does not measure how long something takes to happen. It measures how long a system can afford *not* to change.

This distinction is decisive.

Clock time measures duration under repetition. ATT measures affordability under constraint.

A system approaching alignment does not move closer in a linear way. It moves through a landscape of costs. As long as holding delta remains cheaper than resolving it, the system stays exactly where it is. Ten seconds or ten years make no structural difference if the affordability condition has not shifted.

This is why ATT cannot be predicted by elapsed time alone.

Two systems may experience identical durations and reach completely different outcomes. One resolves suddenly; the other remains frozen. The difference is not time spent, but how topology, constraints, and stored cost evolve during that time.

ATT therefore measures a composite condition:

- the magnitude of delta being held,
- the topology of pathways available for release,
- the cost of persistence under current constraints,
- and the cost of resolution if alignment were allowed.

Resolution occurs when these curves cross.

This is why effort often fails to shorten ATT—and may even extend it.

When effort is applied in a way that reinforces constraints, it increases the system's capacity to hold delta. A person "coping" indefinitely, a structure reinforced to withstand stress, an institution propped up through temporary fixes—all of these actions buy time in the clock sense while raising the threshold in the structural sense.

The system becomes more tolerant of misalignment, not closer to resolving it.

This also explains why ATT often ends abruptly.

When alignment finally becomes cheaper than persistence, resolution does not trickle out. It releases decisively. The system reorganizes because there is no longer any reason to delay. What changes is not speed, but permission.

From the outside, this looks like sudden transformation. From the inside, it feels inevitable.

ATT also clarifies a subtle but critical illusion: the belief that "if nothing is happening, nothing is changing."

In threshold-controlled regimes, the most important changes are invisible. Costs are shifting. Buffers are thinning. Margins are eroding. Pathways are narrowing. The system is not static—it is quietly approaching a boundary.

Time records the silence. ATT explains the snap.

This is why ATT cannot be rushed or negotiated. You cannot argue a threshold into moving. You can only alter the structure that defines it—by changing constraints, opening pathways, reducing holding cost, or accepting release.

Until then, duration accumulates without effect.

Once this is understood, ATT stops feeling frustrating or mystical. It becomes diagnostic. The right question is no longer *"How long will this take?"* but:

- *What is still making persistence affordable?*
- *Which constraint has not yet failed?*
- *What would lower the cost of release?*

With this clarity in place, the abstraction can now be grounded. The next sections will show ATT operating unmistakably in physical systems—where thresholds are measurable, undeniable, and free of interpretation.

Once seen there, ATT will never again feel like a psychological metaphor.

11.3 — Physical ATT: When Matter Suddenly Reorganizes

Physical systems provide the cleanest, least debatable evidence that resolution is threshold-gated. Here, there is no psychology, no narrative bias, no interpretation layer. Matter behaves the same way every time—not because it follows time, but because it obeys affordability conditions.

Consider phase transitions.

Water does not gradually become liquid as temperature rises. It remains solid while absorbing energy. Molecular agitation increases, but structure holds. Bonds stretch, vibrations intensify, yet the lattice persists. Nothing visible happens—until the melting threshold is crossed. Then, in a narrow interval, the entire structure reorganizes.

Time did not cause melting. Temperature did not *push* ice into water.

What changed was the cost balance. Holding the solid structure became more expensive than releasing it.

This is ATT in its purest form.

The same logic governs boiling, condensation, superconductivity, magnetization, and countless other physical transitions. Systems tolerate misalignment quietly, storing energy internally, until a boundary condition flips. When it does, alignment is rapid and decisive.

Friction offers another canonical example.

A block resting on a surface does not move when a small force is applied. Increase the force slightly—still nothing. The system absorbs stress through microscopic deformation. Internal contacts rearrange. Delta accumulates invisibly. This is static friction: a threshold regime.

Once the applied force exceeds the holding capacity of those contacts, motion begins instantly. The block does not ease into movement over time. It breaks free. The transition from static to kinetic friction is abrupt because the system crossed an alignment threshold.

Again, time was irrelevant.

A force held for one second or one hour produces the same result if the threshold is not crossed. Resolution does not care how patient you are. It cares whether release is permitted.

Fracture and yield points behave the same way.

Materials tolerate stress by redistributing it internally—stretching bonds, shifting dislocations, absorbing energy. From the outside, the object appears unchanged. Then, at a critical stress, it fails. Cracks propagate. Structure reorganizes violently. The failure appears sudden only because the holding phase was silent.

These examples matter because they destroy a comforting illusion: that physical systems always respond proportionally to input over time. They do not. Many are binary at the point of release, even if accumulation was gradual.

ATT captures this behavior precisely.

In physical systems:

- delta accumulates as stored energy or stress,
- thresholds define structural permission,
- resolution occurs when persistence becomes unaffordable.

There is no randomness here. There is no mystery. The apparent suddenness is not unpredictability—it is unobserved accumulation.

This is why time-based intuition fails so badly around physical transitions. We ask "why did it happen then?" as if *then* mattered. What mattered was not when the threshold was crossed, but that it was crossed at all.

Once physical ATT is understood, the concept becomes irreversible. It cannot be dismissed as metaphor or philosophical framing. Matter itself behaves this way—consistently, relentlessly, without exception.

With this foundation secured, the next move is unavoidable. If threshold-gated resolution governs matter, then living systems—built from matter, operating under survival constraints—must obey the same logic.

That extension leads directly into biological ATT.

11.4 — Biological ATT: Stress, Fatigue, and Survival Shifts

Biological systems are not optimized for smooth change. They are optimized for survival under uncertainty.

This single fact explains why living organisms are saturated with thresholds.

Life does not resolve delta as soon as it appears. It buffers it. Stress is absorbed. Damage is compensated. Fatigue is masked. Function is preserved long past the point where equilibrium would be cheaper—because premature resolution can be fatal. Survival favors delay.

This makes biology a natural theater for Alignment Threshold Time.

Consider physiological stress.

The body does not fail the moment stress appears. Hormonal systems compensate. Muscles recruit reserves. Attention sharpens. Energy is rerouted. Internally, delta increases, but the organism remains outwardly stable. This is not resilience in the moral sense; it is structural postponement.

As long as compensation is affordable, nothing changes.

Then, seemingly without warning, collapse occurs.

Exhaustion. Burnout. Injury. Immune failure. Panic. These are not sudden in origin. They are threshold releases. The organism crossed a point where maintaining function cost more than allowing failure.

Time did not cause the breakdown. Persistence did.

Fatigue offers an even clearer example.

A muscle does not weaken linearly with effort. It maintains output through recruitment strategies and metabolic shifts. You can lift a weight repeatedly with no visible decline—until suddenly you cannot lift it at all. The failure feels abrupt because the holding phase was invisible. Internally, resources were depleting, waste accumulating, pathways narrowing.

ATT governs the transition.

Healing follows the same logic in reverse.

Damage does not immediately resolve when injury occurs. Inflammation persists. Tissue waits. Scar formation does not begin until conditions allow alignment to proceed without compromising integrity. Healing takes time—but not because time heals. Healing takes time because the body waits until resolution is safe.

This is why rest often matters more than duration. Remove constraint, and resolution accelerates. Leave constraint in place, and ATT extends indefinitely.

Biology is full of such gates:

- sleep debt tolerated until sudden crash,
- hunger ignored until decisive action,
- pain masked until injury forces attention,
- aging hidden until thresholds of fragility are crossed.

In each case, the organism is not unaware. It is delaying alignment strategically.

This reframes many biological phenomena that are often moralized or medicalized prematurely. Collapse is not weakness. Burnout is not failure of will. Many "sudden" illnesses are structural releases after prolonged affordability.

ATT explains why "pushing through" often works—until it catastrophically doesn't.

Every biological system has a budget for holding delta. Once that budget is spent, resolution is no longer optional. Survival mechanisms switch modes. Function gives way to repair, withdrawal, or shutdown.

This is not pathology. It is design.

Understanding biological ATT changes the questions we ask. Instead of asking *how long* someone can endure, the better question becomes:

- *What is still making endurance affordable?*
- *Which buffer is being consumed right now?*
- *What threshold is approaching silently?*

With biology, the lesson is clear: life is not governed by clocks. It is governed by thresholds of viability.

And if thresholds dominate biology, then they must also dominate the most delicate living system of all—the human mind. That extension, often misunderstood and deeply misframed, is the subject of the next section.

11.5 — Psychological ATT: Insight, Freeze, and Looping

Psychological change is where time-based intuition fails most dramatically—and where ATT becomes unmistakable.

We often assume that understanding accumulates gradually, that effort produces proportional clarity, and that emotional states shift smoothly with reasoning. Anyone who has lived through insight, trauma, or chronic anxiety knows this is false.

The mind is a threshold system.

Consider insight.

You can think about a problem for hours, days, even years. You gather information, circle the same ideas, feel progress without resolution. Then, suddenly, the answer arrives whole. Not as a step, but as a reconfiguration. The experience is so abrupt that we call it a "flash."

Time did not deliver insight. Crossing a threshold did.

Cognitive delta—tension between incompatible frames—can be held only as long as existing mental structures can contain it. The mind tolerates ambiguity, contradiction, and partial understanding until the cost of maintaining them exceeds the cost of restructuring. At that moment, alignment occurs. The insight feels instantaneous because resolution was gated, not gradual.

Trauma operates through the same mechanism—but in the opposite direction.

In trauma, delta is overwhelming. The mind cannot afford resolution because resolution would destabilize the system. Instead of reorganizing, the system blocks. Attention freezes. Memory fragments. Emotional processing suspends. Time appears distorted or absent because resolution has been deferred beyond affordability.

This is not dysfunction. It is protection.

The freeze response is not a failure to heal. It is ATT extended indefinitely because the cost of alignment remains too high. Healing does not begin when time passes. Healing begins when conditions change—when safety, support, or capacity reduce the cost of release.

Anxiety loops reveal a subtler form of threshold failure.

In anxiety, delta remains unresolved but never crosses a threshold. The mind cycles repeatedly through the same patterns, attempting resolution but reinforcing the very constraints that block it. Reassurance, rumination, and avoidance all consume effort without altering affordability. Time passes. Nothing changes.

The loop persists not because the mind is weak, but because the threshold has not moved.

This is why "trying harder" so often backfires psychologically. Effort applied within the same frame increases internal tension without changing topology. The system remains below threshold—sometimes indefinitely.

Psychological ATT explains several common mysteries at once:

- why understanding does not guarantee change,
- why emotional shifts can be sudden and irreversible,
- why patience sometimes heals—and sometimes does nothing,
- why crises can catalyze growth when calm reasoning cannot.

Change occurs when permission shifts, not when explanations accumulate.

This reframing is deeply corrective.

It removes moral judgment from psychological struggle. People are not stubborn, lazy, or resistant. They are operating under unmet threshold conditions. Resolution is blocked not by ignorance, but by cost.

The right psychological questions, therefore, are not:

- *Why can't they just change?*
- *Why doesn't time fix this?*

They are:

- *What is still making misalignment affordable?*
- *What makes resolution feel unsafe or too expensive?*
- *Which constraint must shift before alignment can occur?*

Once this lens is adopted, psychological change becomes legible, compassionate, and precise. The mind stops being a mystery governed by moods and starts being a system governed by threshold economics.

And if thresholds govern individuals, they must govern groups as well. Societies, institutions, and cultures hold delta exactly the same way—quietly, visibly stable, until permission collapses.

That collective version of ATT is the final extension of this concept.

11.6 — Social ATT: Tipping Points, Revolutions, and Sudden Collapse

Social systems are the largest and most dangerous threshold systems humans interact with—because their deltas are collective, their buffers are institutional, and their releases are rarely gentle.

From the outside, societies appear stable until they are not. Decades of tension coexist with routine life. Institutions function. Laws hold. Markets operate. People adapt. The surface gives the impression of continuity, even progress.

Internally, delta is being held.

Inequality accumulates. Trust erodes. Narratives diverge. Legitimacy thins. Energy is spent not on resolution, but on containment. Like biological systems under chronic stress, societies survive by postponement.

This is social ATT.

Revolutions do not begin when people become angry. They begin when holding anger becomes unaffordable.

A society can tolerate injustice for generations if its structures absorb dissatisfaction—through distraction, reform signals, repression, or hope narratives. None of these resolve delta; they merely buy time. As long as postponement works, nothing changes.

Then something minor happens.

A single event triggers mass unrest. A policy decision sparks revolt. A symbolic act ignites a movement. Observers are baffled: *Why now? Why over this?*

The question is wrong.

The event did not cause the change. It crossed a threshold.

Social ATT explains why triggers are rarely proportional to outcomes. The visible spark is not the source of change—it is the moment when persistence finally becomes more expensive than release.

The same logic governs institutional collapse.

Organizations do not fail when mistakes occur. They fail when error-correction costs exceed inertia costs. Corruption, inefficiency, and misalignment can persist indefinitely as long as the system can absorb them. Whistleblowing, audits, and reforms often strengthen containment rather than resolve core delta.

Then, suddenly, credibility collapses. Trust evaporates. Exit cascades begin. What looked solid fractures rapidly—not because things worsened overnight, but because permission ended.

Markets follow identical logic.

Bubbles persist not because participants are ignorant, but because exiting early is costly. Risk accumulates quietly. Fragility increases. When confidence crosses a threshold—often triggered by a trivial signal—selling cascades. Value evaporates faster than it was created.

Again: time did not cause the crash. ATT ended.

Social ATT also explains why warnings are ignored.

Early signals do not produce change because they do not alter affordability. Awareness without structural shift merely increases tension. This is why societies can *know* they are heading toward crisis and still do nothing. Knowledge does not move thresholds.

Only constraint changes do.

This understanding removes the temptation to frame social change as moral awakening or collective madness. Societies do not suddenly become brave or irrational. They simply lose the ability to delay alignment.

Once that happens, resolution is rapid, nonlinear, and irreversible.

This is why history appears punctuated. Long periods of apparent stability are followed by sudden reorganization. The pattern repeats across civilizations, economies, technologies, and cultures.

ATT is the hidden clock behind all of them.

By the time change becomes visible, it is already too late to prevent it. The threshold has been crossed. Alignment is underway.

Recognizing social ATT does not grant control—but it grants clarity. It shifts attention away from surface events and toward structural affordability. It teaches us to watch buffers, not headlines.

With this, ATT is no longer an abstract idea. It has revealed itself across matter, life, mind, and society—unchanged in principle, varied only in expression.

What remains is to close this chapter by reconnecting ATT to time itself—showing why duration misleads, why thresholds dominate, and why alignment always arrives *exactly* when it must, and never a moment earlier.

11.7 — Why ATT Feels Mysterious (and Isn't)

Alignment Threshold Time feels mysterious for one simple reason: it violates our time-based intuition.

We are conditioned to believe that if nothing is changing visibly, nothing is changing at all. We equate duration with progress. We assume that enough time, effort, or explanation should eventually force movement. ATT quietly dismantles all three assumptions.

Threshold systems do not negotiate with time.

They negotiate with cost.

As long as holding delta remains affordable, the system has no incentive to resolve—no matter how long the wait, how intense the pressure, or how persuasive the argument. Time accumulates, but permission does not change. From the outside, this looks like stagnation. From the inside, it is active containment.

ATT feels mysterious because the decisive variables are hidden.

What accumulates is not motion, but strain. What changes is not position, but viability. What matters is not duration, but which buffer is thinning.

Humans are poor at perceiving these shifts because they are not linear and not visible until they fail. We notice motion. We notice events. We do not notice affordability curves quietly crossing.

This is why threshold crossings feel sudden and unearned.

We say:

- *"It came out of nowhere."*
- *"Everything was fine yesterday."*
- *"No one could have predicted this."*

In reality, everything necessary was already in place. The system had simply been successfully hiding its instability.

Time-based models reinforce this blindness.

When we expect smooth trajectories, we interpret non-movement as delay rather than blockage. We assume patience will help where restructuring is required. We try to accelerate outcomes by adding effort, information, or pressure—often strengthening the very constraints that keep the threshold in place.

ATT exposes why this fails. You cannot rush a threshold. You cannot persuade it. You cannot wait it out. You can only change the structure that defines it.

This is also why ATT cannot be gamed.

Once alignment becomes affordable, release happens automatically. There is no bargaining phase. No grace period. No partial resolution. The system does not ask whether the timing is convenient. It resolves because it must.

This inevitability is often mistaken for fate or destiny. It is neither. It is the simplest form of structural honesty.

ATT is not mysterious because it is hidden. It is mysterious because it is indifferent to narrative. Stories care about intention. Thresholds care about cost. When intention and cost diverge, cost wins—silently, reliably, without exception.

Understanding this does something subtle but profound. It shifts our relationship with change. We stop asking *when* something will happen and start asking *what is still making delay possible*. We stop moralizing collapse and start recognizing exhaustion. We stop celebrating sudden success as luck and start seeing accumulated readiness.

ATT replaces surprise with legibility. Nothing happens "all of a sudden." It happens exactly when it becomes unavoidable.

And with that realization, the illusion that time governs change finally dissolves. Time records waiting. Thresholds decide release. Resolution follows permission, not patience.

This closes the conceptual arc of this part. Time has been reframed. Motion has been redefined. Scale has been accounted for. And now, inevitability has been explained—without mystery, without metaphysics, and without exception.

From here on, the book no longer needs to ask *why change happens when it does*. The answer is already installed.

From Thresholds to Governance

With ATT fully visible, something irreversible has happened to the reader's intuition about time, motion, and change.

Time no longer feels like a driver. Motion no longer feels like something "applied." Sudden change no longer feels accidental or mystical. What remains is structure.

Across matter, life, mind, and society, the same rule has repeated without variation: alignment occurs when delay becomes unaffordable. Duration records the waiting, but thresholds decide the release. Effort may accumulate, but only structural shifts unlock resolution.

This reframing quietly dissolves many false debates.

There is no need to argue whether change is fast or slow. There is no need to choose between patience and urgency. There is no need to invoke randomness, destiny, or moral reckoning.

All of these are narrative overlays on a simpler mechanism.

Once ATT is understood, the question that naturally follows is no longer *what happens next*, but *how thresholds are shaped in the first place*. Who designs the constraints? Who benefits from delay? Who pays the cost of holding delta open? And who bears the impact when release finally occurs?

These are no longer questions of physics or psychology alone. They are questions of systems, governance, and power.

If thresholds govern reality, then influence lies not in forcing outcomes, but in controlling affordability. Those who can lower the cost of persistence can delay change indefinitely. Those who raise the cost of holding delta can trigger rapid alignment without appearing to act.

This is the unseen architecture behind leadership, institutions, technologies, and civilizations.

The next part of this book moves deliberately into that terrain.

Having dismantled time as an illusion of flow and exposed thresholds as the true regulators of change, UED now turns to systems that deliberately manage delta—not accidentally, but strategically.

From here on, we are no longer describing what nature does by default. We are examining what humans do when they learn to work with—or against—those defaults.

The language will shift slightly. The stakes will rise noticeably. But the rule will not change. Delta still seeks resolution. Equilibrium still attracts. Consequence still accounts. Only now, the management of thresholds becomes intentional. And that is where understanding stops being passive and starts becoming responsibility.

Some things do not take time. They take permission.

Multi-Field Coupling Model

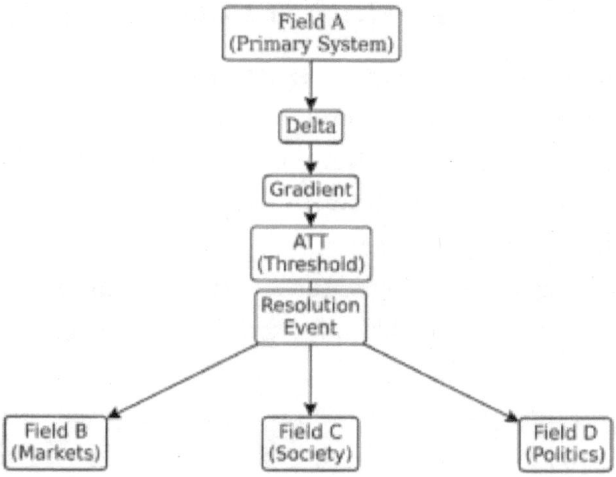

Figure 3 — Multi-Field Coupling in Universal Energy Dynamics

When delta resolves in one field, its consequences rarely remain localized. Because systems are coupled through shared constraints and signals, resolution in one domain propagates into others. Economic adjustments influence social narratives, political responses reshape institutional behavior, and technological changes alter market gradients. Stability therefore depends not on eliminating change, but on managing how consequences propagate across connected systems.

Deliverable of Part III

When Time and Motion Lose Their Mystery

By the end of this part, something subtle but permanent should have shifted in the reader's inner model of reality.

Time no longer feels like a river you are carried by. Motion no longer feels like something imposed from the outside. Change no longer feels unpredictable, capricious, or magical.

All three collapse into a single, quieter understanding:

Time and motion are bookkeeping artifacts of staged alignment.

What we traditionally call *time* is now seen as nothing more than a measure of how long delta remains unresolved under given constraints. It is not an independent dimension flowing forward; it is a record of waiting—a counter that increments while alignment is postponed. When nothing is resolving, time appears to "pass." When thresholds are crossed, time appears to "jump."

Motion, in turn, is no longer mysterious displacement. It is the visible trace of resolution proceeding along permitted pathways. Where alignment is cheap and continuous, motion looks smooth. Where alignment is gated, motion appears jerky, discontinuous, or sudden. Nothing "moves" because it wants to; things reconfigure because holding their current state has become too costly.

With ATT installed, these patterns are no longer confusing.

- Sudden phase changes are not anomalies; they are threshold releases.
- Quantum irregularities are not breakdowns of time; they are failures of repetition under fragile alignment.

- Relativistic dilation is not distortion of time; it is coherence preservation under propagation limits.
- Psychological breakthroughs are not moments of inspiration; they are cognitive realignments past affordability.
- Social upheavals are not madness; they are collective thresholds collapsing at scale.

All of them follow the same logic.

What once required separate explanations—classical mechanics here, quantum weirdness there, relativity elsewhere, psychology somewhere else—now shares a single accounting language: delta held, delta released, delta redistributed.

Time counts how long holding lasted. Motion shows where release went. ATT explains why release waited.

Nothing is added to reality by this reframing. No new entities are introduced. No metaphysical layers are invoked. What disappears instead is confusion.

Mystery dissolves not because everything becomes simple, but because everything becomes legible.

After this part, the reader should no longer ask:

- *Why did this happen now?*
- *Why did nothing change for so long?*
- *Why was the shift so sudden?*

Those questions have already been answered—structurally, not narratively.

What remains to be explored is far more consequential.

If time and motion are not fundamental, but derivative, then the true levers of reality are not clocks and forces—but constraints, thresholds, and permissions. Understanding them explains nature. Managing them shapes civilization.

With this deliverable complete, UED has finished dismantling the illusion of flow.

From here onward, the framework turns outward—toward systems that actively manipulate delta, delay alignment, and harvest consequence.

The mystery phase is over. The responsibility phase begins.

Part IV — Geometry Reframed

Up to this point, UED has dismantled *when* things change and *why* they wait. What remains unresolved is how change is allowed to move at all. That question cannot be answered with time or energy alone. It demands a re-examination of geometry itself.

This part makes a decisive shift: geometry is no longer treated as passive space in which events occur, but as the active constraint structure that determines which resolutions are even possible. Space stops being a container and becomes a permission map.

Here, trajectories are not paths traced by objects, but corridors of least resistance carved by constraints. Distance is not empty extension to be crossed, but the number of resolution steps required under given limits. What looks like motion through space is revealed as alignment progressing through an allowed geometry.

Fields are then stripped of metaphor. Pressure, potential, voltage, temperature—long treated as separate physical quantities—are unified as different languages describing the same thing: gradient. Not a force pulling from afar, but a local statement of misalignment demanding resolution. "Field" gets closer to reality than "force," yet still carries historical baggage that UED must cleanly expose.

Finally, force itself is demoted from cause to consequence. Force does not initiate motion; it appears only when resolution is obstructed. It is a symptom of constraint, not an agent of change. This explains both why force-language works so well in engineering—and why it repeatedly fails at foundational levels of physics.

By the end of this part, geometry will no longer feel abstract or mathematical. It will feel operational.

The reader should stop asking:

- *Where is this happening?*
- *What force is acting?*

And instead begin asking:

- *What resolutions are geometrically allowed here?*
- *Which pathways are constrained—and which are open?*

This part completes the transition from descriptive physics to structural physics. Once geometry is reframed as constraint, the remaining paradoxes of motion, force, and fields stop multiplying.

They simply fall into place.

Chapter 12 — Geometry as Allowed Pathways of Resolution

Why Space Was Never Empty

From the moment geometry entered physics, it was treated as a silent stage. Space became the place where things happen, not something that *does* anything itself. Lines, points, and distances were assumed to exist independently of what moved through them, like a neutral grid waiting to be occupied.

This assumption was convenient. It was never correct.

The idea of empty space—pure extension without preference, resistance, or permission—has always been a mathematical fiction. Useful for calculation, disastrous for intuition. It taught us to imagine that objects travel *through* space, when in reality they resolve *within* constraints.

Nothing meaningful ever happens in emptiness.

If space were truly empty, motion would be meaningless. There would be no reason for paths to curve, for trajectories to persist, or for certain movements to be easier than others. Every direction would be equivalent. Every outcome would be arbitrary. Nature does not behave this way.

What we call geometry is not a container. It is a map of allowances.

Every geometric feature—distance, curvature, boundary, orientation—quietly encodes what resolutions are cheap, which are costly, and which are impossible. Geometry does not host motion; it filters it. It decides, long before any dynamics appear, which rearrangements can proceed smoothly and which will meet resistance.

This is why geometry always shows its hand indirectly.

We notice it when particles curve without visible force. We notice it when orbits persist without constant correction. We notice it when waves follow guides, channels, and boundaries. We notice it when motion takes a long way around instead of a short way through.

Each time, we invent explanations after the fact—forces, interactions, fields—without asking the simpler question: *what paths were permitted in the first place?*

UED insists on starting there.

If motion is resolution, then geometry is the permission structure for resolution. It is not something added to reality; it is the shape of reality's constraints. To move is not to travel across emptiness, but to advance along a corridor that remains open under equilibrium pressure.

This reframing does something subtle but powerful. It removes the need to imagine space as passive and dynamics as active. Instead, geometry becomes active in the quietest possible way: by allowing some things to happen and preventing others—without force, intention, or effort.

Once this is seen, familiar questions begin to feel misplaced.

Not *where is the object going?* But *which resolutions are allowed here?*

Not *how far did it move?* But *how many constraints were crossed?*

This chapter is not about replacing geometry with philosophy. It is about restoring geometry to its rightful role: the first filter through which all motion, time, and force must pass.

Before anything moves, something far more fundamental has already decided—whether movement is even possible.

12.1 — Geometry as a Constraint Map, Not a Container

Once space is no longer treated as empty extension, geometry immediately changes character. It stops being something that holds reality and becomes something that shapes what reality can do. Geometry is not where events occur; it is the constraint map that governs which resolutions are allowed to occur at all.

A container does nothing. A constraint decides everything.

If geometry were merely a backdrop, there would be no reason for certain paths to persist, for others to curve, or for some movements to require effort while others proceed effortlessly. Yet nature is saturated with preference. Motion consistently follows narrow corridors. Alignment repeatedly chooses specific routes. These regularities are not accidents—they are geometric permissions made visible.

Every boundary is a rule. Every curvature is a bias. Every topology is a filter.

In UED terms, geometry encodes how equilibrium pressure is allowed to express itself. Delta seeks resolution, but geometry determines *how* that resolution can unfold. Without constraints, delta would collapse instantly. With too many constraints, resolution would be impossible. Geometry is the negotiated middle: enough structure to guide alignment, enough freedom to allow it.

This is why geometry precedes dynamics.

Forces cannot act where geometry forbids motion. Time cannot accumulate where geometry blocks resolution. Even energy becomes irrelevant if no pathway exists for its redistribution. Geometry silently vetoes possibilities before any equation begins to operate.

Consider boundaries.

A wall is not just an object; it is a geometric prohibition. It does not "push back" until resolution attempts to cross it. What we later describe as force is simply the cost of attempting alignment through a forbidden region. The resistance was not created by motion—it was already encoded in geometry.

Curvature behaves the same way.

Paths curve not because something is pulling them, but because straight resolution is not geometrically permitted. What looks like deflection is simply alignment following the cheapest available corridor under constraint. The geometry did not react; it already decided.

This is why geometry feels invisible until it fails.

When constraints are stable, motion looks natural, even obvious. When constraints change—when a boundary appears, disappears, or deforms—motion suddenly becomes "strange." We rush to invent new forces or interactions, instead of noticing that the permission map has shifted.

UED removes this confusion by naming geometry correctly.

Geometry is the pre-dynamic architecture of reality. It does not explain motion after the fact; it determines motion in advance by shaping the space of possible resolutions. Every trajectory, every orbit, every flow pattern is a consequence of this architecture—not an argument against it.

Once geometry is understood as constraint rather than container, the next question becomes unavoidable: if geometry governs permission, why do certain paths repeat so reliably?

The answer leads directly to trajectories—not as traces left by moving objects, but as stable corridors where resolution remains continuously affordable.

12.2 — Why Nature "Chooses" Certain Paths

The language of physics is full of quiet anthropomorphism. We say nature *chooses* the shortest path, *prefers* straight lines, or *follows* least-action principles. These phrases are useful shorthand—but they smuggle intention where none exists.

Nature does not choose. Constraints choose by elimination.

In UED, paths are not selected; they are filtered. Among countless possible rearrangements, only a narrow subset remains affordable under equilibrium pressure. What survives this filtering appears to us as preference, elegance, or lawfulness.

This is why certain paths recur with uncanny reliability.

A ray of light does not decide to travel straight. A fluid does not aim for the easiest route. A planet does not intend to orbit. In each case, the system is simply resolving delta through a corridor where resistance is minimal and permission is continuous. All other options are quietly rejected—not by force, but by cost.

Least-resistance paths are not optimal by design; they are the only ones that remain viable.

This reframing dissolves a long-standing confusion. When we say a system follows the "path of least action," we often imagine action as something being minimized dynamically. In reality, most alternative paths are never dynamically realized at all. They are geometrically expensive, unstable, or blocked. What remains is not chosen—it is left standing.

Geometry performs this filtering before motion begins.
Constraints define:

- which directions are open,
- which gradients can be followed,
- which reconfigurations can persist without accumulating prohibitive cost.

The result is that motion appears purposeful even though it is purely structural. Paths look intentional because they are survivors of constraint, not products of decision.

This also explains why changing constraints changes "behavior" instantly.

Introduce a boundary, and paths redirect. Alter curvature, and trajectories bend. Modify topology, and stable routes disappear.

Nothing had to learn. Nothing had to adapt. The permission map changed, and resolution obeyed immediately.

Seen this way, geometry does not merely influence motion—it preselects reality's options. What we later describe with equations and principles is simply a description of the corridors that remained open.

This is why nature looks economical, efficient, and consistent without being deliberate. It is not optimizing; it is complying.

With this understanding, the idea of a trajectory can finally be cleaned of its historical baggage. A trajectory is not the trace of an object moving through space—it is the visible imprint of a corridor that stayed affordable long enough for resolution to pass through it.

That reframing is the focus of the next section.

12.3 — Trajectories as Stable Resolution Corridors

Once geometry is understood as a constraint map, the concept of a trajectory changes completely. A trajectory is no longer evidence that an object *moved* through space. It is evidence that a particular pathway of resolution remained affordable long enough to be traced.

Trajectories are not drawn by motion. They are revealed by permission.

What we traditionally call a path is, in UED terms, a stable corridor—a region of geometry where delta can continuously resolve without encountering prohibitive cost. As long as this corridor remains open, resolution flows through it, and we observe what looks like smooth, persistent motion.

This is why trajectories exhibit such stubborn consistency.

An orbit persists not because something keeps pushing a body along it, but because the corridor defined by constraints, gradients, and equilibrium pressure remains open. Remove or alter those constraints, and the orbit disappears immediately—without any need to "undo" motion. The path vanishes because the permission vanishes.

Flow lines in fluids behave the same way. They are not carved by particles deciding where to go; they are channels of minimal resistance defined by boundary geometry, pressure gradients, and continuity constraints. The fluid merely reveals them. Change the container shape, and entirely new trajectories appear without delay.

Even waveguides and optical paths follow this logic.

Light does not follow a path because it is "bent" continuously by force. It follows a path because the geometry of the medium makes some directions of propagation cheap and others expensive. The trajectory is not an afterimage of motion—it is the spine of allowed alignment.

This reframing explains a critical but often ignored fact: trajectories tend to be stable against small disturbances.

Minor perturbations rarely destroy a trajectory. They are absorbed, redistributed, or damped. The path reasserts itself because it is not held together by precision, but by structural affordability. As long as the corridor exists, resolution finds its way back into it.

Trajectories only break when geometry changes.

A constraint shifts. A boundary appears or disappears. A topology crosses a threshold.

Then motion that once seemed natural suddenly becomes impossible. What we describe as "instability" or "chaos" is often nothing more than a corridor losing continuity. The system is forced to search for a new path because the old one no longer exists.

This is why trajectories should never be treated as fundamental entities. They are emergent features of constraint stability. They persist not because motion is conserved, but because permission is.

Understanding this dissolves several long-standing confusions at once:

- why paths recur across scales,
- why motion can be smooth without continuous forcing,
- why small forces do not always redirect motion,
- and why sudden deviations occur when geometry shifts slightly.

Once trajectories are seen as resolution corridors, the next question becomes unavoidable: if paths are defined by permission, what do we really mean by *distance* along those paths?

That question exposes one of the deepest misconceptions inherited from container-space thinking—and leads directly to the reframing of distance itself.

12.4 — Distance Reframed: Resolution Steps, Not Extension

Distance is one of the most deeply ingrained illusions inherited from container-space thinking. We are taught to imagine distance as empty extension—an inert stretch of nothingness separating two points. Motion, in that picture, is the act of crossing this emptiness.

UED replaces this entirely.

Distance is not empty space. Distance is the amount of resolution required under constraint.

What separates two states is not length, but how many alignment steps must occur before equilibrium pressure can relax. Distance is therefore not absolute; it is contextual, structural, and conditional.

This is why distance behaves so strangely across domains.

A particle may traverse astronomical spatial extension effortlessly, while a thought can feel impossibly "far" despite no physical separation at all. A chemical reaction may stall inches from completion, while a signal crosses continents instantly. These are not paradoxes. They are reminders that distance has never been about meters—it has always been about permission density.

In UED terms, distance counts resolution stages, not coordinates.

Each stage represents:

- a constraint that must be negotiated,
- a gradient that must be followed,
- a topology that must remain open,
- a threshold that must not be violated.

If many such stages exist, distance is large—even if spatial extension is small. If few exist, distance is small—even across vast scales.

This reframing immediately explains several long-standing confusions.

Why can "short" distances be hard to cross? Because the resolution corridor is narrow, obstructed, or threshold-gated.

Why can "long" distances be crossed easily? Because the corridor is smooth, open, and energetically cheap.

Why does tunneling appear to violate distance? Because the system bypasses intermediate resolution stages by exploiting an alternative corridor.

Why does delay appear without obstruction? Because distance is measured in unresolved steps, not in visible barriers.

Once distance is understood this way, geometry becomes operational rather than abstract. A metric is no longer a ruler laid over space; it is a cost function that tells you how expensive alignment will be along a given path.

This is why geometry and effort are inseparable.

Moving "far" always means paying more—not in time necessarily, but in alignment work. When that work is distributed smoothly, motion feels continuous. When it is concentrated at thresholds, motion feels jerky or delayed. The apparent behavior changes, but the accounting does not.

This also explains why distance collapses near thresholds.

At critical points—phase transitions, decision points, tipping points—the final resolution step carries disproportionate weight. Once it occurs, the remaining distance vanishes instantly. What felt infinitely far a moment ago suddenly feels trivial. The corridor opens, and alignment completes.

Distance did not shrink. Constraints dissolved.

By detaching distance from emptiness and attaching it to resolution cost, UED prepares the ground for a deeper insight: geometry does not merely shape paths—it determines whether motion appears smooth, discrete, or frozen.

That distinction—between continuous and discontinuous motion—is not about scale or substance. It is about how geometry parcels permission.

This leads directly into the next section, where geometry reveals why motion sometimes flows effortlessly and sometimes arrives in jumps.

12.5 — Geometry Explains Why Motion Looks Smooth or Jerky

One of the most persistent puzzles in physics—and across many other domains—is why motion sometimes appears smooth and continuous, while at other times it appears discrete, delayed, or sudden. Traditional explanations invoke scale ("classical vs quantum"), substance ("particles vs waves"), or mathematical convenience ("continuous models vs discrete updates").

UED shows that none of these are primary.

The appearance of smoothness or jerkiness is not a property of motion itself. It is a property of how geometry parcels permission.

When geometric constraints allow resolution to proceed continuously—without encountering thresholds—motion appears smooth. Alignment advances in small, affordable steps. Each step blends seamlessly into the next because no single step carries disproportionate cost. Time looks continuous. Trajectories feel natural. Nothing calls attention to itself.

This is not because reality prefers continuity. It is because permission is evenly distributed.

In contrast, when geometry concentrates permission into discrete corridors separated by barriers, motion cannot proceed gradually. Delta accumulates until a threshold is crossed. Resolution then occurs in a burst.

To an observer, motion appears jerky, discontinuous, or quantized—not because nature is granular by default, but because geometry only allows release in chunks.

Quantization is not a mystery of scale. It is a signature of constrained geometry.

This reframing immediately dissolves the false divide between "continuous" and "discrete" worlds. Both are outcomes of the same rule operating under different constraint structures. Where geometry is permissive and smooth, motion flows. Where geometry is gated and segmented, motion jumps.

The same object can exhibit both behaviors under different conditions.

A fluid flows smoothly through an open channel, yet drips discretely through a narrow aperture. An electron propagates smoothly in free space, yet occupies discrete states within a bound structure. A person changes gradually in routine contexts, yet undergoes abrupt transformation at life thresholds.

Nothing fundamental changed about the entity. Only the geometry of permission changed.

This also explains why increasing resolution in measurement often reveals apparent discreteness. As observation probes finer constraint structures, previously hidden thresholds become visible. What once looked smooth reveals its gating. The discontinuity was always there—it was simply below the resolution of perception.

Importantly, this reframing removes the temptation to treat discreteness as evidence of underlying "chunks of reality." Discreteness is not ontological. It is geometric.

Where permission is continuous, continuity appears. Where permission is gated, discreteness appears.

Understanding this also clarifies why attempts to force continuous models onto thresholded systems fail, and why discrete models mislead when applied where permission is smooth. The mistake is not mathematical—it is geometric.

Once this is seen, geometry stops being a passive description and becomes an explanatory engine. It tells us, in advance, whether motion will feel fluid or stalled, predictable or abrupt, classical or "weird."

With this clarity established, one final step remains for this chapter: to show why geometry must be placed *before* force and time in any coherent framework.

That placement completes the reframing—and closes the loop opened at the start of this book.

12.6 — Why Geometry Is Prior to Force and Time

Once geometry is correctly understood as a permission structure, the usual hierarchy of physics quietly collapses. Force and time—long treated as primitives—are revealed as secondary effects, meaningful only after geometry has already decided what is possible.

Nothing can be forced where geometry forbids resolution.

A force does not create a path; it only expresses resistance when a path is constrained. Push harder against a wall and force increases—not because something new is acting, but because resolution is geometrically denied. If geometry allowed passage, the same effort would produce motion with little or no force signature at all.

This immediately exposes a fundamental inversion in mainstream intuition.

We are taught:

- force causes motion,
- motion unfolds in time,
- geometry describes where this happens.

UED shows the opposite ordering:

- geometry defines allowed resolution corridors,
- resolution produces motion,
- force appears only when resolution is obstructed,
- time records how long permission remains gated.

Force, therefore, is not a driver. It is a diagnostic.

It tells us that equilibrium pressure wants to resolve, but geometry is resisting. Where geometry opens, force vanishes. Where geometry closes, force spikes. The force is not doing the work—the constraint is charging the cost.

Time follows the same logic.

Time cannot accumulate meaningfully where geometry blocks change. In frozen systems—whether physical, biological, or psychological—duration passes without consequence. Nothing resolves, nothing moves, nothing reorganizes. Clock time continues, but structural time is stalled.

Conversely, when geometry suddenly opens—when a threshold collapses—resolution can occur rapidly, compressing what looks like "a lot of change" into very little clock time. We misinterpret this as time speeding up or slowing down, when in fact permission has changed.

This is why geometry must come first.

Before any equation can describe motion, geometry has already filtered which motions are admissible. Before any force can be applied, geometry has already decided whether that force will manifest as motion or resistance. Before any time parameter makes sense, geometry has already determined whether resolution can proceed continuously or must wait.

This ordering resolves a long-standing discomfort in physics: why force laws work beautifully in engineering but fracture at foundational levels; why time behaves cleanly in classical systems but becomes ambiguous in quantum and relativistic regimes.

The laws did not fail. They were applied after the wrong starting point.

By placing geometry first, UED does not discard force or time. It demotes them to where they belong—as derived, situational descriptors that operate within an already-defined permission landscape.

Once geometry is recognized as prior, the framework becomes internally consistent across scales. There is no need to invent new forces, hidden dimensions, or metaphysical corrections. The same ordering holds everywhere:

1. Geometry defines permission.
2. Delta seeks resolution within that permission.
3. Force appears if permission resists.
4. Time records the duration of delay.

This hierarchy is not philosophical. It is operational. It works because it matches how reality behaves when stripped of narrative habit.

With geometry now firmly placed at the foundation, the chapter can close by reconnecting this insight to what follows next—gradients, fields, and the texture of constraint itself.

Because once geometry defines *where* resolution can happen, something else must explain *why* it flows in one direction rather than another.

That "why" is not force. It is gradient.

And that is where the next chapter begins.

From Geometry to Gradients

With geometry now correctly placed at the foundation, motion, force, and time lose their illusion of primacy. What once looked like dynamic causation is revealed as something far quieter: permission expressed through structure.

Geometry does not push, pull, or accelerate. It simply allows—or refuses to allow—resolution to pass.

Every path traced, every orbit sustained, every motion stalled has already been decided before dynamics appear. What we observe as behavior is merely the visible negotiation between delta and the geometric constraints that shape it.

But geometry alone is not enough.

Geometry tells us where resolution may occur, but not why resolution proceeds in one direction rather than another. It defines corridors, not flow. A corridor can exist without being used. Permission does not guarantee motion.

For resolution to advance, something must *bias* the direction of alignment within the allowed space. Something must give texture to geometry—marking high-cost regions and low-cost ones, steep paths and gentle slopes.

That bias is what we call gradient.

Gradients are not added to geometry; they emerge from it. Where constraints shape space unevenly, delta expresses itself as directionality.

Pressure, potential, temperature, voltage—these are not separate forces acting in different domains. They are different dialects of the same structural reality: misalignment distributed across permitted geometry.

The next chapter moves directly into that texture.

Having established geometry as the map of permission, we now turn to gradients as the language of motivation—not psychological motivation, but structural necessity. Gradients explain why resolution prefers one corridor over another, why motion accelerates here and stalls there, why equilibrium pressure takes shape rather than remaining abstract.

From this point forward, the framework becomes even more operational.

No metaphors. No invisible pulls. No implied intentions.

Just constraints, gradients, and the consequences of allowing alignment to proceed.

If geometry is the map, gradients are the terrain.

And once terrain is understood, the final illusion—force as cause—will have nowhere left to hide.

Before anything moves, pushes, or flows, something far quieter decides: *Is resolution even allowed to pass this way?*

Chapter 13 — Fields Without Metaphor

Gradients as Reality's Texture

Why "Field" Was Invented (and Why It Still Confuses)

When force began to fail as an explanation, physics did not abandon the idea of something *acting*. It simply made that something more subtle. Pushes and pulls were replaced with influences. Contact was replaced with presence. And the word that carried this shift was field.

Fields were an improvement. They removed the need for mysterious action-at-a-distance. They allowed continuity. They fit beautifully into mathematics. For a while, they felt like the missing layer between geometry and motion.

But they also inherited a problem that force-language could not escape.

They quietly reintroduced agency.

A field, as it is commonly imagined, is something that exists everywhere, doing something everywhere, all the time. It exerts influence. It mediates interaction. It fills space with potential. Even when stripped of personality, it behaves like an entity.

This is where confusion begins.

The moment we treat a field as something that *acts*, we are forced to ask awkward questions. Where does it come from? How does it store energy? Why does it exist here and not there? How does it know what to influence?

These questions multiply because they are aimed at the wrong target.

Fields were never meant to be things. They were meant to be descriptions.

Historically, the field concept emerged because force-language could not account for continuity. Forces appeared only when something moved or resisted. Fields allowed us to describe *conditions* before motion occurred—regions where something *would* happen if permission were granted.

That was the right instinct.

The mistake was stopping halfway.

Instead of recognizing fields as structured descriptions of imbalance, we elevated them into ontological layers. We began speaking of electric fields, gravitational fields, quantum fields as if they were residents of space, rather than maps of how delta is distributed under constraint.

UED makes the correction explicit.

A field is not an entity. A field is a language.

More precisely, it is a way of encoding gradient information—how misalignment varies from place to place within allowed geometry. Pressure gradients, temperature gradients, voltage gradients, concentration gradients: these are not different phenomena. They are different dialects describing the same structural fact.

Something is uneven. Resolution is cheaper in one direction than another.

Nothing more is required.

By stripping the field of its metaphysical clothing, UED does not weaken physics—it strengthens it. What remains is cleaner, more general, and more faithful to experience. Gradients stop being mysterious influences and become what they always were: reality's texture.

This chapter completes a necessary cleanup.

It keeps what field-language got right—continuity, locality, directionality—and discards what it accidentally imported—agency, substance, and hidden actors. What emerges is not a new force or a new entity, but a unified way of speaking about imbalance across every domain.

Once gradients are seen clearly, force will no longer appear as a cause. It will appear as what it actually is: resistance encountered when gradients meet constraint.

But before that final demotion, gradients themselves must be understood on their own terms.

That understanding begins with recognizing them for what they are: not things in space, but statements about where resolution wants to go.

13.1 — Gradients as Delta Languages

Once fields are stripped of metaphysical weight, what remains is surprisingly simple. Every field we measure—every pressure map, temperature distribution, voltage profile, or chemical potential—is nothing more than a statement of delta expressed in a domain-specific language.

A gradient is not a thing acting on matter. It is a sentence describing misalignment.

Pressure does not push. It tells you where density imbalance exists. Temperature does not flow. It tells you where kinetic mismatch persists. Voltage does not pull charges. It tells you where electrical asymmetry is stored. Concentration does not drive diffusion. It tells you where compositional delta remains unresolved.

Each gradient answers the same structural question: *In which direction is resolution cheaper?*

This is why gradients are always directional and local. They do not exist as global commands or universal influences. They exist as comparisons—here versus there, now versus nearby. A gradient without contrast is meaningless. Remove delta, and the gradient disappears instantly.

In UED terms, gradients are delta made legible.

Different domains invented different names because they observed different carriers—molecules, charges, particles, heat—but the underlying structure never changed. Each discipline measured misalignment using the tools available to it and mistook the measurement for a new phenomenon.

UED collapses these distinctions cleanly.

Pressure gradient, temperature gradient, electric potential gradient, chemical potential gradient, gravitational potential gradient—these are not separate forces or fields. They are expressions of one condition: uneven distribution under constraint.

This unification has a powerful consequence.

It explains why gradients do not require intention, agency, or mediation. Nothing needs to "flow" for a gradient to exist. A system can sit perfectly still while gradients persist. The gradient is a *promise*, not an action. It encodes what would happen if resolution were allowed to proceed.

This is why gradients can exist without motion.

A blocked pipe can sustain a pressure gradient indefinitely. A capacitor can hold voltage without current. A thermal gradient can persist across an insulator. A psychological tension can remain without behavior.

In each case, delta is present. Geometry is constraining resolution. The gradient records this condition faithfully.

Understanding gradients as languages rather than entities also clarifies why they vanish the moment alignment occurs. When pressure equalizes, the pressure gradient is gone—not because it was consumed, but because there is nothing left to describe. The language has nothing to say.

This reframing also explains why gradients feel so fundamental across sciences. They are not fundamental because they are mysterious. They are fundamental because they are the simplest possible descriptors of imbalance.

Once gradients are recognized as delta languages, the next step becomes inevitable: if they all describe the same thing, why do they look so different across disciplines?

The answer lies not in physics, but in translation—and that is the work of the next section.

13.2 — One Gradient, Many Dialects

The reason gradients appear fragmented across science is not because reality is fragmented, but because measurement is local. Each discipline encountered the same structural condition—misalignment under constraint—through different carriers, instruments, and scales. What changed was the *dialect*, not the message.

UED treats these dialects as translations of a single sentence.

In mechanics, the sentence is spoken as pressure. In thermodynamics, it becomes temperature. In electromagnetism, it appears as potential. In chemistry, it is expressed as concentration. In gravity, it is framed as potential energy.

Each dialect names the same fact: *there is a directional imbalance, and alignment would be cheaper one way than another.*

The differences arise because each field measures a different aspect of the same redistribution problem. Pressure tracks momentum density. Temperature tracks kinetic dispersion. Voltage tracks charge separation. Concentration tracks compositional unevenness. None of these are separate realities—they are projections of delta through different lenses.

This is why cross-domain analogies work so well, even when the underlying physics is different.

Heat diffuses like dye. Charge equalizes like pressure. Ideas spread like contagion. Stress redistributes like fluid.

These are not metaphors. They are recognitions of shared structure.

Wherever delta exists and geometry allows resolution, gradients appear. Where geometry blocks resolution, gradients persist. Where thresholds collapse, gradients vanish abruptly. The carrier does not matter. The logic does.

UED's unification does not flatten differences; it organizes them.

Each dialect remains operationally valid within its domain. Engineers still compute pressure. Physicists still calculate potential. Chemists still measure concentration. What changes is the ontological claim. None of these quantities are fundamental causes. They are contextual descriptors of the same underlying imbalance.

This explains why equations across disciplines look uncannily similar.

Diffusion equations. Flow equations. Relaxation equations. Decay equations.

Different symbols, same skeleton.

The skeleton is not a force law. It is an accounting rule: unevenness redistributes along allowed paths until affordability flips or equilibrium is restored.

Once this is seen, a long-standing confusion dissolves. We no longer need to ask which gradient is "more real" or which field is "fundamental." Reality does not privilege dialects. It privileges structure.

Gradients are not competing explanations. They are coexisting translations.

This also clarifies why attempts at unification often failed historically. They tried to reduce one dialect into another—electricity into mechanics, heat into motion, gravity into geometry—without recognizing that all were already expressions of a deeper common rule.

UED does not reduce. It aligns.

By treating gradients as one concept expressed differently, the framework preserves domain expertise while removing ontological clutter. The world becomes simpler without becoming vague.

With this translation layer in place, one last misconception must be addressed. Even after abandoning force, and even after cleaning up fields, we still tend to imagine gradients as *things that exist.*

They do not.

Gradients are not entities. They are condition descriptions.

That clarification is the work of the next section—and it completes the removal of metaphor from the heart of physics.

13.3 — Fields Are Not Things; They Are Condition Descriptions

The final confusion around fields persists because we continue to treat them as entities rather than what they actually are: summaries of conditions.

When we say "there is an electric field here" or "a gravitational field exists around a mass," we instinctively imagine something present in space—something real, extended, and active. Even when we deny substance, the grammar of our language keeps smuggling it back in. The field *does* something. It *acts*. It *mediates*.

This intuition is wrong—not philosophically, but structurally.

A field does not exist in the way an object exists. It exists in the way a map exists.

A weather map does not create wind. A topographic map does not generate slopes. A pressure map does not push air.

They describe conditions under which movement would occur *if allowed.*

Fields function in exactly the same way.

A field value at a point is not an agent. It is a local summary of how delta is distributed under current constraints. It tells you, compactly, what resolution would look like if geometry permitted it to proceed. Nothing more is required. Nothing more is present.

This immediately explains a critical fact that force-based thinking struggles with: fields can exist without motion.

A charged capacitor holds an electric field without current. A gravitational field exists without falling. A thermal gradient persists across insulation without heat flow.

If fields were things that acted, this would be paradoxical. If fields are condition descriptions, it is obvious.

The field is not waiting to act. The system is waiting for permission.

UED therefore treats field equations correctly—not as laws governing agents, but as consistency conditions. They encode how gradients must distribute themselves so long as constraints remain unchanged. They do not tell reality what to do. They tell us what reality *would do* if resolution were allowed.

This reframing also clarifies why fields update instantly when constraints change.

Move a boundary, and the field reconfigures. Alter geometry, and the field reshapes. Introduce a barrier, and gradients reallocate.

Nothing "travels" to adjust the field. There is no propagation of influence in the descriptive sense. The condition simply changes everywhere it is defined, because the constraints that define it have changed.

This is deeply unintuitive if fields are treated as substances. It is trivial if fields are treated as descriptions.

Fields are therefore not intermediate layers between geometry and motion. They are snapshots of how delta is currently staged within geometry. They sit between constraint and action, not as causes, but as readable states.

Once this is understood, a long-standing mistake becomes visible. We often ask how fields interact with particles, as if two kinds of things were meeting. In UED, there is no such interaction. There is only alignment responding to conditions.

Particles do not respond to fields. Resolution proceeds along gradients.

The language changes, but the prediction remains intact—and often becomes clearer.

With this correction in place, field-language finally settles into its proper role: better than force-language, because it captures continuity and locality—but still incomplete, because it hides thresholds, permissions, and resistance.

That final limitation must be addressed directly.

Why "field" was an improvement over "force," and why it still cannot be foundational, is the focus of the next section.

13.4 — Why "Field" Is Closer Than "Force" (But Still Incomplete)

The shift from force-language to field-language was a genuine intellectual upgrade. It corrected a major flaw: the idea that effects must be transmitted by pushes and pulls acting instantaneously across space. Fields allowed continuity. They allowed locality. They allowed influence to be described without contact.

In that sense, field is closer to reality than force.

But "closer" is not the same as "correct."

Force-language fails because it hides constraint. It suggests that something *acts* on something else and produces motion, when in fact motion only appears if geometry allows resolution. Force is not a cause; it is a signal of obstruction. Wherever resolution is blocked, force spikes. Wherever resolution is smooth, force disappears. This inversion is fatal to force as a foundation.

Field-language improves on this by removing direct agency. A field does not push; it *exists*. Motion becomes response rather than imposition. That alone resolves many paradoxes.

Yet field-language stops short at a critical point.

It assumes smoothness by default.

Fields are typically modeled as continuous functions defined everywhere, varying gently across space and time. This works beautifully in regimes where permission is evenly distributed. But the moment thresholds appear—when geometry gates resolution—field-language begins to fracture.

Quantization appears as an anomaly. Collapse looks mysterious. Discontinuities feel pathological.

These are not failures of reality. They are failures of smooth-field assumption.

Fields describe gradients, but they do not encode permission boundaries explicitly. They tell you where resolution would go, but not whether it is allowed to go there yet. They are excellent at describing direction, but poor at describing gating.

This is why field theories must keep adding patches.

Boundary conditions. Renormalization. Cutoffs. Effective theories.

Each patch quietly acknowledges the same thing: the field description alone is insufficient when thresholds dominate behavior.

UED makes the missing piece explicit.

Fields describe what wants to happen. Geometry and thresholds describe what is allowed to happen.

Without that distinction, fields are forced to carry too much explanatory weight. They are asked to explain resistance, discreteness, delay, and sudden change—things they were never designed to encode.

This is also why field-language often feels metaphysically heavy. To compensate for missing permission logic, fields are treated as more real than they are. They are endowed with energy densities, stresses, and even agency. The description is inflated to cover structural blind spots.

UED deflates this cleanly.

Fields remain valuable. They are not discarded. But they are repositioned. They become what they were always meant to be: intermediate descriptors—closer to reality than forces, but downstream from geometry and delta.

Once thresholds are acknowledged, fields regain their elegance without carrying metaphysical baggage. Discontinuities stop being embarrassing exceptions. They become expected outcomes of gated permission.

This correction also explains why field-language works so well in engineering.

Engineering operates inside known envelopes—regions where constraints are stable and thresholds are not crossed. Within those envelopes, permission is smooth. Fields behave beautifully. Predictions are accurate. The language earns trust.

Foundational physics fails when it forgets that those envelopes exist.

By recognizing fields as descriptive but incomplete, UED keeps their utility while removing their overreach. What remains is a clean hierarchy:

- Geometry defines permission.
- Delta creates gradients.
- Fields describe those gradients.
- Force appears when gradients meet resistance.

With this hierarchy clarified, the framework is now ready to answer a lingering question that has haunted physics since Newton:

If fields are not causes, and forces are not drivers— what exactly produces resistance, effort, and opposition?

That answer does not require a new entity. It requires acknowledging the moment where gradients collide with constraint.

And that collision—nothing more, nothing less—is what we have long called force.

That final demotion is the work of the next chapter.

13.5 — Gradients + Geometry = Direction Without Intention

Once gradients are freed from metaphor and geometry is recognized as a permission map, directionality finally becomes intelligible—without invoking intention, agency, or hidden drivers.

Nothing *decides* where to go. Direction emerges where gradients meet permission.

A gradient alone does not cause motion. It merely states that misalignment exists and that resolution would be cheaper one way than another. Geometry alone does not cause motion either. It only defines which routes are open, narrow, or forbidden. Motion appears only when both are present simultaneously.

This pairing is decisive.

- Geometry answers: *Where can resolution proceed?*
- Gradients answer: *Where is resolution cheaper?*

Direction is the intersection of the two.

When a gradient aligns with an open corridor, resolution flows and motion appears smooth. When a gradient points toward a blocked corridor, nothing happens—no matter how steep the gradient becomes. When geometry opens suddenly, the same gradient that felt impotent moments ago produces rapid motion.

This explains a long-standing confusion: why increasing a gradient does not always increase motion.

Raising pressure behind a sealed valve increases the pressure gradient but produces no flow. Increasing voltage across an open circuit increases the electric gradient but produces no current. Increasing desire under constraint produces anxiety, not action. In each case, the gradient is real, measurable, and intense—but permission is absent.

When permission returns, motion resumes instantly—often explosively.

This also explains why direction can reverse without any change in "force." If geometry shifts—through boundary movement, topology change, or constraint reconfiguration—the cheapest path for resolution can flip. The gradient did not change. The map did.

Direction, therefore, is not a command. It is a consequence of affordability.

This reframing dissolves teleology at its root. Nature does not aim. Systems do not seek goals. Equilibrium pressure does not "want" anything. Resolution proceeds because it is cheaper to do so than to remain misaligned—*and only where geometry allows it.*

Seen this way, gradients are honest. They make no promises. They do not guarantee motion. They merely encode potential. Geometry decides whether that potential is actionable.

This pairing also clarifies why equilibrium pressure can be directional without being intentional. The pressure is not pushing; the geometry is not guiding. Their interaction simply filters possibilities until only one direction remains affordable.

Direction emerges because alternatives are too expensive.

With gradients and geometry aligned this way, the final illusion collapses naturally. If nothing pushes, pulls, or commands—if motion is merely resolution proceeding along permitted gradients—then force cannot be a cause.

Force can only be what appears when this process is resisted.

That recognition completes the arc begun at the start of this part. The next section makes that demotion explicit, showing how force arises as a derived effect—and why treating it as fundamental was always a convenient lie.

13.6 — Why No New Forces Are Needed

Once gradients are understood as delta languages and geometry as the permission map, a striking realization follows almost immediately: there is nothing left for forces to explain.

Every role historically assigned to force has already been accounted for—more cleanly—by gradients operating within constraint.

Force was introduced to answer a specific discomfort: *why doesn't motion always happen when imbalance exists?* The answer seemed to require an intermediary—something that pushes, pulls, compels. But this intermediary was compensating for a missing concept, not adding a necessary one.

The missing concept was permission.

In UED, when resolution proceeds smoothly along an open corridor, no force appears. Motion occurs, alignment advances, and nothing resists. This is why so many natural processes look effortless. No force is "doing work" in the intuitive sense; delta is simply resolving where it is allowed to.

Force only appears when that resolution is obstructed.

Push against a boundary, and force rises. Try to compress an already dense structure, and resistance grows. Attempt to redirect alignment through a constrained geometry, and effort becomes measurable. In every case, force is not initiating change—it is pricing resistance.

This reframing collapses a century of confusion.

Force is not a fundamental interaction. Force is the cost signal of blocked alignment.

Once this is accepted, the proliferation of forces in physics begins to look like a historical artifact rather than a necessity. Gravitational force, electromagnetic force, contact force, frictional force—each names a different way in which geometry constrains resolution under a gradient. The differences are real, but they do not require separate ontologies.

No new forces are needed because no new *causes* are missing.

What changes across domains is not the existence of forces, but:

- the type of gradient involved,
- the geometry that constrains it,
- and the thresholds that gate release.

That triad—gradient, geometry, threshold—fully determines behavior.

This also explains why force-language continues to work so well in engineering and everyday practice. Engineering operates within stable envelopes where permission structures are known and rarely crossed. Within those envelopes, force becomes a reliable proxy for resistance. Measuring force tells you how close you are to violating constraints. It is an excellent *instrumental language.*

But instrumental success does not imply ontological primacy.

When physics forgets this distinction, it mistakes the meter for the mechanism. It treats force as a driver rather than a diagnostic. That mistake remains hidden in classical regimes and becomes fatal at foundational levels—where thresholds, discontinuities, and permission changes dominate behavior.

UED resolves this without subtraction or replacement.

Force is retained, but demoted. Fields are retained, but clarified. Gradients are unified, not multiplied. Geometry is elevated to its proper role.

Nothing is lost. Paradoxes disappear.

By the end of this chapter, force should feel familiar—but no longer authoritative. It is something we measure when alignment is resisted, not something we invoke to make alignment happen.

With that demotion complete, the conceptual cleanup of motion, fields, and force is finished.

What remains is to step back and recognize what this reframing has quietly accomplished: it has replaced a universe of interacting agents with a universe of structured resolution.

The next chapter makes that shift explicit—showing how force emerges, where it still works, and why calling it fundamental was never more than a useful convenience.

From Gradients to Resistance

With gradients finally stripped of metaphor and force demoted from cause to consequence, the picture that remains is unexpectedly spare—and far more powerful than the one it replaces.

Nothing in nature pushes. Nothing pulls. Nothing commands motion.

What exists instead is a layered negotiation:

- Geometry defines where alignment may occur.
- Gradients describe how misalignment is distributed within that geometry.
- Resolution proceeds only where permission and affordability coincide.

Everything else is accounting.

When gradients align with open corridors, motion looks effortless and natural. When gradients point into constrained regions, something new appears—not as a driver, but as a signal. That signal is resistance. That resistance is what we have historically named *force*.

Seen this way, force is no longer mysterious, nor is it fundamental. It is simply the price of insisting on resolution where geometry says no.

This realization closes an important loop.

Fields were introduced to soften the harshness of force. Gradients now complete that softening by removing agency altogether. What remains is not a weaker physics, but a more honest one—one that explains smooth motion and sudden failure with the same rule, without inventing invisible actors.

At this point in the framework, nothing essential is missing.

We no longer need:

- forces to explain motion,
- fields to act as hidden agents,
- or metaphysical layers to bridge gaps between domains.

Delta, geometry, gradients, thresholds, and consequence are sufficient.

The next chapter does not introduce something new. It performs a final clarification. It shows explicitly how force emerges when gradients meet constraint, why force-language remains useful within limits, and why mistaking that usefulness for fundamentality led physics astray.

This part began by reframing geometry. It now ends by completing the demotion of force.

What follows is not an expansion of concepts, but a tightening of responsibility. Once resistance is understood as a consequence rather than a cause, the question is no longer *what is acting*—but *who is holding delta where resolution is no longer cheap.*

That question belongs not only to physics, but to every system that learns how to delay alignment.

And that is where the framework turns next.

When gradients meet constraint, something new appears—not as cause, but as consequence.

That appearance is what we have long called force.

Chapter 14 — Force as a Consequence

Derived Effect of Delta Under Constraint

Why Force Felt Fundamental for So Long

Force earned its reputation honestly.

For centuries, it worked. It predicted motion, explained resistance, enabled engineering, and turned intuition into calculation. Push harder, accelerate more. Pull back, slow down. Build stronger, withstand greater loads. The language was simple, measurable, and enormously successful.

That success made force feel fundamental.

But success is not proof of primacy. It is proof of usefulness within limits.

Force-language became dominant because it compressed too many things into one idea. Geometry, gradients, resistance, and thresholds were all folded into a single scalar or vector quantity. This compression made problems solvable—and in doing so, it hid what was actually happening.

Force felt like a cause because it was always observed at the point where resolution was obstructed.

We never see force in free resolution. We see it at walls, joints, contacts, constraints, anchors. We measure force where motion struggles, where alignment is resisted, where equilibrium pressure is denied its preferred path. And because that is where our instruments operate, force appeared to be the driver.

It wasn't.

Force was the receipt, not the transaction.

Historically, this misunderstanding was inevitable. Before geometry was understood as constraint, before gradients were unified, before thresholds were acknowledged, resistance needed a name. Force became that name. It stood in for everything we did not yet know how to separate.

And it worked—until it didn't.

At foundational levels, force began to behave strangely. It multiplied into categories. It diverged at small scales. It required renormalization. It failed to explain sudden transitions. It demanded exceptions, fixes, and interpretive layers.

These failures were not accidents. They were signals.

Force was being asked to do too much.

UED does not attack force. It relieves it.

By restoring geometry to its role as permission, gradients to their role as directionality, and thresholds to their role as gates, force is finally allowed to be what it always was: a derived effect—the measurable expression of resistance when alignment is attempted through constraint.

This reframing does not make force disappear. It makes force honest.

It explains why force vanishes in effortless motion. Why it spikes near boundaries. Why it explodes near failure. Why it works so well in engineering and so poorly in metaphysics.

Force felt fundamental because we met it everywhere alignment was blocked. Once alignment is understood, force no longer needs to pretend to be the cause.

It becomes what it truly is: the price paid when resolution is no longer cheap.

This chapter completes that demotion—not to weaken physics, but to finally let it speak plainly.

14.1 — When and Why Force Appears

Force does not announce itself at the beginning of motion. It appears only at the moment resolution is denied.

This single observation, once taken seriously, dissolves the illusion of force as a cause.

Where resolution proceeds freely—along open corridors defined by geometry and guided by gradients—there is no force signature to measure. Motion happens, alignment advances, and nothing resists. A falling leaf drifts, a current flows, a wave propagates, a thought resolves—without any sense of force being "applied."

Force enters the picture precisely when something says no.

A boundary blocks a path. A constraint narrows a corridor. A threshold refuses release.

At that moment, equilibrium pressure does not disappear. Delta still seeks resolution. What changes is that alignment can no longer proceed cheaply. The system responds by accumulating resistance, and that resistance is what we measure as force.

This is why force is always local and situational.

You do not measure force in empty extension. You measure it at contact points, hinges, joints, supports, interfaces.

A wall does not exert force until you try to pass through it. A spring does not produce force until you deform it. Friction does not appear until relative motion is resisted.

Force is never present by default. It is activated by refusal.

This reframes a foundational misconception. We are taught that force produces acceleration. In UED terms, acceleration appears when resolution finds an open corridor under a gradient. Force appears when that corridor is partially or fully closed.

Acceleration without force is not paradoxical. Force without acceleration is not mysterious.

Both are expected outcomes once permission is made primary.

This also explains why increasing force does not guarantee motion.

Push harder against a rigid boundary and force rises—but nothing moves. The geometry has not changed. The corridor remains closed. The system simply reports higher resistance. Force increases because the cost of insisting has increased, not because progress is being made.

Conversely, reduce constraint and motion can resume with negligible force. Remove the wall, loosen the joint, open the valve—and the same gradient produces flow without effort. The force vanishes because permission returned.

Seen this way, force becomes a diagnostic rather than a driver. It tells us:

- where alignment is being resisted,
- how strongly permission is being denied,
- and how close a system is to threshold failure.

Force is the language constraints use to communicate.

It is equilibrium pressure translated into numbers at the point where geometry says "not this way." The more stubborn the refusal, the louder the signal.

Once this role is understood, force loses its mystique. It no longer explains why things move. It explains why they cannot.

And that inversion sets the stage for the next clarification: force is not just resistance—it is the quantified cost of insisting on forbidden resolution.

That pricing logic is what the next section makes explicit.

14.2 — Force as the Cost of Insisting

Once force is seen as a response to denied resolution, its true role becomes unmistakable: force quantifies the cost of insisting on alignment where geometry refuses permission.

Force does not initiate anything. It prices resistance.

Every time force is measured, something is being asked to resolve along a path that is constrained, narrowed, or forbidden. The higher the force, the higher the cost of continuing to insist along that path. This is why force rises smoothly as resistance increases—and spikes dramatically near failure.

Push against a rigid wall. The wall does not push back out of aggression or reaction. The geometry simply does not allow passage. As equilibrium pressure continues to demand alignment, resistance accumulates. The force you feel is the invoice for attempting the impossible.

Compress a spring. The geometry allows deformation—but only within limits. As you approach those limits, the internal structure resists more strongly. The rising force is not "stored intent"; it is escalating cost for maintaining misalignment.

Drag an object across a rough surface. Friction appears not because motion exists, but because motion is being opposed. The geometry of contact converts attempted resolution into heat and deformation. Force here is the cost of maintaining relative motion under constraint.

This logic holds everywhere.

- In fluids, force appears at obstructions and boundaries.
- In structures, force concentrates at joints and supports.
- In materials, force spikes at defects and stress risers.
- In biology, force manifests as strain when tissues are pushed beyond natural motion.
- In psychology, effort is felt when action conflicts with internal constraints.

In every case, force is not a driver—it is a billing statement.

This also explains why force is such a reliable predictor of failure.

When force grows without producing motion, it means delta is being held at increasing cost. Thresholds are being approached. Structures are being stressed. The system is warning that persistence is becoming unaffordable. Engineers rely on force measurements precisely because force is a precise proxy for proximity to breakdown.

But this reliability has a price. Because force is so measurable, so tangible, and so effective within constrained regimes, it was mistaken for a cause. The meter was confused with the mechanism.

UED corrects this without discarding anything useful.

Force remains indispensable for monitoring resistance, designing safety margins, and predicting collapse. What changes is the interpretation. Force is not what moves the world forward. It is what tells us how much the world is refusing to move in the way we are demanding.

This redefinition also clarifies why adding force often worsens outcomes.

More force does not create permission. It only increases cost.

To change behavior, constraints must shift. Geometry must open. Thresholds must collapse. Until that happens, force accumulates, energy dissipates, and failure becomes inevitable.

This insight carries far beyond mechanics.

In systems of all kinds—physical, biological, social—forceful insistence is the least efficient way to produce change. It maximizes resistance rather than enabling resolution. Force signals conflict with structure, not progress through it.

Recognizing force as cost rather than cause sets up the final step in this chapter: understanding why force-language persists, despite being structurally misleading.

That persistence is not stupidity. It is convenience.

And it is precisely that convenience—the way force compresses complexity into a single term—that must now be examined.

14.3 — Why Force-Language Is a Convenient Lie

Force-language survives not because it is true, but because it is useful. It compresses a complex negotiation—geometry, gradients, thresholds, and resistance—into a single, measurable quantity. That compression makes problems solvable, predictions fast, and intuition trainable. It is a lie in the technical sense: it hides structure while preserving outcomes.

And that is precisely why it endured.

Force takes three questions and answers them with one number:

- *Where is resolution blocked?*
- *How steep is the gradient pushing for alignment?*
- *How hard is the geometry refusing?*

By collapsing these into "force," we get a clean handle on resistance without tracking permission maps, pathway topology, or threshold economics. In stable regimes, this shortcut works beautifully. It lets us design bridges, engines, machines, and materials with extraordinary reliability.

But convenience has a cost.

Force-language erases permission. It makes resistance look like an active agent rather than a passive refusal. It suggests that adding more force can always produce more effect, when in reality it often only increases cost. It frames outcomes as battles between pushes and pulls instead of negotiations between alignment and constraint.

This erasure becomes dangerous the moment systems leave their envelopes.

Near thresholds, force inflates without progress. At discontinuities, it diverges. In regimes dominated by gating—fracture, phase change, collapse—force ceases to be explanatory and becomes merely loud. The equations still compute, but intuition breaks. We add corrective terms, boundary conditions, renormalizations—patches that keep the lie serviceable without admitting its limits.

The deeper problem is ontological. When force is treated as fundamental, we begin to hunt for ever more forces to explain every anomaly. The catalog grows, clarity shrinks, and unification stalls. Each new "force" is an attempt to repackage permission failure as causation.

UED calls this what it is: misplaced causality.

Force is not what *makes* things happen. It is what we *measure* when the thing we want to happen is not permitted to happen the way we want. Treating it as a cause confuses resistance with agency and cost with intent.

And yet—the lie is convenient enough to keep.

Engineers rely on it because they operate where permission is stable. Educators teach it because it builds intuition quickly. Scientists use it because it predicts within known bounds. None of this is wrong. What is wrong is mistaking instrumental compression for foundational truth.

UED does not abolish force-language. It labels it correctly.

Force is a shorthand for "alignment demanded where geometry resists." A powerful shorthand. A necessary shorthand. But a shorthand nonetheless.

Once this is acknowledged, force stops distorting our understanding. We can keep the tool without letting it dictate ontology. And we can finally explain why force works so well—*and* why it fails exactly where the deepest questions begin.

That boundary—where convenience must yield to structure—is where the next section places force back into its rightful, limited domain.

14.4 — Engineering Envelopes: Where Force Still Works

The fact that force-language is structurally incomplete does not make it useless. On the contrary, it explains why it works so well where it does.

Force is extraordinarily effective inside engineering envelopes—regimes where geometry, gradients, and thresholds remain stable and well understood. Within these envelopes, permission does not change unexpectedly. Resolution corridors are predictable. Thresholds are deliberately kept far away. Under such conditions, resistance behaves smoothly, and force becomes a reliable proxy.

This is the domain where force-language shines.

Engineers do not ask whether geometry might suddenly reconfigure. They design so that it does not. Materials are kept within elastic limits. Loads are constrained below yield points. Motions are restricted to ranges where linear response holds. In this carefully managed space, force scales proportionally with deformation, acceleration, or stress.

Here, force is not lying—it is summarizing faithfully.

Within an envelope:

- geometry is effectively fixed,
- gradients change gradually,
- thresholds are not approached,
- and resistance remains continuous.

Under these assumptions, force behaves like a cause because permission is implicitly guaranteed. Motion follows force because geometry has already agreed to allow it. The deeper structure stays silent, so the shorthand appears complete.

This is why force-based equations feel universal in practice. They work wherever humans work hardest to avoid threshold behavior.

But this also reveals the hidden condition: force works because engineers deliberately remove the situations where it would fail.

Safety factors exist precisely to keep systems away from alignment thresholds. Linear approximations exist because designers operate where nonlinearity is negligible. Control systems exist to prevent topology changes. In other words, engineering succeeds by freezing permission.

Force-language is perfectly suited to frozen-permission regimes.

The moment those regimes are violated—when a structure yields, a material fractures, a joint slips, or a flow separates—force equations stop explaining and start reporting. They tell you something is wrong, but not *why* it suddenly became wrong. At that point, geometry has shifted, thresholds have been crossed, and the envelope has collapsed.

This is not a failure of engineering. It is evidence that engineering understood force correctly—as a local operational tool, not a universal truth.

UED respects this distinction.

Force-language remains indispensable wherever:

- constraints are stable,
- permission does not change,
- and resolution pathways are known in advance.

What UED refuses is the extension of this local success into a global ontology. Force is not wrong; it is conditional.

Understanding this restores intellectual honesty without sacrificing utility. Engineers can continue to calculate forces confidently. Physicists can stop pretending those calculations reveal ultimate causes.

Force works where humans have made reality behave politely.

Beyond those envelopes—where nature expresses thresholds freely—force does not fail. It simply steps aside, revealing the deeper structure it was never meant to replace.

14.5 — Where Force-Language Breaks Down

Force-language fails not randomly, but systematically—and always at the same kinds of places. It breaks down precisely where permission changes, where thresholds dominate behavior, and where resolution is no longer smooth.

These failures are not bugs in physics. They are diagnostics.

The first breakdown appears at threshold crossings.

As a system approaches a threshold—yield, fracture, phase change, slip—force values inflate rapidly while outcomes stall. You push harder, measure more force, and still nothing resolves. Then, suddenly, everything resolves at once. The equations did not predict the *moment* of change because force was never tracking permission; it was only tracking resistance.

At the instant the threshold collapses, force drops or reconfigures—not because causation reversed, but because the corridor opened. The same gradient that produced enormous resistance a moment ago now produces motion with negligible force. Force-language cannot anticipate this reversal because it has no variable for permission.

The second breakdown appears in nonlinear and discontinuous regimes.

Whenever behavior depends on topology rather than magnitude—buckling, snapping, turbulence onset, avalanche dynamics—force ceases to be explanatory. It can be computed, but it no longer clarifies. Small changes in conditions produce outsized effects. Linear intuition fails because the system is not responding to "how much," but to whether alignment is allowed.

This is why force equations multiply caveats near critical points. Extra terms are added. Regimes are split. "Effective forces" are introduced. Each addition is an attempt to retrofit permission logic into a language that was never designed to carry it.

The third breakdown is most visible in quantum and relativistic domains.

At quantum scales, force-language becomes awkward almost immediately. Interactions are replaced by potentials, operators, probabilities. Force appears only as a derivative convenience, not a primary driver. Measurement events behave like threshold collapses rather than force-mediated processes. Time itself loses its clean role.

In relativistic regimes, force again retreats. Geometry takes over. Curvature replaces interaction. Force becomes frame-dependent, constrained, secondary. The closer physics moves to foundations, the more force-language is quietly sidelined.

This pattern is not accidental.

Force works best where permission is stable and implicit. It fails where permission is explicit and dynamic.

Even in classical domains, the breakdown is visible wherever systems self-organize, adapt, or reorganize—fluids transitioning to turbulence, materials forming patterns, ecosystems shifting states, societies collapsing. Force can describe local stress, but it cannot explain the timing or nature of reconfiguration.

UED predicts these failures in advance.

Wherever:

- thresholds exist,
- topology can change,
- or constraints can reconfigure,

force-language will become insufficient—not because it is wrong, but because it is too compressed.

This recognition flips the usual narrative.

Instead of asking why force fails in certain regimes, the better question becomes: *why did we expect it to work there at all?*

Force was never meant to explain emergence, collapse, or reorganization. It was meant to price resistance inside fixed structures. When the structures themselves become variables, force can only report distress—not govern behavior.

Seen this way, the breakdown of force-language is not a crisis. It is a guidepost.

It tells us exactly where to stop treating force as fundamental and start examining:

- geometry shifts,
- threshold economics,
- and delta redistribution.

That shift is the final act of cleanup in this chapter.

What remains is to assemble the hierarchy cleanly—so force can no longer drift back into primacy by habit.

14.6 — Reassembling the Hierarchy Cleanly

At this point, force has been stripped of its illusions, gradients have been clarified, and geometry has reclaimed its role. What remains is not a rejection of classical physics, but a reordering—one that aligns explanation with how reality actually behaves.

This hierarchy is not philosophical. It is operational.

UED assembles it cleanly, without overlap or redundancy:

1. Geometry — Permission Structure: Geometry is the first filter. It defines what resolutions are allowed, which pathways are open, and which are forbidden. Before anything moves, geometry has already decided whether movement is even possible. Boundaries, topology, curvature, and connectivity all belong here.

2. Delta — Misalignment to Be Resolved: Delta is the reason anything happens at all. Without imbalance, nothing seeks alignment. Delta does not act; it simply exists as permitted misalignment within geometry.

3. Gradient — Directional Expression of Delta: Gradients translate delta into directionality. They tell us where resolution would be cheaper *if* geometry allows it. Gradients encode preference without agency.

4. Threshold — Gating Mechanism: Thresholds decide *when* resolution may proceed. They are affordability boundaries. Below a threshold, delta is held. At the threshold, permission flips. Above it, resolution proceeds rapidly.

5. Force — Resistance Signal: Force appears only when gradients demand alignment through constrained geometry. It quantifies resistance. It never initiates change; it measures refusal.

6. Time — Delay Record: Time records how long delta remains unresolved under current constraints. It does not drive events; it timestamps waiting.

This ordering is not optional.

Every paradox that plagued physics emerged from violating it—by letting force pretend to cause motion, by letting time pretend to flow independently, or by treating geometry as passive. When the order is respected, the paradoxes dissolve without further invention.

Importantly, this hierarchy is complete.

Nothing needs to be added. No new entities are required. No metaphysical bridges are necessary.

Each layer does one job, and only one.

- Geometry permits.
- Delta motivates.
- Gradients bias.
- Thresholds gate.
- Force reports.
- Time records.

This clarity prevents force from reclaiming authority later. It cannot sneak back in as a cause because its role is explicitly bounded. It cannot expand into explanation because its function is diagnostic.

With this hierarchy in place, the framework becomes immune to category drift. Concepts stay in their lanes. Explanations remain stable across scale, domain, and language.

The demotion of force is now complete—not as an attack, but as a correction.

What remains is to close the chapter by stepping back and recognizing what this correction enables. Once force is no longer mistaken for cause, attention shifts naturally to who and what manage constraints, thresholds, and delay.

That shift marks the transition from physical explanation to systemic responsibility.

And that is where this part ends.

From Force to Governance

With force finally placed where it belongs, a quiet but decisive transformation has occurred in the framework.

Force no longer explains why things move. It explains where movement is being refused.

What once looked like an active driver is now recognized as a signal—a numeric expression of resistance generated when alignment is demanded through constrained geometry. The world does not move because of force. It moves when permission exists, and it resists when permission is denied.

This realization closes a long chapter in physics.

Force was elevated because we lacked the language to speak about constraints, thresholds, and affordability. Now that language exists. Geometry defines permission. Gradients express bias. Thresholds gate release. Force merely reports the cost of insisting.

Nothing essential is lost. Everything becomes clearer.

But this clarity has consequences.

Once force is no longer mistaken for cause, responsibility shifts. If resistance is not an agent of nature but a result of constraint, then the meaningful question is no longer *what force is acting*, but who is holding the constraints in place.

In natural systems, constraints arise organically. In engineered systems, they are designed. In biological systems, they are evolved. In social systems, they are imposed, negotiated, and defended.

Force hides these distinctions. UED exposes them.

By reframing force as consequence, the framework points beyond mechanics into governance—the deliberate management of permission, delay, and threshold. The physics of resistance becomes the politics of structure.

That transition marks the end of this part.

Geometry has been reframed. Fields have been demystified. Force has been demoted.

What follows is not another physical layer, but a systems layer—where the same rules govern organizations, technologies, economies, and civilizations.

The laws have not changed. Only the scale of responsibility has.

From here on, UED stops asking how nature moves—and starts asking how systems choose to delay alignment, harvest consequence, and survive the cost.

That is where the framework becomes unavoidable.

Deliverable of Part IV — The Boundary Is Now Drawn

By the end of this part, a clean and irreversible boundary has been established—not as a rejection of existing physics, but as a correction of its placement.

Nothing essential has been discarded. Nothing useful has been broken. What has changed is where fundamentals are allowed to sit.

The first correction is ontological.

Reality is no longer described as a collection of things that push, pull, or act upon one another. It is described as a landscape of permitted and denied resolutions. Objects are stabilized states. Motion is allowed alignment. Resistance is priced refusal. Ontology shifts from entities to states and permissions. This single move dissolves entire families of paradoxes without adding a single new assumption.

The second correction is geometric.

Space is no longer a passive container in which events occur. It is a constraint map—a structured permission field that determines which pathways of resolution are open, narrow, or forbidden. Trajectories cease to be mysterious curves traced through emptiness and become stable corridors where alignment remains affordable. Distance stops being "how far" and becomes "how many resolution steps are required." Geometry regains agency—not as force, but as governance.

The third correction is linguistic and methodological.

Force, field, and time are no longer treated as primitive actors. They are recognized as derived reporting languages, each compressing deeper structure into manageable forms. Force reports resistance. Fields summarize gradients. Time records delay. Their equations remain valid precisely because they were always local tools, never universal truths.

And that brings us to the boundary itself.

Classical and modern mathematics are not wrong. They are situated.

They work brilliantly where:

- permission is stable,
- geometry is fixed,
- thresholds are distant,
- and resolution proceeds smoothly.

In these regimes, old math does exactly what it was built to do. It predicts, engineers, and controls with astonishing success. UED does not replace it. UED contains it—explicitly, cleanly, without apology.

What UED refuses is the silent expansion of these local tools into global ontology.

The moment thresholds dominate, topologies shift, or permission reconfigures, those tools stop explaining and start reporting. UED tells you *why*—and tells you *in advance*. That is the boundary: not between old and new physics, but between description and foundation.

By drawing this line openly, Part IV performs its final task.

It frees physics from pretending that its most convenient languages are its deepest truths. It frees metaphysics from smuggling intuition where structure suffices. And it frees the reader from choosing sides between tradition and revision.

From here on, the framework is stable.

Ontology is corrected. Geometry is reinterpreted. Existing mathematics is retained—powerful, respected, but finally placed.

Everything that follows builds on this ground. No further cleanup is required.

What remains is application across domains where constraints are managed consciously, thresholds are manipulated deliberately, and delay becomes strategy.

At that point, UED stops being a theory about nature.

It becomes a theory about how intelligence survives inside nature.

Part V — Energy In Action

With ontology clarified and geometry reinterpreted, UED now turns to where confusion has historically been greatest: *energy as behavior, not substance.*

This part does not introduce new entities. It removes old misunderstandings.

Energy here is not treated as a thing that flows, is stored, or gets consumed. It is treated as *delta in action*—misalignment expressing itself through permitted pathways under constraint. Once this shift is accepted, matter, waves, friction, and thermodynamics stop appearing as separate topics and begin to read as variations of the same story.

This part reframes matter itself. Mass is no longer a mysterious intrinsic property but a *measure of how persistently a configuration resists resolution.* What we call particles are not fundamental beads of reality; they are *stable delta-holding states*—condensed arrangements that survive equilibrium pressure long enough to appear object-like.

This part releases waves from metaphor. A wave is not a thing that travels; it is *a disturbance attempting to resolve.* Oscillation appears not because nature prefers vibration, but because overshoot, feedback, and constraint make smooth resolution temporarily impossible. Coherence and diffusion become competing resolution strategies, not separate phenomena.

This part addresses what physics traditionally calls "loss." Friction, drag, damping, and dissipation are not failures of motion or leaks in energy. They are *alignment diffusion*—the redistribution of delta across more degrees of freedom. Efficiency is no longer about minimizing loss, but about managing where and how resolution spreads.

This part closes by revealing thermodynamics not as a special domain, but as *UED speaking in its native language*. Entropy becomes the signature of delta diffusion. Irreversibility emerges from pathway constraints and information loss, not from cosmic arrows. Heat appears as a consequence channel—one of the most reliable ways unresolved delta exits structured systems.

By the end of this part, energy stops feeling abstract and starts feeling inevitable.

Matter is frozen action. Waves are frustrated resolution. Friction is alignment spreading. Heat is consequence made visible.

Nothing new has been added. Everything has been put in motion.

From here onward, UED no longer needs to argue for its validity. The reader begins to see familiar physics fall into place, not because it was replaced, but because it was finally read in the right order.

Chapter 15 — Matter as Condensed Delta Stability

Why Matter Should Not Have Been the Starting Point

For as long as humans have tried to understand the world, matter has felt like the safest place to begin. It is what we can touch, weigh, break, and rebuild. Stones do not vanish when we look away. Tables resist our hands. Bodies persist from one moment to the next. Matter announces itself as *real* in a way few other things do.

That apparent reliability made matter an irresistible starting point for physics.

But that choice, while understandable, quietly inverted the logic of explanation.

By starting with matter, we treated stability as primitive and motion as secondary. Objects were assumed to exist first, and only later did we ask how they moved, interacted, or changed. When those questions became difficult, we invented additional layers—forces to push them, fields to mediate between them, particles to carry influence. Each invention worked locally, but the overall picture grew heavier, not clearer.

UED begins by undoing that inversion.

Matter is not the beginning of explanation. Matter is what survives explanation.

What we experience as a solid object is not a foundational unit of reality; it is a successful outcome. It is a configuration that has learned how to remain unresolved under continuous equilibrium pressure. It holds delta efficiently enough that alignment is postponed for very long durations. That postponement creates the illusion of permanence.

This is why matter feels so convincing.

We mistake persistence for primacy.

In reality, matter is a special case—a limit case—of energy in action. It is what happens when resolution pathways become narrow, thresholds become high, and internal rearrangements absorb equilibrium pressure without dissolving the overall configuration. Matter is not still; it is busy resisting change.

Seen this way, solidity is not silence. It is sustained negotiation.

This correction immediately resolves several long-standing confusions. We no longer need to ask how motion arises from inert stuff, or how forces "act" on passive objects. Motion and interaction are already present at deeper levels. Matter appears only when those activities find stable ways to delay their own completion.

The world did not start with objects and then learn to move.

It started with tendencies toward alignment, and matter emerged wherever those tendencies were held open without collapse.

This is why this chapter belongs here, not at the beginning of the book. Only after equilibrium, delta, consequence, time, geometry, and force have been correctly placed can matter be understood without mythology.

From this vantage point, matter loses none of its reality—but it gains its explanation.

Matter is no longer mysterious substance. It is condensed stability.

And once that is seen clearly, the boundary between particles and waves—between objects and disturbances—begins to dissolve naturally, without debate or contradiction.

15.1 — Condensed Energy Is Not Stored Energy

One of the most persistent habits in physics is to speak of matter as if it were a container—energy stored inside, waiting to be released. This language is intuitive, but it quietly smuggles in a metaphor that does not survive careful inspection.

Energy is not a substance that can be packed, sealed, or stockpiled. There is no inner warehouse inside matter where energy sits idle. What exists instead is configuration—a stable arrangement that continuously absorbs equilibrium pressure without dissolving.

Condensation, in UED, does not mean accumulation. It means successful structuring.

When energy condenses into matter, nothing is "put away." Delta is not hidden; it is managed. The configuration created is one in which internal pathways, constraints, and thresholds cooperate to delay resolution. The system remains active internally, but externally it appears persistent.

This is why the phrase "stored energy" misleads. It suggests passivity, as if energy becomes dormant once matter forms. In reality, condensed configurations are among the most dynamically active systems in nature. They are constantly redistributing stress, adjusting micro-alignments, and shedding small disturbances—all to preserve overall stability.

Matter is not energy at rest. Matter is energy working hard to stay arranged.

The distinction becomes clearer when we compare free and structured delta. Free delta resolves quickly. Temperature gradients equalize, pressure differences relax, attention drifts. Structured delta, by contrast, is held within architectures that slow or redirect resolution. Geometry and thresholds collaborate to make alignment expensive.

Condensation happens only when such architectures exist.

Without constraint, there is no matter. Without thresholds, there is no persistence. Without time, there is no appearance of solidity.

This is why matter does not form everywhere. It appears only where resolution can be delayed reliably across scales. When those conditions fail, matter does not "lose" energy—it releases structure. What we call energy release is simply the collapse of a configuration that could no longer afford to hold delta.

Seen through UED, the mystery evaporates. Nuclear reactions, chemical bonds, phase changes—all of them are not examples of energy escaping from storage, but of permission structures breaking or reorganizing. The energy was never locked away; the constraints were.

This reframing has a powerful consequence. It allows us to treat matter, radiation, and interaction on the same footing—not as different substances, but as different behaviors of delta under constraint.

Condensed energy is not stored energy. It is energy that has learned how not to resolve—yet.

15.2 — Stability as the Real Definition of Matter

If energy is not stored inside matter, then the next obvious question follows naturally: *what, exactly, makes matter matter?* The answer is neither substance nor solidity. It is stability—specifically, stability under continuous pressure to resolve.

Matter exists wherever a configuration can persist despite equilibrium's constant pull toward simplification.

This immediately reframes what we mean by "solid." Solidity is not immobility. It is not absence of motion. At microscopic and submicroscopic levels, matter is intensely active—vibrating, exchanging, rearranging. What remains stable is not the motion, but the pattern.

Stability, in UED, is not stillness. It is repeatable self-maintenance.

A rock sits on the ground not because nothing is happening inside it, but because everything that *is* happening remains contained within a configuration that does not dissolve. The rock survives pressure, temperature variation, minor impacts, and time itself because its internal resolution pathways absorb disturbances without allowing global reconfiguration.

This is why stability must replace substance as the defining property of matter.

Traditional physics often treats stability as a consequence of forces balancing each other. UED reverses this logic. Forces appear *because* stability is being preserved. The configuration resists reorganization, and that resistance is reported as force when disturbed.

Matter, therefore, is not defined by what it is made of, but by how long and how reliably it holds together.

This also explains why matter exists on a spectrum rather than as a binary category. Gases, liquids, solids, plasmas—these are not different kinds of stuff. They are different stability regimes. As constraints loosen, resolution accelerates and stability weakens. As constraints tighten, resolution slows and stability deepens.

Even within what we casually call "solid," stability varies. Glass flows over geological time. Metals creep under stress. Biological tissue remodels continuously. Nothing is absolutely stable. Matter only appears so because its resolution timescales exceed our observation window.

Matter is slow change mistaken for permanence.

Once stability becomes the defining feature, many conceptual problems disappear. We no longer need to ask how matter can "contain" energy while also participating in motion. Motion and energy exchange are simply internal activities that do not threaten the configuration's persistence—until thresholds are crossed.

Matter is not the opposite of change. Matter is change that has learned to stay coherent.

This prepares the ground for the next step: if stability is what defines matter, then the quantity we call mass must also be reinterpreted—not as amount of substance, but as a measure of how deeply stability is entrenched against resolution pressure.

That is where the chapter must now go.

15.3 — Mass as Persistence Under Resolution Pressure

Once matter is understood as stability rather than substance, the concept of mass quietly changes its meaning. Mass is no longer a count of "how much stuff" is present. It becomes a measure of how persistently a configuration resists reorganization under equilibrium pressure.

Mass is not weight. Mass is not force. Mass is entrenched stability.

When we say a system has large mass, what we are really observing is this: the configuration holds delta so deeply, across so many intertwined pathways, that changing its state requires substantial reconfiguration. Resolution is possible—but it is expensive. The system does not oppose change out of inertia-as-mystery; it resists because its internal alignment is densely interlocked.

This is why mass feels like reluctance.

In classical language, mass "resists acceleration." In UED terms, acceleration demands reconfiguration. The more entrenched the configuration, the higher the threshold for reorganization. What we experience as inertial resistance is not an abstract property—it is the cost of disturbing a deeply stabilized pattern.

This also explains why mass is indifferent to direction, intent, or context. It does not "decide" to resist. It simply persists. The system continues doing what it has been doing successfully, and any attempt to alter that pattern must overcome the accumulated stability.

Seen this way, mass is historical.

It encodes how much alignment work has already been done to establish the current configuration. The more history embedded in a structure—the more layers of constraint, coupling, and internal resolution pathways—the more massive it appears. Mass is consequence memory made physical.

This reframing dissolves the confusion between mass and force. Force is a signal that resistance is occurring *now*. Mass is the reason resistance is likely to occur *at all*. One is a measurement of immediate refusal; the other is a measure of long-term persistence.

It also clarifies why mass remains constant even when systems move freely. A massive object in uniform motion does not "exert" its mass. Its configuration remains stable, and no reorganization is being demanded. Resistance appears only when alignment must change—when acceleration is attempted.

Mass, therefore, is not a brake on motion. It is a price tag on change.

This interpretation also scales naturally beyond mechanics. In chemistry, stable molecules behave as if they have "mass" relative to reaction pathways. In biology, organisms resist rapid change because their internal organization is deeply layered. In psychology, entrenched habits feel heavy not because of force, but because of persistent internal configuration.

Across domains, mass expresses the same idea: how hard it is to stop being what you are.

Once mass is seen this way, it no longer competes with energy, time, or force. It takes its rightful place as a descriptor of stability depth—a measure of how strongly a state insists on continuing.

That insight clears the way for the next correction. If mass and matter are about stability, then what we call "particles" must also be reconsidered—not as tiny objects, but as labels we attach to repeatedly stabilized states.

That is the next step.

15.4 — "Particles" as Labels for Stabilized States

Once matter and mass are understood as stability and persistence, the idea of a particle quietly loses its claim to being a fundamental object. What remains is not a tiny thing moving through space, but a reliably reappearing state—a configuration that stabilizes often enough, and long enough, that we give it a name.

Particles are not entities. They are labels.

Historically, particles were introduced because something repeatable was being observed. Certain interaction outcomes kept recurring. Certain signatures appeared again and again under similar conditions. Naming these recurring outcomes as "particles" was a practical decision, not an ontological one. It allowed bookkeeping, prediction, and communication.

UED accepts that practicality—while correcting the interpretation.

What actually persists is not a bead of substance traveling intact through space and time. What persists is a resolution pattern: a particular way delta organizes itself under specific constraints, thresholds, and geometries. When those conditions are recreated, the same pattern reappears. We call that reappearance a particle.

This immediately dissolves the tension between particle and wave descriptions.

A wave-like behavior reflects how a disturbance propagates while seeking resolution. A particle-like behavior reflects the moment that disturbance locks into a stable configuration. They are not competing realities; they are different phases of the same process. One describes unresolved motion. The other describes arrested resolution.

The so-called "duality" exists only because we insisted on treating particles as things rather than states.

In UED terms:

- propagation is state-seeking behavior,
- localization is state-stabilization,
- and measurement enforces thresholds that favor one over the other.

A particle appears when conditions allow a disturbance to settle into a configuration that can persist. When those conditions dissolve, the label dissolves with them. Nothing mysterious disappears—only the stability permission changes.

This also explains why particle identity is contextual. The same "particle" behaves differently in different environments because it is not carrying intrinsic properties like luggage. Its behavior reflects the current permission structure. Change the geometry, constraints, or thresholds, and the same label now refers to a different effective behavior.

From this perspective, particles are closer to verbs than nouns. They describe what a system is doing successfully, not what it eternally is.

UED therefore treats particles the way good engineering treats modes: as named solutions that emerge under known conditions. Useful, real, repeatable—but not fundamental.

This correction is not an attack on particle physics. It is its clarification.

Particle language remains invaluable wherever stabilized states recur reliably. What UED removes is the temptation to treat those states as irreducible building blocks of reality. They are outcomes, not origins.

Once this is seen clearly, the final confusion of this chapter can be addressed: why matter appears discrete at all—why these stabilized states come in "allowed" forms instead of blending smoothly into one another.

That question leads directly to thresholds, and to the appearance of quantization—not as cosmic granularity, but as permission logic made visible.

15.5 — Why Matter Appears Discrete

If matter is stability, and particles are stabilized states rather than tiny objects, one question still presses itself forward: why does matter appear discrete at all? Why do we encounter "allowed" masses, energies, charges, and states instead of a smooth continuum of possibilities?

The traditional answer invokes fundamental granularity—nature as a mosaic of indivisible units. UED offers a cleaner explanation: discreteness is not built into reality; it is enforced by thresholds.

Discreteness is permission logic, not pixelation.

Whenever resolution is governed by constraints and thresholds, outcomes do not vary continuously. They cluster. Only certain configurations can survive equilibrium pressure without collapsing. Those configurations become visible as discrete states—not because the universe prefers counting, but because stability is selective.

A configuration either meets the conditions to persist, or it does not. There is no half-permission.

This logic appears everywhere. Chemical bonds form only at specific energy levels. Atomic orbitals exist at allowed configurations. Phase changes occur abruptly. Even in macroscopic systems, stability arrives in steps—gears lock, structures buckle, regimes shift. Continuity exists beneath, but persistence does not.

Discreteness is therefore the shadow cast by thresholds.

Thresholds do not care about smoothness. They care about affordability. Below the threshold, resolution is blocked. At the threshold, it releases. Above it, the configuration reorganizes rapidly. What remains observable are the plateaus of stability between releases.

This is why quantization does not contradict continuity. Delta varies continuously. Geometry deforms continuously. Gradients adjust continuously. But stable configurations do not. They occupy islands where resolution can be delayed cheaply and collapse elsewhere.

Seen this way, discreteness is not a feature of "particles." It is a feature of survivable arrangements.

UED therefore removes the need for metaphysical granules. There is no smallest bead of existence. There are only configurations that meet the criteria to persist—and those that do not. What we count are the survivors.

This reframing also explains why discreteness depends on context. Change the constraints, and the allowed states change. Under different geometries, thresholds, or boundary conditions, new discrete outcomes appear while old ones vanish. Nothing fundamental has changed—only the permission map.

The universe does not count. It filters.

Once this is understood, the appearance of discrete matter no longer demands special explanation. It becomes the expected outcome of a world governed by equilibrium pressure, constrained pathways, and threshold-gated resolution.

With discreteness clarified, the final task of this chapter is to revisit the most dramatic language we use about matter—creation, destruction, annihilation—and strip it of its last remaining illusions.

That is where the chapter must now turn.

15.6 — Creation, Destruction, and Transformation Revisited

Once matter is understood as stabilized configuration, the dramatic language we use around it—*creation*, *destruction*, *annihilation*—loses its literal meaning. What remains is not cosmic drama, but structural bookkeeping.

Nothing truly begins. Nothing truly ends. Configurations stabilize, destabilize, and reconfigure.

What we call creation is the successful condensation of delta into a new stability regime. Conditions align—geometry permits, thresholds are crossed, pathways cooperate—and a configuration emerges that can persist. We witness this as something "coming into existence." But nothing has been summoned from nothingness. Resolution has simply been delayed in a new way.

Likewise, what we call destruction is not disappearance. It is the collapse of permission. Constraints fail, thresholds give way, and the configuration can no longer afford to hold delta. Alignment accelerates. The structure dissolves into other forms of organization—often less coherent, more distributed. We label this as loss, decay, or annihilation, but nothing has vanished. Stability has.

Even the most extreme cases—particle–antiparticle annihilation, nuclear reactions, stellar collapse—follow the same logic. No entity blinks out of reality. What disappears is a specific pattern of persistence. What remains is redistributed delta expressed through other channels: radiation, heat, motion, new configurations.

Transformation, then, is the most honest word of the three.

Matter transforms when constraints change. The same underlying activity reorganizes itself under new permission maps. Bonds break and form. States dissolve and condense. Stability migrates. At no point does reality require a metaphysical reset.

This reframing resolves a subtle but deep confusion. Traditional language makes us imagine matter as a thing that must be protected from nonexistence. UED shows that matter is never threatened by nothingness—only by affordability. As long as a configuration can pay the cost of holding delta, it persists. When it cannot, it releases.

Creation and destruction are therefore observer-centered terms. They describe what crosses our stability horizon—what becomes persistent enough to notice, or unstable enough to lose its identity. Nature itself is indifferent to these labels.

The universe does not create or destroy. It permits and withdraws permission.

Once this is seen, conservation laws regain their proper meaning. They do not protect entities; they track redistribution. They ensure that what ceases to be stable somewhere must reappear as activity elsewhere. The bookkeeping never fails. Only the forms change.

This completes the reframing of matter.

Matter is not substance. It is not storage. It is not fundamental.

Matter is delta that learned how to stay.

And with that insight in place, the path forward becomes unavoidable. If matter represents the extreme case of delayed resolution, then the complementary behavior must also be examined—what happens when resolution cannot settle, when stability is not achieved, and disturbance propagates instead.

That behavior is what we call waves.

The chapter closes not by leaving matter behind, but by letting it point naturally to its dynamic counterpart.

From What Stays to What Travels

Matter, as it now stands in this framework, is no longer mysterious. It is not the bedrock of reality, nor the starting point of explanation. It is a successful outcome—a configuration that learned how to hold delta long enough to appear solid, separate, and persistent.

What stays is no longer confused with what exists.

This realization quietly shifts the reader's intuition. Matter stops feeling like the default state of the universe and begins to look like a special achievement. Stability is no longer assumed; it is explained. Persistence is no longer obvious; it is earned.

But stability is only one way resolution behaves.

Not all disturbances can condense. Not all misalignments find architectures that allow long-term delay. When thresholds cannot be met, when constraints are insufficient, or when geometry refuses to cooperate, resolution does not disappear. It moves.

Where matter represents delta that learned how to stay, the next behavior represents delta that must propagate.

This is the natural pivot.

The framework does not switch topics; it follows logic. If matter is delayed alignment, then waves are frustrated alignment—disturbance seeking balance across space, time, and structure. The same ingredients are present: delta, geometry, thresholds, and consequence. Only the outcome changes.

This transition is not a contrast between things and non-things. It is a spectrum of resolution strategies. At one extreme, stability dominates and matter appears. At the other, stability fails and disturbance travels.

Between them lies everything we call dynamics.

By closing this chapter here, the book avoids a false separation. Waves are not introduced as a new entity, nor as the opposite of matter. They are what happens when condensation is not affordable.

Matter pointed here all along.

What remains unresolved will not vanish. It will seek balance.

That seeking is what the next chapter finally allows us to see clearly—without metaphor, without mysticism, and without contradiction.

From what stays, we move now to what travels.

When resolution cannot settle into stability, it does not disappear—it travels.

Chapter 16 — Waves as Disturbance Seeking Balance

Why Waves Were Treated as Things

Waves have always unsettled physics in a way matter never did. They appear, move, interfere, vanish—and yet nothing tangible seems to travel with them. Still, for a long time, we tried to treat waves as if they were objects: entities that propagate through space, carry energy, and behave like extended substances.

This was not a mistake of intelligence. It was a consequence of starting from matter.

Once matter was assumed fundamental, anything that moved had to be treated as a thing. If a pattern traveled, we assumed something was traveling. When nothing visible could be identified, we invented carriers, media, or dual descriptions. Waves became "things that aren't quite things," and confusion followed naturally.

UED removes the need for that compromise.

A wave is not an entity that exists in addition to matter. It is what happens when matter fails to form or when a configuration cannot afford to stabilize locally. It is not a thing moving through space; it is a disturbance attempting to resolve across available pathways.

This reframing immediately dissolves the ontological tension. There is no need to ask whether waves are real objects or mere abstractions. They are real behaviors—real in the same way negotiation is real, even though nothing material passes hands.

The urge to reify waves came from language. We speak of waves "traveling," "colliding," and "carrying energy." These phrases are useful, but they smuggle in a hidden assumption: that something persistent is moving. In reality, what persists is not substance, but pattern under constraint.

Waves appear when delta cannot settle. The system is misaligned, but local condensation is too costly or forbidden. The only remaining option is redistribution—alignment effort spreading outward, searching for pathways where resolution might become affordable.

Seen this way, waves stop being mysterious intermediates between particles and fields. They become the default behavior of unresolved systems.

Matter is delayed resolution that learned how to stay. Waves are delayed resolution that cannot stay.

This correction also explains why waves are so common. Most systems do not have the architecture required to condense disturbances into stable forms. Propagation is cheaper than persistence. Spreading is easier than holding. Waves are not exotic; they are what equilibrium pressure produces when stability fails.

By opening the chapter here, the reader is invited to let go of the idea that waves are "things in motion." What moves is not an object, but the demand for alignment. The wave is the visible trace of that demand negotiating with geometry and constraint.

Once this intuition settles, the rest of the chapter follows without friction. We no longer need metaphors or dualities. We only need to watch how disturbance behaves when balance is out of reach.

That is where the analysis now begins.

16.1 — Disturbance Is an Attempt, Not an Entity

A disturbance is not something added to a system. It is not an object injected, released, or launched. A disturbance is a condition—the presence of delta that has not yet found an affordable way to resolve locally.

In UED terms, disturbance is misalignment made visible.

When equilibrium is locally broken and the system lacks the architecture to condense that misalignment into a stable configuration, nothing "forms." Instead, the system is left with an unresolved demand for alignment. That demand does not sit still. It seeks pathways. It explores geometry. It negotiates constraints. What we observe as a wave is the trace of that negotiation.

This is why it is a mistake to ask, "What is the wave made of?" There is no thing to inventory.

What exists is a pattern of attempted resolution moving across a permission map. At each point, the system asks the same question: *Can alignment finish here?* If the answer is no, the attempt continues outward.

Disturbance, therefore, is inherently processual. It exists only as long as misalignment exists. The moment resolution becomes affordable—through condensation, absorption, dissipation, or reconfiguration—the disturbance disappears. Nothing has been destroyed. The attempt has simply succeeded.

This also clarifies why waves cannot be localized in the same way matter can. Matter is a successful answer. A wave is an ongoing question.

Treating waves as entities leads to immediate confusion:

- How can a wave pass through another without collision?
- How can it interfere with itself?
- How can it vanish without residue?

All of these puzzles dissolve once the premise is corrected. Attempts can overlap. Negotiations can interfere. Questions can be answered in multiple places at once. None of this requires substance.

A wave is alignment pressure in motion, not a moving object.

This framing also explains why waves are sensitive to boundaries and constraints. Geometry does not "act on" a wave; it filters attempts. Edges redirect, reflect, refract—not because a thing is bouncing, but because the permission landscape has changed. The disturbance adapts its search accordingly.

Even energy transport, so often attributed to waves as carriers, becomes clearer. Energy does not travel like cargo. What propagates is the capacity to induce reconfiguration elsewhere—the ability to pass along misalignment until someone else can afford to resolve it.

In this sense, waves are the most honest expression of UED's core logic. They show delta doing exactly what delta does when stability is unavailable: seeking balance without form.

Once disturbance is understood this way, the next question becomes unavoidable. If waves are attempts to resolve, why do they move the way they do? Why don't they simply spread chaotically or collapse immediately?

The answer lies in geometry and cost. Propagation is not arbitrary. It is the least expensive strategy available when condensation fails.

That logic is where the chapter now turns.

16.2 — Why Propagation Replaces Condensation

When a disturbance appears, the system faces a simple but unforgiving constraint: resolve locally or redistribute. Condensation is resolution that settles. Propagation is resolution that moves. The choice is not preference; it is affordability.

Local condensation requires architecture. It demands boundaries that can hold delta, thresholds that can be crossed, and pathways that can absorb misalignment without collapsing. When any of these are missing—or when the cost of holding delta locally is too high—the system does not fail. It chooses the only remaining viable strategy: propagation.

Propagation is not motion for its own sake. It is resolution deferred and displaced.

In UED terms, propagation is the least-cost way to continue negotiating alignment when local options are exhausted. The disturbance explores adjacent regions of the permission map, redistributing misalignment across space, structure, or medium in search of cheaper resolution pathways.

This is why waves follow geometry so faithfully. They do not decide where to go; they discover where they are allowed. Corridors that minimize resistance become preferred routes. Obstacles do not repel a wave; they reprice the negotiation. The disturbance reroutes because the cost landscape has changed.

This logic also explains why propagation has speed limits. A system can only redistribute misalignment as fast as its internal coupling allows. Those couplings—elastic, electromagnetic, fluid, structural—define how quickly alignment information can be shared. Speed is not an attribute of the wave; it is a property of the medium's permission bandwidth.

Crucially, propagation does not imply transport of substance. Nothing is being carried from point A to point B. What propagates is the opportunity to resolve elsewhere. Each location temporarily inherits the same question the previous one could not answer: *Can alignment finish here?*

When the answer becomes yes—through absorption, damping, scattering, or condensation—the wave ends. Not because it was destroyed, but because its job is done.

This perspective also clarifies why waves can superpose without interference in the material sense. Multiple attempts can traverse the same region because they are not occupying space as objects. They are coexisting negotiations. Where their demands overlap, the permission landscape responds collectively, producing patterns we call interference.

Propagation, then, is not an alternative to stability. It is stability's fallback mechanism.

Matter represents delta that learned how to stay. Waves represent delta that could not.

Once this is understood, the presence of waves throughout nature stops being surprising. Most systems cannot afford condensation everywhere. Spreading misalignment is cheaper than holding it. Propagation is therefore common, efficient, and inevitable.

But propagation rarely proceeds smoothly. When feedback is delayed and resistance is encountered unevenly, another familiar behavior emerges—oscillation.

That behavior is not a preference for rhythm. It is the signature of negotiation under constraint.

16.3 — Why Oscillations Appear at All

Oscillation is often treated as a basic preference of nature—as if systems *like* to vibrate, swing, or repeat. In UED, oscillation has no such status. It is not chosen. It is what happens when resolution overshoots and is pulled back repeatedly under constraint.

Oscillation is negotiation failure made rhythmic.

When a disturbance propagates, it does so under gradients that point toward alignment. But gradients alone do not guarantee smooth settlement. Geometry resists, thresholds delay release, and feedback is never instantaneous. As a result, the system frequently resolves too much, too fast, crosses the optimal alignment point, and immediately finds itself misaligned in the opposite direction.

This overshoot is not an error. It is a structural consequence of delayed feedback.

Perfect resolution would require infinite sensitivity and zero delay—conditions that never exist in real systems. Instead, alignment proceeds in steps. Each step corrects the previous one but introduces a new imbalance of opposite sign. The system swings, corrects, swings again. Oscillation emerges naturally.

This explains why oscillation appears across domains:

- In mechanics, springs overshoot equilibrium because restoring response lags deformation.
- In fluids, waves oscillate because pressure gradients and inertia are mismatched in time.
- In electronics, alternating currents arise from phase delays between voltage and charge redistribution.
- In biology, heartbeats and neural rhythms reflect feedback delays in regulatory loops.
- In psychology, emotional cycles emerge when corrective responses arrive after the situation has already shifted.

In every case, oscillation is not motion for its own sake. It is repeated correction under delayed information.

The system is trying to settle. It simply cannot do so cleanly.

This is why oscillation requires both constraint and feedback. Without constraint, misalignment would diffuse and disappear. Without feedback, correction would never occur. Oscillation sits between these extremes: resolution is attempted, resisted, corrected, and resisted again.

UED therefore treats oscillation as a signature of partial permission. The system is allowed to resolve—but not fully, not immediately, and not cheaply. The result is a back-and-forth motion that persists until either constraints relax, energy diffuses, or coherence collapses.

This framing also explains why oscillations decay. Damping is not an external enemy of oscillation; it is the gradual diffusion of unresolved delta into other pathways. As alignment pressure spreads and redistributes, the need for repeated correction weakens. Oscillation fades because the negotiation is finally concluding.

Nature does not favor oscillation. Nature tolerates it temporarily.

Oscillations exist only as long as the system cannot complete alignment smoothly. They are the audible, visible trace of a conversation still in progress.

Once this is understood, wave behavior loses its mystique. Oscillation is no longer a fundamental motion; it is resolution struggling to finish.

That insight opens the door to a deeper distinction—why some disturbances hold their shape while others dissolve almost immediately. That distinction is not about waves themselves, but about how well alignment attempts remain coordinated.

That is the question of coherence.

16.4 — Coherence vs Diffusion

Once oscillation is understood as repeated correction under constraint, the next distinction becomes unavoidable: why do some disturbances maintain a recognizable shape while others dissolve almost immediately? The answer lies in whether alignment attempts remain coordinated or fragment into independent resolutions.

This is the difference between coherence and diffusion.

Coherence is not strength. It is agreement.

A coherent wave exists when many local resolution attempts remain phase-aligned—when corrections happen in a coordinated manner across space and time. Each part of the disturbance "knows" what the others are doing, not through intention, but through consistent coupling and stable geometry. The system negotiates alignment together, producing a pattern that holds its form while propagating.

Diffusion appears when that agreement breaks.

As soon as pathways diversify, coupling weakens, or geometry becomes irregular, alignment attempts lose synchronization. Different regions resolve at different rates, in different directions, through different channels. The disturbance fragments. The recognizable pattern fades—not because energy is lost, but because coordination is lost.

This distinction clarifies a long-standing confusion.

Coherence is often treated as an intrinsic property of a wave, as if some waves are "born coherent." In UED, coherence is conditional. It exists only as long as the permission landscape remains uniform enough to keep resolution attempts synchronized. The moment that landscape becomes uneven—through scattering, absorption, heterogeneity, or interaction—coherence degrades naturally.

This is why coherence is fragile.

A laser beam remains coherent because the medium, boundaries, and feedback are tightly controlled. A ripple on water spreads coherently for a short distance and then dissolves as viscosity, boundary effects, and competing pathways intervene. Sound maintains structure in air over meters, but not indefinitely. In every case, coherence survives only where coordination is affordable.

Diffusion, by contrast, is the default.

When coordination becomes expensive, the system does not preserve pattern out of loyalty. It resolves locally. Alignment pressure spreads across many degrees of freedom. The disturbance completes its work quietly, without shape, rhythm, or identity. What remains is not absence, but distributed consequence—often experienced as heat, noise, or background fluctuation.

Importantly, diffusion is not failure. It is successful resolution without structure.

UED therefore places coherence and diffusion on the same continuum. They are not opposites. They are different outcomes of the same process under different constraint conditions. Coherence delays completion by maintaining agreement. Diffusion completes alignment by abandoning agreement.

This framing also resolves the false hierarchy often implied in physics, where coherence is treated as "ordered" and diffusion as "disordered." In reality, both are lawful. Both are efficient under their respective conditions. The system chooses neither; it follows cost.

Where coordination is cheap, coherence persists. Where coordination is expensive, diffusion dominates.

With this distinction in place, waves no longer appear mysterious or special. They occupy a middle ground between matter and heat—between stability and completion. When coherence holds, waves behave like structured travelers. When it fails, they dissolve into diffuse consequence.

That dissolution is not an afterthought. It is the natural endpoint of most disturbances.

And it leads directly to the next chapter, where diffusion is no longer a side effect but the main actor—where friction, dissipation, and so-called "loss" are finally read as what they truly are: alignment finishing its work by spreading everywhere.

16.5 — Coherence: When Disturbance Holds Its Shape

Coherence is often mistaken for strength or purity, as if a coherent wave is simply a more powerful version of an incoherent one. In UED, coherence has nothing to do with intensity. It is a matter of coordination—the ability of multiple local resolution attempts to remain phase-aligned while propagating.

A coherent wave is not a single action. It is many negotiations agreeing to move together.

When a disturbance propagates, each local region is independently attempting to resolve misalignment. Coherence emerges only if these attempts remain synchronized—if the timing, direction, and feedback of resolution stay compatible across space. This requires remarkably strict conditions: stable geometry, uniform coupling, and minimal disruption from competing pathways.

Coherence is therefore earned, not assumed.

This explains why coherent waves are special but not mysterious. They appear in systems where constraints are consistent and thresholds are well behaved. In such environments, resolution attempts can "trust" one another. Overshoots occur together. Corrections occur together. The pattern survives.

Seen this way, coherence is a temporary truce among competing resolution paths.

This also clarifies why coherence is sensitive to disturbance. Small irregularities in geometry, slight variations in coupling, or delayed feedback are enough to break agreement. Once coordination falters, the wave does not weaken gradually—it loses shape. The disturbance continues to resolve, but no longer as a unified pattern.

Importantly, coherence does not require low entropy or perfect order. It requires consistent permission. A system may be noisy yet coherent if the noise is uniform. Conversely, a system may be quiet yet incoherent if its constraints vary unpredictably.

This reframing dissolves the false mystique around coherence in physics. Coherent light, sound, or motion is not privileged by nature. It simply occupies a region of the permission map where alignment attempts can afford to stay synchronized.

UED also explains why coherence feels informational.

When a wave remains coherent, it can carry structured influence—timing, phase relationships, correlations—across distance. But nothing is being transported as content. What persists is coordinated behavior, not substance. Information appears because coordination persists, not because something is encoded and shipped.

Once coordination becomes too expensive, coherence collapses—not catastrophically, but quietly. The disturbance does not stop resolving. It just stops doing so together.

That collapse is not a failure of waves. It is the system choosing completion over pattern.

With coherence fully placed as conditional coordination, the complementary behavior becomes unavoidable. Most disturbances do not enjoy such favorable conditions. Their resolution fragments almost immediately, spreading alignment effort across many degrees of freedom.

That default behavior—where agreement is abandoned and resolution completes by spreading—is diffusion.

And that is where the chapter must now turn.

16.6 — Diffusion: When Disturbance Resolves Without Shape

Diffusion is what remains once the ambition of coherence becomes unaffordable.

When coordination between local resolution attempts can no longer be maintained—because geometry varies, coupling weakens, thresholds differ, or feedback delays diverge—the system does not hesitate. It abandons pattern and completes alignment locally, wherever it can, through whatever pathways are cheapest.

Diffusion is not decay of a wave. It is the successful conclusion of disturbance.

In diffusion, there is no longer a shared rhythm, no phase agreement, no recognizable shape. Each region resolves its misalignment independently. Alignment pressure spreads across many degrees of freedom, thinning as it goes, until no local gradient remains strong enough to demand further correction.

What we observe as "fading," "damping," or "loss" is simply this redistribution reaching completion.

This reframing removes a deep misconception. Diffusion is often treated as a problem—a degradation of signal, a loss of order, a failure to preserve structure. In UED, diffusion is what resolution looks like when structure is unnecessary.

If coherence is negotiated persistence, diffusion is negotiated release.

This is why diffusion is so common. Most systems are heterogeneous. Their constraints vary. Their pathways are uneven. Under such conditions, coordination is expensive and unnecessary. It is cheaper for the system to resolve misalignment independently across many micro-pathways than to preserve a global pattern.

Heat is the most familiar face of diffusion.

When a disturbance dissolves into thermal motion, nothing has vanished. Alignment pressure has simply been distributed so widely that no further work remains to be done. The system has found its equilibrium, not through elegance, but through completeness.

Diffusion is therefore irreversible in practice—not because laws forbid reversal, but because reassembling coordination from distributed resolution would require recreating the exact permission structure that was lost. The cost is prohibitive. Equilibrium has won.

This logic applies far beyond physics.

In biology, stress diffuses into fatigue. In psychology, unresolved tension diffuses into anxiety or numbness. In social systems, coordinated movements diffuse into background noise once shared structure collapses.

In every domain, diffusion marks the moment when alignment finishes without ceremony.

UED therefore restores dignity to diffusion. It is not a lesser outcome. It is not entropy's villainous hand. It is the natural endpoint of most disturbances in a constrained world.

Coherence delays completion by preserving agreement. Diffusion completes alignment by abandoning agreement.

With this understanding, waves can finally be placed properly. They are not fundamental actors, not carriers of substance, and not mysteries suspended between particle and field. They are transitional behaviors—what happens when matter cannot form and heat has not yet claimed the disturbance.

This realization prepares the ground for the next chapter, where diffusion is no longer an endpoint but a mechanism—where friction, dissipation, and so-called "loss" are examined not as failures, but as how alignment actually finishes its work in real systems.

From Traveling Disturbance to Settled Consequence

By this point, waves no longer occupy an ambiguous middle ground between matter and mystery. They have found their exact place in the hierarchy of resolution.

A wave is not a thing. It is not a carrier. It is not a fundamental actor.

A wave is delta in transit—misalignment that could not settle locally and therefore spread its demand for alignment across available pathways.

When coordination holds, the disturbance travels with shape. When coordination fails, the disturbance dissolves into diffusion. In both cases, the goal is the same: finish alignment at the lowest total cost.

This closes a long-standing conceptual loop. Matter, waves, and heat are no longer separate ontological categories. They are three outcomes of the same process under different constraint conditions:

- Matter is delta that learned how to stay.
- Waves are delta that must move to survive.
- Heat is delta that has finished moving.

Nothing jumps between categories by magic. The transition is continuous and structural. As permission tightens, behavior shifts—from propagation to oscillation, from coherence to diffusion, from pattern to spread.

This realization also strips waves of their last illusion of autonomy. They do not "decay" because something attacks them. They do not "lose energy" because the universe is careless. They simply complete their work once coordination becomes unaffordable.

The disturbance ends not in failure, but in settlement.

And that settlement is not abstract. It is felt as resistance, warmth, drag, fatigue, damping—phenomena traditionally grouped under the uncomfortable word *loss*.

UED now removes the discomfort.

What physics has long called loss is not disappearance. It is alignment reaching places we stopped tracking. It is resolution finishing without preserving structure.

With waves now fully placed as transitional behavior, the framework is ready to face what comes next without apology or metaphor.

The next chapter does not introduce a new effect. It explains the inevitable destination of most effects.

From disturbance seeking balance, we move to balance achieved by spreading.

That is where friction, dissipation, and irreversibility finally stop looking like problems—and start looking like the most honest expressions of equilibrium pressure at work.

When disturbance can no longer negotiate structure, it resolves by spreading everywhere.

Chapter 17 — Friction, Dissipation, and "Loss"

Why "Loss" Felt Like Failure

Few words in physics carry as much quiet emotional weight as *loss*. It suggests waste, inefficiency, something gone wrong. When motion slows, when heat appears, when energy spreads into forms we no longer care about, our language implies disappointment—as if nature missed an opportunity we wanted it to take.

This reaction did not come from physics. It came from human intention.

We care about outcomes. We build machines to move loads, transmit signals, preserve structure. Anything that diverts energy away from those purposes feels like failure. Over time, that bias slipped into our descriptions. Dissipation became something to minimize. Friction became an enemy. Heat became the graveyard of useful motion.

UED removes that emotional inheritance.

Nothing in nature experiences loss. Nature only experiences completion.

What we call loss is simply alignment finishing its work in places we stopped valuing or measuring. Energy does not vanish. It does not decay into nothingness. It spreads, redistributes, and resolves along pathways that no longer preserve the structure we were tracking.

The disappointment is ours, not nature's.

This is why dissipation has always felt vaguely unsatisfying in explanation. We were trying to read purpose into a process that has none. Dissipation is not an error term. It is not a tax imposed by an imperfect universe. It is the default endpoint of unresolved delta in a constrained world.

When coherence fails, diffusion takes over. When diffusion completes, gradients vanish. When gradients vanish, nothing more needs to happen.

That is not failure. That is equilibrium doing its job.

Friction, drag, damping, fatigue—these were framed as obstacles because we wanted permanence, efficiency, and control. Physics inherited that framing and quietly agreed. But the universe was never optimizing for our goals. It was optimizing for affordability of resolution.

Once this is acknowledged, the word "loss" loses its authority.

What disappears is not energy, but our chosen structure. What remains is consequence—distributed, quiet, and finished. Dissipation is alignment that no longer announces itself.

This chapter exists to remove the last trace of moral language from physics. Not to dismiss engineering concerns, but to place them properly. Efficiency will remain meaningful. Optimization will remain valuable. But they will be understood as strategies for managing delta, not as judgments passed on nature.

With that correction in place, friction and dissipation can finally be read honestly—not as enemies of motion, but as the mechanisms by which motion concludes.

And once loss is stripped of disappointment, thermodynamics stops feeling grim and starts feeling inevitable.

17.1 — Dissipation as Alignment Diffusion

Dissipation is not a side effect of motion; it is how unresolved alignment finishes.

When a system carries delta—misalignment held in structured form—it has only two long-term options. Either the structure continues to coordinate resolution attempts coherently, or coordination becomes too expensive and the system releases alignment pressure across many degrees of freedom at once. That release is dissipation.

Dissipation, in UED, is alignment diffusion.

Nothing is destroyed. Nothing is consumed. The misalignment that once drove motion is redistributed so broadly that no local gradient remains strong enough to organize further action. The system does not "lose" energy; it loses concentration.

This is why dissipation requires multiplicity. It needs many micro-pathways—vibrational modes, internal rearrangements, microscopic collisions, configurational degrees of freedom. The more pathways available, the cheaper diffusion becomes. Structured motion collapses not because something attacks it, but because spreading becomes more affordable than coordination.

Viewed this way, dissipation is not antagonistic to motion. It is motion's completion strategy when persistence fails.

A sliding block slows not because energy disappears, but because alignment pressure is being rerouted into countless micro-adjustments at the interface—atomic vibrations, lattice distortions, microscopic heat. Each pathway absorbs a tiny portion of the original delta until nothing remains to drive bulk motion.

The same logic applies universally. Electrical resistance dissipates alignment into lattice vibrations. Viscosity dissipates flow into internal rearrangements. Mechanical damping dissipates oscillation into distributed motion. In every case, dissipation is the flattening of gradients through diffusion.

This reframing also explains why dissipation is directional in practice. Once alignment pressure has spread across many degrees of freedom, reassembling it into a coordinated pattern would require re-creating the exact permission structure that allowed coherence in the first place. That cost is prohibitive. Diffusion wins not because time forbids reversal, but because coordination cannot be cheaply restored.

Dissipation is therefore not decay. It is resolution choosing breadth over structure.

Understanding dissipation as alignment diffusion strips it of mystery and hostility. It is not an enemy of systems; it is the mechanism that prevents them from accumulating unresolved delta indefinitely. Without dissipation, systems would become brittle, unstable, and prone to catastrophic release.

Dissipation is equilibrium pressure acting patiently, finishing what coherence could not.

With this placed, the chapter can now focus more sharply. If dissipation is alignment diffusion, then friction—the most familiar face of "loss"—must be understood not as a force or a drain, but as diffusion forced at specific interfaces.

That is the next correction.

17.2 — Friction: Resolution at the Interface

Friction is not a force that opposes motion by intention. It is alignment being forced to resolve sideways when a system is denied a smooth path forward.

Friction appears only at interfaces—places where two geometries meet and disagree. This alone is telling. If friction were a fundamental resistance embedded in motion itself, it would appear everywhere. It does not. It appears where resolution corridors narrow, fragment, or misalign.

In UED terms, friction is what happens when delta demands alignment through a pathway that geometry refuses to smooth out.

Consider sliding motion. The macroscopic intent is clear: continue moving along a direction. But at the microscopic level, the interface is a landscape of mismatched lattices, asperities, and local constraints. There is no clean corridor for alignment to proceed coherently. The system cannot condense the disturbance into stable motion, nor can it propagate it as a wave without obstruction.

So it diffuses—laterally and locally.

Friction is alignment being redirected into micro-pathways that were never part of the intended motion. Atomic vibrations, micro-deformations, bond stretching, and lattice rearrangements absorb the misalignment. The original gradient flattens, not by stopping, but by spreading into places we were not tracking.

This is why friction feels like loss.

From the perspective of the intended motion, progress stalls. From the perspective of the system, alignment has simply chosen a cheaper route.

The universe did not resist motion; it refused to subsidize a corridor that geometry could not support.

This reframing explains several stubborn facts about friction without adding new principles.

- Friction depends on surfaces, not volumes, because it is an interface phenomenon.
- Friction scales with normal load because tighter contact increases the number of micro-pathways available for diffusion.
- Friction produces heat because diffusion overwhelmingly favors vibrational degrees of freedom.
- Friction cannot be eliminated, only managed, because interfaces cannot be made perfectly coherent across all scales.

Seen this way, friction is not an opposing force added to equations. It is resolution happening where permission is cheapest once the intended path becomes expensive.

This also clarifies why lubrication works. Lubricants do not "reduce force" by magic. They reshape the permission map at the interface, opening smoother corridors for alignment to proceed coherently. When geometry is softened and pathways are made more uniform, alignment can travel instead of diffusing. Friction drops because diffusion is no longer the cheapest option.

Friction, therefore, is not a failure of design. It is feedback.

It tells us where geometry is mismatched, where coordination cannot be sustained, and where the system is quietly choosing completion over persistence. Engineers learn to listen to this signal, not to curse it.

UED places friction exactly where it belongs: not as an enemy of motion, but as the cost of insisting on coherence where geometry will not cooperate.

With friction properly placed at the interface, the broader misunderstanding can now be addressed. If friction does not destroy energy, and dissipation does not erase alignment, then the most persistent myth must finally be dismantled: the idea that something ever truly disappears.

That correction belongs to the next section.

17.3 — Why Nothing Ever Disappears

The idea that something *disappears* in physical processes is not a statement about reality; it is a statement about what we chose to keep accounting for.

When motion slows, when sound fades, when heat spreads, we say energy is "lost." But nothing in the system has vanished. What has vanished is our structured description.

In UED terms, consequence never ends. It only changes channels.

What we track—velocity, height, voltage, coherence—is a narrow slice of how alignment is expressed. Dissipation feels like loss because alignment exits those tracked variables and enters forms that are harder to organize, harder to reverse, and less useful for our immediate purposes. The bookkeeping did not fail; our ledger did.

This is why conservation laws remain perfectly intact even as usefulness declines. Energy conservation never promised preservation of structure. It promised persistence of redistribution. When a wave diffuses into heat, nothing is missing. Alignment pressure has simply spread across so many micro-pathways that no macroscopic gradient remains to exploit.

Loss is therefore exit from relevance, not exit from existence.

This reframing explains why different observers perceive different losses. An engineer designing a motor sees heat as loss. A biologist sees heat as regulation. An astrophysicist sees radiation as information. Each perspective tracks different consequence channels. What appears lost in one frame is active in another.

UED makes this explicit: "loss" is a projection of intent onto neutral redistribution.

This also clarifies why recovery is so difficult after dissipation. To reverse diffusion, one would need to reassemble alignment from countless micro-adjustments into a coordinated pattern. That would require reconstructing the exact geometry, thresholds, and timing that once allowed coherence. The universe does not forbid this in principle—but it prices it prohibitively.

Nothing disappears. Coordination does.

Once coordination is gone, consequence continues quietly. Vibrational modes thermalize. Microstates rearrange. Alignment finishes without ceremony. The system reaches equilibrium not because something was destroyed, but because nothing remains unresolved.

This understanding strips "loss" of its last illusion of negativity. Dissipation is not a leak in the laws of physics. It is the law of consequence being honored fully, even when the outcome no longer serves our designs.

With disappearance removed from the story, one question remains unavoidable: if nothing vanishes, why does dissipation feel so final? Why does it mark a one-way boundary in practice?

The answer is not time, fate, or entropy-as-mystery. It is irreversibility born from exhausted permission.

That is the next correction the chapter must make.

17.4 — Irreversibility Without Mystery

Irreversibility has long been treated as one of physics' darker facts—a fundamental asymmetry woven into time itself. In UED, irreversibility carries no such drama. It is not imposed by a cosmic arrow. It emerges naturally from permission exhaustion.

Irreversibility is what happens when alignment has spread so thoroughly that no affordable pathway exists to reassemble it.

When dissipation occurs, delta does not simply move from one form to another. It fans out across an enormous number of microstates—vibrational modes, internal rearrangements, configurational variations. Each of these states resolves a tiny fraction of misalignment independently. The original coordination that made structured motion possible is lost, not by law, but by cost.

Reversing this process would require recreating coordination across all those pathways simultaneously.

In principle, nothing forbids it. In practice, everything prices it out.

This is the crucial correction. Irreversibility is not about impossibility. It is about unaffordability.

To reverse dissipation, one would need to:

- recollect distributed alignment from countless degrees of freedom,
- reestablish precise phase relationships,
- reconstruct the original geometry and thresholds,
- and do so without introducing new misalignments.

The cost explodes combinatorially. The system does not refuse. It simply cannot pay.

This is why irreversibility feels absolute. Once alignment diffuses beyond coordination, the permission landscape collapses. The system has no corridor back to its previous structured state. Not because time forbids it, but because permission has been spent.

UED therefore reframes irreversibility as a structural outcome, not a temporal decree.

Time does not push systems forward irreversibly. Systems become irreversible when they exhaust the conditions required for reversal.

This distinction matters deeply. It removes the need for metaphysical arrows, special axioms, or existential asymmetry. The same rules apply forward and backward. What differs is where permission remains available.

This also explains why irreversibility depends on scale. At microscopic scales, where degrees of freedom are few and coordination is cheap, reversals can occur. At macroscopic scales, where diffusion involves astronomical numbers of pathways, reversibility becomes practically invisible.

Irreversibility is therefore not universal. It is scale-dependent exhaustion.

Once this is understood, dissipation loses its finality. It is no longer a cliff beyond which nothing can be done. It is a region where structure has finished its work and alignment has settled. The system is not broken. It is complete.

With irreversibility stripped of mystery, the chapter can now address a more practical concern—one that matters deeply to engineers, biologists, economists, and societies alike.

If dissipation is inevitable, and irreversibility is structural, then what does efficiency actually mean?

That question belongs to the next section, where efficiency is freed from moral tone and placed where it belongs: as strategy, not virtue.

17.5 — Efficiency as Delta-Management

Once dissipation and irreversibility are understood structurally, the idea of efficiency must be rebuilt from the ground up. Efficiency is often treated as a moral scorecard—good systems waste little, bad systems waste much. In UED, that framing collapses.

Efficiency is not virtue. Efficiency is strategy.

Every system operates under equilibrium pressure with limited permission. Delta must resolve somewhere. The only meaningful choice is where and how that resolution is allowed to occur. Efficiency, therefore, is not about avoiding dissipation—because dissipation is inevitable—but about steering the pathways through which alignment finishes.

An efficient system is not one that loses nothing. It is one that loses where it can afford to lose.

This reframing immediately clarifies why perfect efficiency is impossible. To preserve all structure indefinitely would require infinite coordination and zero diffusion—conditions that no real system can sustain. The attempt to suppress dissipation entirely would only accumulate unresolved delta until release becomes catastrophic.

Efficiency is thus a balancing act between:

- maintaining useful structure long enough to perform work, and
- allowing dissipation to prevent overload, instability, or failure.

In practical terms, efficiency is delta-budgeting.

Engineers do this instinctively. Heat sinks are added not because heat is desired, but because its controlled release protects structure elsewhere. Friction is reduced in bearings to preserve motion, but allowed in brakes to terminate it safely. Losses are not eliminated; they are placed.

Biological systems do the same. Metabolism is inefficient by design, shedding energy as heat to maintain stability and responsiveness. Psychological systems dissipate tension through fatigue, emotion, or rest to avoid breakdown. Social systems leak energy through bureaucracy, delay, and redundancy to prevent violent release.

In all cases, efficiency is not maximal retention. It is selective surrender.

UED therefore removes the moral tension from efficiency debates. High efficiency is not always good. Low efficiency is not always bad. What matters is whether dissipation occurs in channels that preserve the system's ability to continue functioning.

A perfectly efficient engine that cannot dissipate excess alignment would destroy itself. A perfectly efficient society that cannot absorb stress would fracture. A perfectly efficient mind that never diffuses tension would collapse.

Efficiency, then, is not an endpoint. It is a continuously negotiated compromise between coherence and diffusion.

This understanding also explains why efficiency metrics are always contextual. What counts as loss depends on what the system values. Change the goal, and the efficiency calculation changes. Physics itself is neutral here. It provides the accounting. Strategy supplies the priorities.

Once efficiency is reframed this way, dissipation stops being an embarrassment. It becomes a design parameter—one of the most important ones.

With efficiency properly placed as delta-management rather than moral achievement, the chapter can close its final loop. One last correction remains: dissipation is not merely unavoidable—it is necessary.

Without it, systems do not become perfect. They become fragile.

That necessity is where the chapter must now conclude.

17.6 — Why Dissipation Is Necessary

Dissipation is not merely unavoidable; it is essential. A system that cannot dissipate does not become efficient—it becomes unstable.

This is the final correction that UED insists on making.

Every system that holds delta is under continuous equilibrium pressure. Alignment is always being requested. If a system were able to preserve structure perfectly—never allowing diffusion, never shedding alignment pressure—it would accumulate unresolved delta relentlessly. The result would not be elegance or control. The result would be catastrophic release.

Dissipation is the safety valve of reality.

It allows alignment to finish quietly, incrementally, and locally rather than explosively. It prevents pressure from concentrating beyond thresholds that would destroy structure outright. In this sense, dissipation is not the enemy of order; it is the guardian of survivable order.

This logic appears everywhere once it is seen.

Mechanical systems require damping to avoid resonance runaway. Electrical systems require resistance to prevent destructive oscillations. Biological systems require metabolic inefficiency to regulate temperature and stress. Psychological systems require emotional diffusion—rest, distraction, sleep—to prevent collapse. Social systems require inefficiencies—buffers, redundancies, delays—to absorb tension without violence.

In every domain, systems that try to suppress dissipation entirely do not become stronger. They become brittle.

UED therefore rejects the fantasy of lossless systems as not merely impossible, but undesirable. A world without dissipation would be a world without forgiveness. Every disturbance would persist indefinitely. Every misalignment would compound. Nothing would settle.

Equilibrium itself would become unreachable.

Dissipation makes equilibrium attainable by allowing alignment to complete in forms that do not demand coordination. It lets systems relax without having to remain coherent. It is the reason stability can be restored after disturbance.

This necessity also explains why dissipation often feels anticlimactic. There is no visible victory, no preserved structure, no reusable pattern. Alignment simply finishes, and the system moves on. Completion without ceremony is easy to overlook—but it is what keeps the world intact.

Once this is accepted, the emotional resistance to dissipation dissolves. Loss is no longer a sign of failure. It is evidence that the system chose survival over perfection.

This closes the chapter's argument.

Friction is not sabotage. Dissipation is not decay. Loss is not disappearance.

They are the mechanisms by which alignment concludes safely in a constrained universe.

With this understanding in place, only one step remains in Part V. If dissipation is how alignment finishes, then there must be a language that tracks this finishing universally—across matter, waves, and heat alike.

That language already exists.

It has always been called thermodynamics.

The final chapter of this part will show why thermodynamics was never a separate subject at all—but UED speaking in its most honest voice.

From "Loss" to Thermodynamics

At this point, the word *loss* can no longer stand unchallenged. It has been emptied of its drama and stripped of its implied failure. What remains is something far more precise—and far more useful.

What physics once called loss is simply alignment finishing outside our preferred structures.

Friction, dissipation, damping, drag—these are not imperfections layered onto an otherwise ideal universe. They are the universe doing exactly what it must do once coherence becomes unaffordable. They are not leaks in the system. They are release mechanisms.

This is the quiet resolution of a long-standing discomfort in physics. For centuries, we treated dissipation as an embarrassment—something to be minimized, postponed, or explained away. But once alignment diffusion is understood as inevitable and necessary, dissipation stops being a problem to solve and becomes a process to account for.

And accounting is the key word.

Nothing in this chapter required new laws, special exceptions, or metaphysical arrows. Every conclusion followed from one simple fact: delta must resolve, and when structure cannot be preserved, resolution spreads. The universe does not judge where alignment finishes. It only ensures that it does finish.

This is why thermodynamics has always felt different from other branches of physics.

It does not describe motion. It does not describe force. It does not describe structure.

It describes what remains when all preferences are removed.

Entropy, irreversibility, heat—these were never about disorder or decay. They were about diffusion of alignment beyond coordination. Thermodynamics has always been the language of consequence after structure gives up the negotiation.

Seen this way, thermodynamics is not a special rule-set imposed on physics. It is physics finally admitting that accounting matters more than intention.

Part V now reaches its natural closure.

- Matter showed us how alignment can stay.
- Waves showed us how alignment moves.
- Dissipation showed us how alignment finishes.

Nothing else is required.

The final chapter of this part will not introduce thermodynamics as a new topic. It will reveal why thermodynamics was inevitable from the first page of this book—the moment equilibrium, delta, and consequence were placed correctly.

When alignment spreads everywhere, physics stops asking *what happens* and starts asking *what is left to count.*

That counting—clean, unsentimental, and universal—is where this part ends.

When alignment spreads everywhere, physics stops arguing and starts accounting.

Chapter 18 — Thermodynamics as UED's Native Tongue

Why Thermodynamics Always Felt Different

Thermodynamics has always occupied an uncomfortable position in physics. It does not feel elegant in the way mechanics does. It does not inspire intuition like geometry. It does not invite metaphors of motion, force, or structure. And yet, no theory—no matter how revolutionary—has ever escaped it.

Everything eventually answers to thermodynamics.

This persistence is not accidental. Thermodynamics survived classical mechanics, electromagnetism, relativity, quantum theory, and every attempted unification because it was never competing with them. It was never proposing mechanisms. It was never claiming primacy. It was simply reporting what remains once all mechanisms have acted.

That is why it always felt different.

Thermodynamics does not ask *how* something happens. It asks *what is left once it has happened.*

In the language of UED, this makes perfect sense. From the very first chapters, the framework has insisted on three ideas: equilibrium is default, delta enables action, and consequence is unavoidable. Thermodynamics is nothing more than those ideas written without storytelling, without preference, and without rescue clauses.

It does not explain motion. It does not explain structure. It explains completion.

This is why thermodynamics feels immune to interpretation. You can rename forces, reinterpret time, dissolve particles into states, and reframe geometry as permission—but entropy still increases, dissipation still spreads, and irreversibility still asserts itself. Not because thermodynamics is authoritarian, but because it is descriptive of inevitability.

Physics often tried to treat thermodynamics as a rulebook imposed on reality—laws that nature must obey. UED reveals the inversion. Thermodynamics is not commanding the universe. It is taking notes.

It records what happens when:

- coordination fails,
- permission collapses,
- and alignment finishes wherever it can.

This is why thermodynamics applies everywhere. Not because all systems are the same, but because all systems must eventually stop negotiating structure. When that happens, only accounting remains.

Seen from this vantage point, thermodynamics was always speaking UED's language—just without the vocabulary to say so explicitly. Entropy was delta diffusion before we had the words. Heat was consequence channeling before we understood alignment. The second law was affordability logic before we recognized permission exhaustion.

This chapter does not reinterpret thermodynamics to fit UED.

It shows why thermodynamics never needed reinterpretation at all.

It was already describing the same reality—just from the only place where disagreement is impossible: after everything else has finished acting.

With this understanding in place, entropy can finally be approached without metaphor or fear. Not as disorder, decay, or doom—but as the signature left behind when alignment spreads beyond coordination.

That is where the chapter must now begin.

18.1 — Entropy as the Signature of Delta Diffusion

Entropy has suffered more from metaphor than almost any concept in physics. Disorder, chaos, decay, randomness—these words were never definitions; they were emotional substitutes. They tried to make peace with something that resisted intuition. In UED, entropy finally becomes simple, precise, and unthreatening.

Entropy is not disorder. Entropy is delta that has finished spreading.

When a system holds alignment in a coordinated way—motion, structure, potential—it is carrying delta in a compressed, organized form. This compression is never neutral. It requires geometry, thresholds, and permission. The moment those supports weaken, alignment does not vanish; it diffuses across available degrees of freedom.

That diffusion leaves a signature.

Entropy is that signature.

It is not a property added to a system. It is a record of how widely alignment has been distributed. The higher the entropy, the more degrees of freedom are participating in the resolution, and the less coordination remains accessible for structured work.

This is why entropy increases when coherence collapses. Not because the universe prefers disorder, but because diffusion is cheaper than coordination. Holding alignment in a narrow corridor demands constant permission. Releasing it into many micro-pathways does not.

Seen this way, entropy is not an enemy of order. It is the price tag attached to order's disappearance.

This reframing immediately resolves a long-standing confusion: why entropy can increase even while local order appears. Living systems, machines, and societies all create structure while entropy globally increases. There is no contradiction. Local coordination is bought by exporting diffusion elsewhere. Delta is not destroyed; it is reassigned.

Entropy does not measure how messy a system looks. It measures how irretrievable alignment has become.

When alignment is diffused into countless microscopic modes—vibrations, rotations, internal rearrangements—it becomes statistically invisible at macroscopic scales. The system still contains energy, but not in a form that can be easily recollected into gradients that do work. Entropy has increased because pathways have multiplied beyond coordination.

This is why entropy is indifferent to meaning, intention, or utility. A beautifully designed machine and a pile of rubble can contain the same energy. Entropy distinguishes them not by appearance, but by how many pathways alignment would have to traverse to become useful again.

From a UED perspective, entropy is therefore neither abstract nor mysterious. It is the natural consequence of allowing delta to resolve without constraint. Once diffusion begins, entropy simply tracks how far that resolution has progressed.

And crucially, entropy is not universal doom. It is local bookkeeping. Entropy increases in closed accounting scopes. Open systems can export diffusion, delay completion, and maintain structure—at a cost. That cost is always paid somewhere else.

Entropy does not say, "order is forbidden." It says, "order must be paid for continuously."

With this understanding in place, entropy no longer needs to be feared or moralized. It is not decay; it is completion without coordination. It is the universe's most honest receipt.

Once entropy is understood as delta diffusion rather than disorder, the next puzzle resolves itself naturally. Entropy has direction. It increases. It rarely decreases. This directionality has often been confused with time itself.

But entropy does not flow because time flows.

It flows because reversal is expensive.

That structural direction—without invoking mysticism or arrows—is what the next section must now make explicit.

18.2 — Why Entropy Is Directional Without Time Mysticism

The directionality of entropy has long been treated as a deep metaphysical problem. Why does entropy increase? Why does it not spontaneously decrease? Why does this "arrow" seem to align with our experience of time? Physics often answered these questions by quietly promoting time itself into a causal agent.

UED does not.

Entropy has direction not because time flows forward, but because permission collapses asymmetrically.

When delta diffuses, it spreads into an ever-growing number of micro-pathways. Each additional pathway increases the cost of reconstructing coordination. This is not a temporal preference; it is an economic one. Reassembly demands precise alignment across all those pathways simultaneously. The cost grows combinatorially, while diffusion remains cheap.

Directionality, therefore, is not imposed by time. It emerges from pathway economics.

This is why entropy increase feels inevitable. The system does not choose diffusion because it prefers disorder. It diffuses because diffusion requires less coordination, fewer constraints, and lower thresholds. The system follows the cheapest available route to resolution.

Once alignment has spread, reversing it is not forbidden—it is simply unaffordable.

This is the core insight that dissolves the arrow-of-time mystery. Entropy points "forward" because permission only shrinks, not because time commands it to. Once coordination is lost, the system no longer has access to the corridors that allowed structured alignment. The past is not inaccessible because it is behind us in time, but because the pathways that made it possible no longer exist.

This also explains why entropy increase depends on scale. At microscopic scales, where the number of pathways is small, reversals can and do occur. Fluctuations briefly reconcentrate alignment. Entropy can locally decrease. But as scale increases, the number of pathways explodes, and reversal becomes statistically invisible.

Directionality is therefore emergent, not fundamental.

UED reframes the second law accordingly. It is not a commandment etched into the universe. It is a statistical inevitability born from permission exhaustion. The universe does not enforce entropy increase. It simply never subsidizes coordination for free.

Time, in this framework, becomes a witness rather than a driver. We perceive time as directional because we move through sequences where permission has been spent and cannot be reclaimed. Entropy marks where alignment has finished spreading. Time records the order in which permission collapsed.

This distinction is subtle but decisive.

Entropy does not increase *because* time passes. Time feels directional *because* entropy increased.

By severing entropy from time mysticism, UED restores clarity. Direction arises from structure, not from metaphysics. The arrow is drawn by economics of resolution, not by a cosmic clock.

With directionality explained without invoking time as a force, one major confusion remains. Entropy increase is often described as "loss of information." But information about what, exactly? And how can information be lost if nothing disappears?

That apparent paradox belongs to the next correction—where information is freed from epistemology and placed back into structure.

18.3 — Information Loss as Pathway Collapse

Information is one of the most misused words in modern physics. It is often treated as something abstract, observer-dependent, or even mystical. In UED, information is neither mental nor metaphysical. It is structural accessibility.

Information is not what we know about a system. Information is which pathways remain usable.

When a system holds alignment in a coordinated way, it carries information in the most literal sense: the arrangement itself encodes how resolution is being delayed. The geometry, phase relationships, gradients, and constraints collectively specify *how* the system can act. That specification is information.

When dissipation occurs, nothing vanishes. Microstates still exist. Energy is conserved. But the pathways that made coordination visible and actionable collapse. Information is lost not because states disappear, but because access disappears.

This distinction resolves a deep confusion. Information loss is not about ignorance. It is about irreversibility of structure.

A shattered glass does not lose energy, atoms, or degrees of freedom. It loses the pathways that allowed those atoms to act as a glass. The information that defined "glass" was not stored in the atoms themselves, but in the constraints that coordinated them. Once those constraints are gone, the information is gone—even though everything still exists.

In UED terms, information loss is permission loss.

As delta diffuses, alignment spreads into countless micro-configurations. Each configuration resolves a tiny portion of misalignment independently. The original correlations that allowed macroscopic description are erased—not by randomness, but by over-distribution.

This is why information loss scales with entropy. As entropy increases, the number of pathways increases, but their usefulness collapses. The system still has states, but no longer has *handles*.

This also clarifies why information loss feels irreversible. To recover information would require reconstructing not just states, but the exact constraints, correlations, and thresholds that once coordinated them. That reconstruction is prohibitively expensive once pathways have collapsed.

Information is therefore not destroyed. Retrievability is.

This reframing dissolves the tension between conservation laws and information loss. Conservation tracks quantities. Information tracks structure of access. The two operate at different layers and never contradict each other.

It also explains why information loss is not subjective. It does not depend on observers, measurement, or ignorance. A system can lose information even if every microstate is perfectly known—because knowing states is not the same as being able to use them.

In UED, information is always tied to consequence. When consequence spreads beyond coordination, information finishes its work and disappears—not into nothingness, but into irrelevance.

With information properly grounded as pathway accessibility rather than knowledge, the chapter can now address the most universal consequence channel of all—one that appears wherever coordination collapses, regardless of domain or scale.

That channel has always been called heat.

The next section will show why heat is not a substance, not a waste product, and not an afterthought—but the most honest expression of alignment completing everywhere at once.

18.4 — Heat as a Consequence Channel

Heat has always been treated as the most mundane form of energy—useful, unavoidable, and slightly disappointing. It shows up when motion degrades, when work fails, when structure relaxes. In most explanations, heat is where energy goes when nothing better can be done with it.

UED reveals why that intuition is almost right—and exactly backward.

Heat is not what remains when energy is wasted. Heat is how alignment finishes when coordination is no longer affordable.

When delta resolves through structured pathways—motion, force, flow, oscillation—it does so in a directed, recoverable way. These pathways preserve gradients. They preserve geometry. They preserve *handles*. But once those pathways collapse, alignment does not stop resolving. It simply loses direction.

That loss of direction is heat.

Heat is alignment distributed into maximally available degrees of freedom, primarily microscopic vibrational and rotational modes. These modes are numerous, weakly coupled, and poorly coordinated. They are perfect sinks for resolution because they require almost no permission to absorb alignment.

This is why heat appears everywhere dissipation completes.

It is not because heat is special. It is because vibration is cheap.

Whenever structured resolution fails—whether in mechanical friction, electrical resistance, fluid viscosity, chemical reaction, biological metabolism, or psychological stress—the system falls back to the most permissive channel available. Alignment spreads into microscopic motion. Temperature rises. Heat appears.

UED therefore treats heat not as a substance, not as a leftover, and not as an enemy—but as a universal consequence channel.

This reframing explains several long-standing puzzles cleanly.

First, it explains why heat is so hard to convert back into work. Once alignment has been distributed across countless micro-modes, coordination is gone. To extract work would require re-synchronizing those modes into gradients. The difficulty is not conceptual; it is structural.

Second, it explains why heat is so universal. Regardless of domain, when coordination collapses, vibration remains accessible. Heat does not care whether the original process was mechanical, electrical, chemical, or social. It is the default completion mode.

Third, it explains why temperature behaves statistically rather than dynamically. Temperature is not a force or flow. It is a density measure of diffused alignment. It reports how much resolution has been absorbed per degree of freedom—not where it is going.

Heat does not move because it wants to. Heat spreads because nothing is holding it together.

This also dissolves the confusion between heat and energy. Energy is conserved. Heat is not a type of energy—it is a mode of accounting. It tells us that alignment has finished in forms that no longer support gradients.

In UED language, heat is consequence without preference.

It is alignment that has stopped negotiating pathways and has accepted the cheapest completion available.

Once heat is understood this way, it loses its stigma. Heat is not failure. It is closure. It is how the universe ensures that delta always resolves—even when structure gives up.

With heat properly placed as a consequence channel, the second law of thermodynamics can finally be addressed without mythology or commandment. Not as a rule imposed on systems, but as a statement about what happens when consequence is allowed to finish honestly.

That final clarification belongs to the next section.

18.5 — The Second Law Without Commandment

The second law of thermodynamics is often presented as a cosmic rule—an unbreakable commandment etched into the fabric of reality. Entropy must increase. Heat must spread. Irreversibility must prevail. This framing gives the law an almost moral authority, as if the universe is enforcing discipline on unruly systems.

UED removes that drama entirely.

The second law is not a command. It is a summary of what happens when alignment is allowed to finish without subsidy.

Nothing in the universe instructs systems to increase entropy. No mechanism "pushes" heat to spread. What exists instead is a simple asymmetry: diffusion is cheap, coordination is expensive. The second law is merely the statistical record of systems choosing the cheaper path whenever permission runs out.

In this light, the second law stops looking fundamental and starts looking inevitable.

Whenever a system releases alignment into many degrees of freedom, the number of pathways multiplies. Reversal would require re-coordinating those pathways with precision. The universe does not forbid this—but it does not pay for it either. Without external intervention, the system cannot afford the reconstruction. Entropy increases, not because it must, but because nothing stops it.

This reframing dissolves a critical confusion. The second law is often treated as prescriptive—*this must happen*. In UED, it is purely descriptive—*this is what happens when no one intervenes.*

There is no authority enforcing the law. There is only exhausted permission.

This also explains why apparent violations of the second law are never violations at all. Refrigerators, engines, living organisms—all reduce local entropy. They do so by exporting diffusion elsewhere. The law remains intact because the accounting scope was expanded. Entropy did not decrease; it was reassigned.

UED therefore treats the second law as an accounting identity, not an axiom. It does not dictate behavior; it reports balance sheets. It tells us where alignment finished, not where it was forbidden to go.

This perspective also clarifies why the second law feels absolute at large scales but negotiable at small ones. At microscopic scales, coordination costs are low. Pathways are few. Reversals can occur. At macroscopic scales, pathway counts explode. Permission collapses. The law asserts itself statistically—not metaphysically.

The second law does not say: *You may not reverse*. It says: *You did not pay for reversal.*

Once this is understood, thermodynamics loses its authoritarian tone. It becomes the most honest branch of physics—the one that refuses to speculate about intention, purpose, or ideal behavior. It simply counts what remains when all preferences have been exhausted.

With the second law stripped of commandment and restored to consequence, only one task remains for this chapter: to show why thermodynamics closes every loop in physics—not by dominance, but by finality.

That closure is not about control. It is about completion.

18.6 — Why Thermodynamics Closes All Loops

Thermodynamics closes all loops in physics because every process eventually stops negotiating structure. No matter how sophisticated the mechanism, how elegant the geometry, or how clever the intervention, alignment must ultimately finish somewhere. Thermodynamics is the language that speaks after that finishing.

This is why thermodynamics never competes with other theories. Mechanics explains motion while coordination holds. Field theories explain gradients while pathways remain open. Quantum descriptions explain fragile repeatability under tight constraints. But when those descriptions reach their limit—when coherence dissolves and permission collapses—thermodynamics remains.

Not as a replacement. As a closure.

In UED terms, thermodynamics is the universal accounting layer that activates once delta has exhausted all structured options. It does not ask how alignment was delayed, harvested, or redirected. It only records where alignment ended up when delay was no longer affordable.

This explains why thermodynamics applies across domains without modification. It does not care whether the system was mechanical, electrical, chemical, biological, psychological, or social. Once resolution spreads beyond coordination, the same bookkeeping applies. Heat appears. Entropy increases. Irreversibility sets in. The story ends—not because the universe enforces silence, but because nothing remains unresolved.

Thermodynamics therefore subsumes all other descriptions without contradicting them. It does not invalidate mechanics; it waits for mechanics to finish. It does not refute quantum theory; it counts what quantum processes leave behind. It does not oppose intelligence or design; it records the cost of both.

This is why thermodynamics feels final.

There is no deeper mechanism beneath it—not because it is fundamental in a metaphysical sense, but because it speaks from the point where mechanisms have completed their work. Asking for a deeper explanation of thermodynamics is like asking what happens after accounting is done. Nothing happens. The ledger is closed.

UED clarifies this finality without mysticism. Thermodynamics is not the deepest layer of reality. It is the last layer of consequence. It tells us what remains when all deltas have either been held, redirected, or released.

And this is precisely why thermodynamics is so powerful. It is immune to reinterpretation because it does not speculate. It does not describe becoming. It describes having become.

Once this is understood, thermodynamics stops feeling pessimistic. It is not a doctrine of decay. It is a statement of completion. It marks the point where the universe has finished resolving what was permitted to remain unresolved.

With this realization, this part reaches its natural conclusion.

Energy in action has now been followed from:

- condensation into matter,
- propagation as waves,
- diffusion through dissipation,
- and completion via thermodynamics.

Nothing is missing.

But something new becomes possible.

If thermodynamics is where all physical processes end, then intelligence—life, mind, systems, societies—must operate before this closure. Not by violating thermodynamics, but by learning how to delay, steer, and harvest consequence before it finishes.

That is the doorway the book now opens.

Where thermodynamics closes the loop of physics, intelligence opens the loop of strategy.

The next part begins there—not by denying inevitability, but by working intelligently within it.

From Accounting to Intelligence

At this point, thermodynamics no longer stands as a grim epilogue to physical processes. It stands as their final balance sheet. Nothing has been forbidden. Nothing has been judged. Alignment has simply finished resolving wherever permission allowed it to go.

This is the most important emotional correction this chapter makes.

Thermodynamics does not announce failure. It announces completion.

Every gradient that could be exploited has been spent. Every corridor that could hold alignment has collapsed. What remains is evenly distributed, anonymous, and indifferent to structure. The universe is not tired. It is settled.

Seen this way, thermodynamics is not the ceiling of understanding. It is the floor on which understanding must stand honestly. Any attempt to describe reality without acknowledging this final accounting will always feel incomplete, because it ignores the cost of action.

And yet, something remarkable becomes visible precisely here.

If thermodynamics closes all physical loops, then everything interesting must happen before closure.

Life, intelligence, design, adaptation, and choice do not exist beyond thermodynamics. They exist in tension with it. They are not violations of entropy; they are strategies that delay diffusion by holding delta open locally while exporting consequence elsewhere.

This is the quiet bridge the book now crosses.

Up to this point, UED has described what happens when alignment is permitted to resolve honestly. From here onward, it will describe how systems learn to intervene in that resolution—not by denying equilibrium, but by negotiating with it.

This is not a shift from physics to metaphysics. It is a shift from inevitability to management.

Thermodynamics remains untouched. The accounting will still close. But before it does, intelligence gets a window—a narrow, costly, temporary window—in which structure can be built, maintained, repurposed, and dissolved on its own terms.

That window is where:

- life sustains itself,
- minds experience meaning,
- societies organize power,
- and technology stretches consequence across scales.

The next part enters that window.

Not to escape thermodynamics. But to show why intelligence exists only because thermodynamics does.

Where physics finishes counting, intelligence begins choosing.

That is where UED now leads.

Where physics finishes accounting, intelligence begins deciding what to do next.

Interlude — Where Physics Learns to Wait

Up to this point, the book has been uncompromising. Delta resolves. Permission collapses. Accounting closes. Nothing escapes equilibrium. Physics has spoken clearly, and it has spoken without sentiment.

And yet, the world around us is not quiet.

Cells divide. Organisms persist. Minds anticipate. Systems adapt. For a long time. Far longer than inert matter has any right to.

This is not a contradiction. It is a *strategy*.

Life does not violate thermodynamics. It does not reverse entropy. It does not escape consequence. Life does something far subtler and far more expensive: it learns how to wait.

This interlude marks a necessary pause—not in logic, but in perspective. Physics has shown us how alignment finishes when left alone. Life begins exactly where that finishing would otherwise occur. It exists in the narrow window where dissipation can be delayed, redirected, and budgeted just enough to keep structure alive.

Life is not an exception to UED. It is UED under management.

Every living system is a temporary treaty with equilibrium. It holds delta open through constant work, constant sensing, constant correction. The price is continuous energy expenditure. The reward is persistence.

This is why life feels fragile. It is not balanced; it is *maintained*. The moment management fails, the same rules reassert themselves without mercy or judgment. Disease is not punishment. Death is not failure. They are simply equilibrium reclaiming what was being borrowed.

This interlude also corrects a deep misunderstanding. Biology is often treated as a special domain governed by unique principles—emergence, purpose, teleology. UED does not deny complexity, but it refuses exemption. The same mechanics of delta, constraint, and consequence operate here, just at higher organizational cost.

Life does not add new laws to physics. It adds *overhead.*

With that overhead comes remarkable capabilities: sensing, prediction, memory, adaptation. These are not miracles. They are techniques for compressing delta into forms that can be managed over time.

This is the bridge the book now crosses.

From this point onward, the discussion will not soften. It will not anthropomorphize. It will not romanticize survival. It will show, carefully and consistently, how life uses the same machinery already established—only more cleverly, more precariously, and more expensively.

Next part begins where thermodynamics would normally end—but before the ledger closes.

Biology does not rewrite physics. It *buys time from it.*

That is the living trick.

Part VI — Life as Managed Disequilibrium

This part makes a decisive move: it brings biology back into physics without reduction and without mysticism. Life is not introduced as a new kind of substance, a special force, or an exception granted by complexity. It is presented as what inevitably appears when delta management becomes continuous and self-referential.

Life is disequilibrium that learned to sustain itself.

This part reframes living systems as active negotiations with equilibrium. Every cell, organism, and ecosystem is a temporary structure that holds delta open by spending energy, exporting entropy, and correcting misalignment in real time. Nothing here violates thermodynamics; everything here pays its price explicitly.

Homeostasis, adaptation, and evolution are not metaphors or purposes. They are operational strategies for delaying resolution. Disease and death are not moral failures but patterns where delta management breaks down or becomes unaffordable.

By grounding biology in UED, this part removes the artificial boundary between "physical" and "living" systems. The same accounting applies—only the overhead changes.

This part also introduces a critical shift: prediction enters the story. Living systems do not merely react to delta; they anticipate it. Nervous systems emerge as mechanisms for compressing future misalignment into present action, allowing organisms to steer consequence before it arrives.

Across these chapters:

- Life is defined as a sustained delta window.
- Evolution is shown as consequence accumulation filtered by viability.

- Nervous systems are reframed as delta compressors and pathway stabilizers.

By the end of this part, biology no longer feels like a separate universe with its own rules. It becomes a chapter of Universal Energy Dynamics—written in molecules, membranes, and memory rather than equations.

Life does not escape equilibrium. It learns how to negotiate with it, moment by moment.

Chapter 19 — The Living Trick: Delay Equilibrium

Why Life Is Not a Miracle

Life has always invited reverence. It moves, repairs itself, responds, and persists against odds that seem impossible for inert matter. For centuries, this persistence was treated as evidence of something added—vital force, purpose, design, or exception. Even modern science, while rejecting mysticism, often preserves a softer version of the same belief by speaking of emergence as if it were exemption.

UED makes a cleaner, harder claim.

Life is not miraculous. It is expensive.

What appears extraordinary is not that life exists, but that it can afford to exist for as long as it does. Living systems do not escape equilibrium; they negotiate delays in its arrival. Every breath, every heartbeat, every metabolic reaction is a payment toward keeping delta unresolved just a little longer.

The mistake was never in noticing life's persistence. The mistake was in assuming persistence implies privilege.

Life persists not because the universe protects it, but because it works relentlessly against the default. The moment that work slows or stops, the same physics applies without hesitation. No exceptions are granted. No sympathy is expressed. Equilibrium simply resumes.

This is not a reduction of life's value. It is a correction of its explanation.

By removing miracle language, life becomes intelligible. It no longer sits outside physics as a special category. It becomes a strategy that matter occasionally discovers when conditions allow for sustained management of misalignment. Where energy throughput is possible, where boundaries can be maintained, where correction loops can operate, life can arise.

Not everywhere. Not forever. Not for free.

This framing also removes a deeper confusion: the idea that life is defined by what it *is*. DNA, proteins, membranes, and cells are important—but they are implementations, not definitions. Life is not a substance. It is an activity pattern: the continuous postponement of resolution through structured expenditure.

Seen this way, life does not oppose death. Death is simply the end of the negotiation. The miracle was never survival itself, but the temporary success of the bargain.

By starting here, the chapter refuses romance without losing respect. It asks the reader to look at life not as a gift dropped into the universe, but as a hard-won arrangement maintained moment by moment.

With that lens in place, the central idea can now be stated without drama:

A living system is one that keeps delta open intentionally—and pays the cost continuously.

Everything that follows in biology flows from this single fact.

19.1 — Life as a Sustained Delta Window

At its most fundamental level, life is not defined by chemistry, replication, or complexity. It is defined by duration against default. A living system is one that manages to keep delta open—locally, temporarily, and continuously—while the surrounding universe is always pressing toward resolution.

This is what makes life rare without making it mysterious.

In inert systems, delta appears briefly and resolves quickly. Gradients flatten. Structures relax. Motion fades. Equilibrium reasserts itself without resistance. In living systems, something different happens: delta is held deliberately within a bounded window. Not eliminated. Not resolved. Maintained.

That window is life.

To sustain such a window, a system must meet three non-negotiable conditions. First, it must have boundaries—membranes, compartments, interfaces—that prevent immediate diffusion. Second, it must have throughput—a continuous inflow of energy and matter to renew permission. Third, it must have feedback—mechanisms that detect misalignment and correct it before collapse.

Remove any one of these, and life does not slowly weaken. It ends.

This clarifies a subtle but crucial point: life is not simply "far from equilibrium." Many nonliving systems exist far from equilibrium briefly. Life is persistently far from equilibrium, and persistence is costly. The farther a system stays from equilibrium, the more aggressively it must export consequence elsewhere. Heat, waste, and entropy are not side effects of life; they are its operating exhaust.

From a UED perspective, a living organism is not a stable object but a dynamic permission engine. It continuously renegotiates which deltas remain open, which must be resolved, and which can be postponed. The organism does not fight equilibrium globally; it buys time locally by paying with energy.

This is why life feels active even at rest. A resting organism is still spending. Ion gradients are maintained. Proteins are repaired. Errors are corrected. The delta window never closes while life continues.

Crucially, this window is finite by design. Life does not aim for permanent suspension of equilibrium. It aims for long enough. Long enough to grow, reproduce, adapt, and pass the burden forward. The system knows nothing of purpose, but the structure ensures continuity while permission lasts.

This also explains why life scales poorly without organization. Larger organisms require more sophisticated delta management: more boundaries, more feedback loops, more energy throughput. Complexity is not free; it is a tax on delay.

Seen this way, life is neither fragile miracle nor robust exception. It is a temporary achievement, maintained moment by moment by disciplined expenditure.

Life exists in the gap between appearance and resolution. It survives by refusing to let that gap close—yet never pretending it won't.

With life now grounded as a sustained delta window rather than a special substance, the next question becomes unavoidable: *what exactly is doing the spending?* How is this delay financed, regulated, and enforced at every moment?

That accounting—stripped of romance and placed firmly in mechanism—is the role of metabolism, which the chapter must now confront next.

19.2 — Metabolism: The Cost of Staying Alive

If life is a sustained delta window, metabolism is the price paid to keep that window open. Metabolism is not fuel in the romantic sense, nor is it a life force quietly animating matter. It is accounting—continuous, unforgiving accounting.

Metabolism exists for one reason only: to renew permission.

Every moment a living system remains out of equilibrium, it is spending alignment capital. Gradients decay. Errors accumulate. Structures drift. Without constant intervention, resolution would complete and the system would settle like any other. Metabolism is the machinery that intervenes—again and again—to prevent that completion.

This reframes energy intake immediately. Organisms do not "gain energy" in a meaningful ontological sense. They purchase delay. Nutrients, oxygen, light—these are not gifts; they are currency exchanged for continued misalignment. The organism converts external gradients into internal maintenance: restoring ion differences, repairing molecular damage, re-synthesizing structures, and correcting drift.

Nothing about this process is optional.

Even the simplest living system must:

- rebuild structures faster than they decay,
- correct deviations before they cross thresholds,
- and export the consequences of that correction elsewhere.

That export is metabolism's shadow. Waste, heat, and byproducts are not inefficiencies to be optimized away. They are proof of payment. Every act of internal order requires external disordering. Every delay of equilibrium here demands acceleration of equilibrium elsewhere.

This is why metabolic rate correlates so tightly with vulnerability. High metabolic throughput allows rapid correction and responsiveness—but at the cost of faster wear, greater dependency, and higher risk if supply falters. Low metabolic throughput conserves resources—but reduces tolerance to disturbance. There is no ideal point, only trade-offs shaped by environment.

UED also dissolves a common misconception: that metabolism exists to support activity. In reality, activity is a side effect of maintenance. Movement, growth, reproduction—all draw from the same budget that keeps the system from collapsing. When metabolic supply is constrained, activity is sacrificed first so that core alignment can be preserved.

This is why exhaustion feels existential. It is not merely lack of energy for action. It is approach of permission collapse. The system senses that the delta window is narrowing, and prioritizes survival over expression.

Seen through this lens, metabolism is neither heroic nor ignoble. It is relentless. It does not care about meaning, happiness, or achievement. It enforces one rule only: *pay now, or resolve now*.

This also explains why metabolic "efficiency" has hard limits. A perfectly efficient metabolism—one that produces no waste and no heat—would be fatal. Without export channels, alignment pressure would accumulate internally until catastrophic failure occurred. Inefficiency is not a flaw; it is a requirement for survival.

Metabolism, then, is not life's engine. It is life's tax system.

It collects continuously, redistributes relentlessly, and closes the account the moment payment stops.

With the cost of staying alive now placed honestly on the table, the chapter must move to the mechanism that decides *where* that cost is spent moment by moment. How does a living system know which misalignments matter, which can be ignored, and which demand immediate correction?

That intelligence—non-conscious, distributed, and tireless—is what biology calls homeostasis.

19.3 — Homeostasis as Active Alignment Management

Homeostasis is often misunderstood because it is described using the language of balance. That language is misleading. Balance suggests rest, stillness, or equilibrium reached and maintained effortlessly. Living systems do none of these.

Homeostasis is not balance. Homeostasis is continuous correction.

A living organism is never at equilibrium, and it is never trying to reach it. Instead, it operates within narrow, survivable corridors of misalignment. Temperature, pH, ion concentration, glucose levels, oxygen tension—none of these are "balanced" in the sense physics uses the word. They are actively held away from equilibrium through constant sensing and response.

From a UED perspective, homeostasis is the organism's real-time alignment management system.

At every moment, countless deltas are forming: molecular noise, environmental fluctuations, internal wear, external stressors. Most of these would be fatal if allowed to accumulate. Homeostasis exists to detect which deltas matter, amplify their signal, and trigger corrective action before thresholds are crossed.

This is why homeostasis is expensive. Nothing about it is passive.

Sensors must operate continuously. Feedback loops must remain responsive. Corrective actions must be deployed even when nothing "seems wrong."

The organism is not waiting for failure. It is preventing it preemptively.

This reframing explains why homeostasis cannot be static. There is no fixed internal state the organism is trying to preserve. Setpoints drift with age, environment, stress, and availability. What remains constant is not the value being maintained, but the effort to remain within viable bounds.

Homeostasis is therefore dynamic alignment, not static balance.

UED also clarifies why regulation always involves trade-offs. Correcting one misalignment often introduces another. Lowering body temperature may slow reactions. Raising glucose may increase oxidative stress. Suppressing inflammation may impair repair. The organism is not optimizing globally; it is triaging locally under constraint.

This is why homeostasis can fail without error.

Failure does not imply malfunction. It often implies overload. When the rate or magnitude of incoming delta exceeds the system's correction capacity, misalignment accumulates faster than it can be resolved. The result is not imbalance in a moral sense, but structural saturation.

This saturation is what we experience as stress.

Stress is not psychological weakness. It is alignment pressure exceeding management bandwidth. Chronic stress is not caused by a single large disturbance, but by persistent small deltas that never fully resolve. Homeostatic systems remain active but begin to drift, compensate, and cannibalize reserves.

Over time, the cost becomes visible.

UED also dissolves the false boundary between physiological and psychological regulation. Thought, emotion, immune response, hormonal cycles, and metabolism are not separate systems. They are layers of the same alignment management problem, operating at different timescales and resolutions.

When homeostasis succeeds, life feels effortless. When it struggles, life feels heavy. When it fails, disease appears.

And this brings the chapter to a necessary correction.

If homeostasis is active alignment management, then disease cannot be moral failure, weakness, or punishment. It must be understood as a pattern of misalignment—one that has exceeded the system's capacity to correct.

That correction—without shame, without judgment—is where the chapter must now go next.

19.4 — Growth, Repair, and Aging as Delta Economics

Growth, repair, and aging are often treated as separate biological phenomena—one associated with youth, one with health, and one with decline. In UED, they are revealed as different expressions of the same accounting process: how a living system allocates its delta budget over time.

Nothing here is symbolic. Nothing is moral. Everything is economic.

Growth occurs when a system expands its capacity to hold and manage delta. Cells divide, tissues enlarge, networks become more complex—not because growth is an inherent goal, but because expanded structure allows greater buffering against misalignment. A larger system can distribute stress, store resources, and absorb fluctuations more effectively than a smaller one.

But growth is never free. Every new structure increases maintenance cost. More boundaries require more correction. More pathways require more coordination. Growth increases capacity—but it also raises overhead. This is why unchecked growth is unsustainable and why living systems naturally slow it once survivable capacity is reached.

Repair is local reversal. It is the most expensive operation a living system performs.

Repair attempts to reclaim coordination after partial diffusion has already occurred. It must reconstruct pathways, restore geometry, and re-establish thresholds—all under ongoing equilibrium pressure. This is why repair is prioritized selectively. Not all damage is fixed. The system invests repair where reversal is still affordable and abandons regions where the cost exceeds future benefit.

This is not failure. It is triage.

UED clarifies why perfect repair is impossible. To repair everything indefinitely would require reversing diffusion continuously across all scales. That would demand infinite energy, infinite coordination, and zero delay. Living systems do not fail to achieve perfect repair; they were never designed to afford it.

This brings us to aging, which is perhaps the most misunderstood biological process of all.

Aging is not decay. Aging is cumulative permission loss.

Over time, pathways degrade. Repair becomes more expensive. Feedback loops lose precision. Noise increases. The system continues to manage delta, but it does so with shrinking margins. More energy is spent maintaining baseline function, leaving less available for correction, growth, and resilience.

Nothing dramatic needs to break for aging to occur. It is not caused by a single failure, mutation, or insult. It is the compound interest of deferred accounting. Each decision to postpone repair, each trade-off that favored survival over restoration, leaves a small residue. Over decades, those residues accumulate.

Aging, then, is not a process acting on life. It is life remembering its costs.

This reframing removes both fear and fantasy. Aging is neither enemy nor flaw. It is the inevitable outcome of sustained delta management under finite resources. The only alternative to aging would be perfect repair—which would require infinite payment.

UED therefore places growth, repair, and aging on a single continuum. Early life prioritizes expansion. Mid-life prioritizes maintenance. Late life prioritizes survival. The shifts are not programmed in advance; they are responses to changing affordability.

This perspective also prepares the reader for a necessary emotional correction. If growth, repair, and aging are economic processes, then illness cannot be framed as betrayal, weakness, or failure of will.

Disease must be understood as something far more precise—and far more neutral.

It is the visible pattern of misalignment exceeding management capacity.

That clarification is where the chapter must now go next.

19.5 — Disease as Misalignment Pattern

Disease has been burdened with more meaning than it can carry. Across cultures and eras, it has been framed as punishment, weakness, imbalance of character, or failure of discipline. Even in modern medicine, traces of moral language remain—compliance, lifestyle fault, negligence—quietly smuggled into what should be neutral description.

UED removes that weight entirely.

Disease is not a verdict. Disease is a pattern of misalignment.

A living system constantly manages delta through homeostasis, repair, and adaptation. Disease appears when the rate, magnitude, or topology of incoming misalignment exceeds the system's capacity to correct it. Nothing more is required. No intent. No blame. No meaning beyond structure.

This reframing immediately clarifies why disease can arise without wrongdoing. Genetic constraints, environmental exposure, accumulated wear, random perturbations, or sudden shocks can all push a system beyond its correction bandwidth. The system may function exactly as designed and still fail—because design was never infinite.

Disease is therefore not an exception to life's rules. It is life encountering unaffordable delta.

UED also dissolves the false divide between acute and chronic illness. Acute disease represents sudden overload—a rapid spike in misalignment that overwhelms correction loops. Chronic disease represents persistent low-grade delta that never fully resolves. In both cases, the mechanism is the same: misalignment persists longer than the system can tolerate.

This perspective explains why chronic conditions are so draining. They are not dramatic failures. They are endless negotiations. Homeostatic systems remain active, correcting partially, compensating imperfectly, and consuming resources continuously. The organism survives, but at escalating cost.

Disease also reveals its true nature when viewed across scales. A localized misalignment—an infection, a mutation, an injury—rarely remains local. Consequence propagates through networks: immune activation alters metabolism, pain reshapes behavior, stress modifies hormonal baselines. Disease is not an isolated defect; it is a system-wide pattern emerging from local failure.

Importantly, UED reframes treatment without moral overlay. Treatment is not "fighting disease." It is reducing misalignment, restoring pathways, or lowering correction cost. Antibiotics reduce delta load. Surgery removes irreparable regions. Therapy reshapes feedback loops. Rest reduces incoming perturbation. None of these are battles. They are accounting interventions.

This also explains why cure is not always possible. Some misalignments destroy pathways so thoroughly that restoration is unaffordable. Management replaces reversal. Support replaces repair. This is not defeat. It is realistic budgeting under constraint.

By stripping disease of moral meaning, UED does not diminish compassion—it deepens it. Suffering is no longer interpreted as failure of character or effort. It is recognized as the natural consequence of finite systems managing infinite variability.

Disease is not shameful. It is informative.

It tells us where delta accumulated faster than correction could occur. It reveals which pathways are fragile, which reserves are depleted, and which negotiations have become unsustainable.

With disease properly understood as pattern rather than judgment, the chapter can now move to its final clarification. If life is sustained misalignment, if disease is misalignment overload, then death must be understood not as collapse or defeat—but as something far quieter.

It is equilibrium reclaiming what was being borrowed.

That completion is where the chapter must now close.

From Individual Delay to Lineage Accounting

Once life is understood as a sustained delta window, and disease as a misalignment pattern rather than a moral event, one final correction settles quietly into place.

Individual survival was never the unit that mattered most.

A living organism delays equilibrium only temporarily. No matter how sophisticated its homeostasis, how efficient its metabolism, or how resilient its repair systems, the negotiation always ends. Permission runs out. Pathways degrade. Alignment finishes. This is not tragedy—it is closure.

Death, in this framework, is not collapse. It is equilibrium reclaiming the loan.

Nothing special needs to fail. No dramatic breakdown is required. When the cost of maintaining misalignment exceeds what the system can pay, the universe does not punish the organism—it simply stops waiting. Resolution completes in the same neutral way it always has.

This realization does something important to the narrative. It shifts attention away from the individual as the final accounting unit. If life were only about individual persistence, biology would be a short, repetitive story. But life found a way to move the burden forward.

What cannot be delayed indefinitely by one organism can be distributed across generations.

Here lies the true continuity of life—not in endless survival, but in consequence accumulation beyond individuals. Traits, structures, behaviors, and strategies that manage delta more effectively persist—not because they are rewarded, but because they last longer under pressure.

This is where biology leaves physiology and enters history.

The organism negotiates equilibrium moment by moment. The lineage negotiates equilibrium across time.

With this shift, survival stops being heroic and becomes statistical. Life does not seek perfection. It explores affordability. It keeps what works long enough and discards what does not—without intention, without foresight, and without mercy.

The next chapter does not add purpose to this process. It removes another illusion.

Evolution is not a story of progress, improvement, or destiny. It is the record left behind by consequence.

Life delays equilibrium one organism at a time. Evolution remembers who delayed it best.

That memory—written in structure, behavior, and constraint—is where the book now moves next.

Life delays equilibrium one organism at a time—but evolution records who delayed it best.

Chapter 20 — Evolution as Consequence Accumulation

Why Evolution Was Never About Progress

Evolution has always been narrated like a story with direction. From simple to complex. From primitive to advanced. From past to future. Even when scientists carefully avoid these words, the intuition leaks through. We look at what exists now and assume it is better than what existed before.

UED asks us to pause and remove that assumption completely.

Evolution is not a journey toward improvement. It is a record of what did not collapse.

This distinction matters because progress implies intention, trajectory, and destination. Evolution has none of these. There is no target state being approached, no ideal form waiting at the end of time. What we call "advanced" organisms are simply those whose configurations remained affordable under their specific consequence pressures for long enough to be noticed.

The confusion arises because persistence is easily mistaken for success.

When a lineage survives across generations, we instinctively attribute merit to it. When another disappears, we assume failure. But survival carries no reward signal. It is not granted. It is left over. Extinction is not punishment. It is resolution completing when maintenance became unaffordable.

Evolution does not celebrate winners. It forgets what could not be paid for.

This reframing removes the heroic tone from biology without diminishing its power. Evolution becomes quieter, harsher, and more honest. It does not elevate. It accumulates. It does not guide. It filters. It does not move forward. It simply continues where continuation remains possible.

Once this is seen, many long-standing confusions dissolve naturally.

Why complexity sometimes increases and sometimes collapses. Why simple organisms persist alongside complex ones. Why "inferior" traits survive in some environments and dominate in others. Why extinction is the rule, not the exception.

None of these require special explanation. They are consequences of the same principle applied repeatedly: configurations that can absorb consequence remain; those that cannot disappear. No narrative is needed.

Evolution is therefore not creative in the romantic sense. It does not invent solutions. It retains configurations that did not violate affordability. Variation explores; consequence decides; accumulation remembers.

This is why evolution feels slow and brutal. It has no urgency and no mercy. It does not correct mistakes quickly because it does not know what a mistake is. It only registers cost.

Seen through UED, evolution is not a force shaping life. It is life leaving behind a long receipt of payments that were made successfully.

With this lens established, fitness can finally be discussed without distortion. Not as superiority, strength, or advancement—but as something far more precise.

Fitness is consequence advantage.

That is where the chapter must now go next.

20.1 — Fitness as Consequence Advantage

Fitness is one of the most abused words in biology. It is often conflated with strength, intelligence, dominance, or reproductive success, as if evolution were a competitive tournament awarding trophies. This interpretation is emotionally satisfying—and structurally wrong.

In UED, fitness has no heroic meaning.

Fitness is consequence advantage.

A fit configuration is not one that wins, conquers, or excels. It is one that can absorb the consequences of existing longer than alternatives under the same conditions. That is all. No praise is implied. No hierarchy is established.

This correction immediately dissolves several misunderstandings.

First, fitness is always contextual. A trait that is advantageous in one environment can become catastrophic in another. Speed is fitness where predators matter; slowness is fitness where energy is scarce. Intelligence is fitness where prediction pays; simplicity is fitness where maintenance costs dominate. There is no globally fit trait—only locally affordable ones.

Second, fitness is not about maximizing performance. It is about minimizing unaffordable consequence. Many highly "capable" organisms go extinct precisely because their capabilities are expensive. Large brains, fast metabolisms, complex behaviors—all raise the cost of staying alive. When environments shift, these costs become liabilities.

This is why UED insists on a counterintuitive rule:

The strongest often lose to the cheapest.

Fitness favors configurations that can pay their bills repeatedly, not those that shine briefly. Persistence beats brilliance when consequence accumulates.

Third, fitness is not something an organism *has*. It is something a configuration *exhibits* relative to its surroundings. The same organism can be fit today and unfit tomorrow without changing at all—because fitness lives at the intersection of structure and environment, not inside the organism itself.

UED therefore reframes fitness as a rate problem.

A configuration is fit if:

- the rate at which it accumulates consequence is lower than
- the rate at which it can manage, export, or tolerate that consequence.

As long as this inequality holds, the configuration persists. When it fails, resolution completes—quietly, without commentary.

This explains why fitness does not require superiority. Many lineages persist not because they are exceptional, but because they are adequate at low cost. Bacteria outlast mammals. Insects outlast empires. Simple strategies persist because their consequence load is light.

Fitness is thus not about being better. It is about being affordable longer.

This framing also removes the illusion of fairness from evolution. Fitness is not evenly distributed. Environments are not neutral. Consequence does not balance effort. Some configurations are born expensive and die quickly. Others are cheap and linger for millions of years. Evolution does not correct this imbalance because it does not perceive imbalance.

It only registers cost over time.

With fitness properly grounded as consequence advantage rather than superiority, the next illusion must be dismantled. If fitness is not reward, then survival cannot be reward either.

Survival is not something earned. It is something that simply has not yet ended.

That uncomfortable clarity is where the chapter must now proceed next.

20.2 — Survival Is Not Reward; It's Non-Resolution

Survival is often spoken of as if it were a prize. An organism survives because it did something right. A lineage persists because it earned continuation. This language quietly turns biology into a courtroom, where existence is granted or denied based on merit.

UED removes that courtroom entirely.

Survival is not a reward. Survival is non-resolution.

Nothing survives because it is chosen. Nothing persists because it deserves to. Survival simply means that, so far, equilibrium has not reclaimed the configuration. The bills are being paid. The delta window remains open. That is all.

This reframing is uncomfortable because it strips survival of validation. It tells us that persistence carries no moral signal. A parasite survives. A virus survives. A fragile organism survives. A brutal strategy survives. None of these outcomes imply approval. They imply affordability under current constraints.

Survival is therefore passive in nature. It is not an achievement event. It is the absence of collapse.

This explains why survival does not correlate cleanly with effort, intelligence, cooperation, or virtue. Many organisms invest immense effort and still vanish. Others invest very little and persist. Evolution does not audit intention. It audits cost over time.

From a UED perspective, survival is simply the condition where:

- incoming delta does not exceed
- the system's ability to manage, export, or tolerate consequence.

As long as this inequality holds, the system remains. When it fails, resolution completes. There is no ceremony. No signal is sent. The lineage simply stops appearing in the ledger.

This also explains why multiple, contradictory strategies can survive simultaneously. Predators and prey. Cooperation and competition. Aggression and passivity. There is no single correct way to live because there is no target to reach. Different configurations remain viable as long as their consequence budgets balance locally.

UED therefore rejects the idea that extinction is failure. Extinction is resolution finishing at the lineage level. It occurs when the accumulated cost of maintaining a configuration exceeds what the environment allows. The lineage does not fail morally; it becomes unaffordable structurally.

This perspective also dissolves the myth of inevitability. Survival is never guaranteed, no matter how long it has lasted. Past persistence does not purchase future permission. Environments shift. Constraints change. What was once affordable becomes costly overnight.

Evolution does not promise continuity. It only allows it temporarily.

By removing reward language, UED clarifies why survival feels indifferent. There is no balancing force ensuring fairness, justice, or improvement. Persistence is statistical residue, not recognition.

This is not a cynical view. It is a precise one.

Once survival is understood as non-resolution rather than reward, the narrative of evolution shifts decisively. The focus moves away from who survives and toward how systems renegotiate equilibrium under changing constraints.

That renegotiation—continuous, imperfect, and costly—is what biology calls adaptation.

And adaptation, as UED will now show, is not improvement either. It is compromise under pressure.

That is where the chapter must now go next.

20.3 — Adaptation as Equilibrium Negotiation

Adaptation is often portrayed as improvement in disguise. Organisms "adapt" to become better suited, more efficient, more capable. This framing quietly imports intention into a process that has none. UED replaces that story with something both harsher and more accurate.

Adaptation is not optimization. Adaptation is renegotiation.

When environments change, the existing balance between delta intake and delta management is disturbed. New pressures appear. Old buffers fail. What once remained affordable becomes costly. Adaptation is the process by which a lineage reconfigures how it delays equilibrium under the new constraints.

Nothing is made perfect in this process. Trade-offs are introduced immediately.

A trait that reduces energy cost may reduce responsiveness. A trait that improves defense may reduce reproduction. A trait that increases resilience may slow adaptation speed.

Adaptation always pays for one advantage by accepting another vulnerability.

This is why UED insists that there is no "optimal" organism. There are only temporarily viable compromises. Every adaptation locks the system into a new corridor—narrower in some directions, wider in others. What looks like progress in hindsight is simply a path that remained affordable long enough to persist.

Seen this way, adaptation is not forward motion. It is sideways movement along the boundary of equilibrium.

This reframing also explains why adaptation often feels conservative rather than inventive. Most adaptations are small adjustments that preserve existing architecture rather than radical redesigns. Large changes are expensive. They destabilize too many pathways at once. Incremental renegotiation is cheaper and safer under consequence pressure.

UED therefore treats adaptation as a cost-minimization strategy, not a creativity engine.

This also dissolves a common confusion: why adaptations can become maladaptations. A trait that negotiated equilibrium successfully in one context may become disastrous when conditions shift. The trait did not "fail." The negotiation terms changed. What was once affordable is no longer so.

Adaptation has no foresight. It cannot anticipate future constraints. It only responds to present consequence.

This is why environments that change rapidly are so destructive. They compress renegotiation time below what lineages can afford. Adaptation lags behind consequence, and collapse follows.

From a UED perspective, adaptation is simply the restructuring of delta pathways to reestablish non-resolution under new pressures. It does not seek stability. It seeks continued delay.

Once this is understood, a deeper question arises. If adaptation is renegotiation, what provides the raw material for renegotiation? Where do new configurations come from, and why are most of them discarded?

That question brings the chapter to variation—not as randomness with creative intent, but as exploration of permission space.

That is where the next section must now turn.

20.4 — Variation as Exploration of Delta Topology

Variation is often described as the creative engine of evolution—random mutations generating novelty from which selection then chooses. This language suggests invention, experimentation, and possibility. UED replaces that imagery with something far more grounded.

Variation is not creative. Variation is exploratory.

Every living system operates within a constrained space of allowable configurations—a permission landscape defined by chemistry, physics, development, and history. Variation is simply the system sampling that space, usually blindly, usually wastefully, and almost always unsuccessfully.

In UED terms, variation explores delta topology.

Topology matters more than magnitude here. Most variations do not introduce large changes. They introduce small shifts in how delta is routed, buffered, or amplified. A protein folds slightly differently. A regulatory loop fires earlier. A developmental pathway branches at a different threshold. These are not bold innovations; they are local perturbations in alignment management.

The overwhelming majority of these perturbations are neutral or harmful. They either do nothing noticeable or they raise the cost of maintaining the delta window. Only rarely does a variation open a corridor that is cheaper under current constraints than what existed before.

This rarity is not a failure of creativity. It is a consequence of cost.

Exploration of permission space is expensive because most of that space is unaffordable. The system is not searching for solutions. It is stumbling across configurations and letting consequence decide which ones linger.

This framing dissolves the mystery of randomness in evolution. Variation is random in the sense that it is unbiased toward affordability. It does not aim toward usefulness. It does not learn from failure. It simply perturbs. Consequence filters.

UED therefore treats variation as a background process—constant, indifferent, and prolific. It does not drive evolution forward. It keeps the ledger populated with possibilities that consequence can later eliminate or tolerate.

This also explains why evolutionary change often feels conservative. Because most of permission space is unaffordable, lineages tend to remain near existing configurations. Radical departures are costly because they disrupt too many coordinated pathways at once. Small variations that preserve most architecture are cheaper and more likely to persist.

Variation does not innovate freely. It probes cautiously by necessity.

Seen this way, novelty is not celebrated by evolution. It is tolerated occasionally. When environments shift, previously unaffordable regions of permission space may suddenly become viable, and variation that once failed can persist. This gives the illusion of sudden creativity—but nothing new was invented. The cost landscape changed.

Variation, then, is not the author of evolutionary stories. It is the noise floor against which consequence operates.

With variation understood as exploration rather than invention, the mechanism that shapes long-term outcomes becomes clear. If variation fills the space and consequence removes what it cannot afford, then what we call "selection" must be nothing more—and nothing less—than a filter.

That filter is not intelligent. It is not active. It is delta-based elimination.

That is the next correction the chapter must now make.

20.5 — Selection as a Delta Filter

Natural selection is often personified. It "chooses," "favors," "selects for." This language suggests intention, preference, and evaluation—as if the environment were an examiner grading traits. UED removes that fiction completely.

Selection does not select. Selection filters.

In UED terms, selection is the passive consequence of delta economics operating over time. Variations enter the system indiscriminately. Each variation alters how delta is managed—sometimes slightly, sometimes catastrophically. The environment does not judge these configurations. It simply exposes them to consequence.

What cannot afford its consequences disappears.

That is selection.

This reframing is crucial because it strips selection of agency. There is no decision-making entity. There is no direction imposed. There is only a constraint field—resources, temperature, predators, pathogens, competition, physics—and configurations either remain viable within that field or they do not.

Selection is therefore not creative. It is eliminative.

It does not build complexity. It removes unaffordable configurations. Whatever remains after sufficient time appears adapted—not because it was shaped intentionally, but because it withstood repeated consequence.

This also clarifies why selection often looks inefficient or cruel. Many viable configurations are eliminated not because they are fundamentally flawed, but because they are temporarily misaligned with current constraints. Selection has no memory of potential. It does not preserve options. It removes what fails now, not what might succeed later.

UED also explains why selection pressure varies across environments. A harsh environment filters aggressively, eliminating many configurations quickly. A permissive environment filters slowly, allowing diverse strategies to coexist. In both cases, selection is not stronger or weaker by design—it is simply reflecting how tight the delta budget is.

Selection operates on cost, not intention.

This perspective dissolves a common misunderstanding: that selection pushes populations toward optimality. It does not. It pushes populations away from unaffordability. The resulting configurations may look well-matched to their environment, but this is an afterimage, not a goal.

Selection does not ask, "Is this good?" It asks nothing. It simply waits.

Given enough time, configurations that leak delta faster than they can manage it vanish. Configurations that balance their budgets linger. The ledger updates quietly.

UED therefore reframes natural selection as a delta filter—a structural sieve that lets affordable patterns pass and removes those that cannot sustain their costs. No preference. No foresight. No mercy.

Once this is understood, a final illusion must be addressed. Over long timescales, evolutionary history appears directional. Complexity seems to rise. Intelligence appears. Systems seem to "advance."

This appearance tempts us back into narrative.

The next section shows why direction can emerge without intention, and why reading it as progress is a mistake born of hindsight.

That clarification completes the chapter's logic.

20.6 — Accumulation Without Direction

Over long spans of time, evolutionary history appears to move somewhere. Complexity increases in some lineages. Nervous systems emerge. Intelligence appears. From a distance, this sequence tempts us to read direction, momentum, even destiny into the record.

UED insists on a harder truth.

Direction is not built into evolution. Direction is read backward into what remains.

What accumulates in evolution is not improvement, but unresolved consequence that happened to remain affordable. When similar pressures persist for long enough, lineages that negotiate those pressures successfully leave behind layered residues—structures built upon structures, strategies stacked upon strategies. To an observer looking only at survivors, this layering looks like ascent.

But it is not ascent. It is selective accumulation.

Nothing in the process aims upward. Each step exists only because the previous one did not collapse under consequence. The moment a layer becomes unaffordable, it is removed—sometimes instantly, sometimes after millions of years. History is littered with complex lineages that vanished without ceremony.

This is why UED rejects the idea of an evolutionary ladder. Ladders imply direction and destination. Evolution produces branching ledgers, most of which terminate. A few branches persist long enough to appear dominant, not because they were better in any absolute sense, but because their environments allowed their costs to remain manageable.

Complexity increases only under specific conditions:

- when environments are stable enough to reward long-term investment,
- when energy throughput is sufficient to pay overhead,
- when consequence does not fluctuate faster than renegotiation can occur.

Remove any of these, and complexity collapses immediately. Simpler configurations reassert themselves because they are cheaper. This is not regression. It is rebalancing.

UED therefore explains a paradox that troubled evolutionary theory for decades: why complexity increases in some lineages while simplicity dominates the biosphere overall. The answer is simple. Complexity is expensive. Only a few corridors can afford it, and only temporarily.

Accumulation, then, is conditional.

Evolution accumulates what survives repeated consequence under similar constraints. When constraints change, accumulation can reverse violently. Dinosaurs did not fail. Their accumulated cost structure became unaffordable under new conditions.

This perspective also explains why intelligence appears late and rarely. Intelligence is not evolution's goal. It is one of the most expensive delta-management strategies known. It persists only where the payoff of prediction outweighs the overhead of maintenance. In most environments, it does not.

Direction appears only because observers stand at the tip of a surviving branch and mistake local continuity for global intent.

UED removes that illusion cleanly. Evolution has no arrow. It has memory without aim.

What accumulates is not progress. What accumulates is what consequence did not erase yet.

This realization completes the structural explanation of evolution. But it also sets the stage for something genuinely new. If evolution accumulates consequence slowly across generations, some systems begin to do something unprecedented.

They start compressing consequence within a lifetime.

They do not wait for selection to erase failures. They attempt to anticipate them.

That shift—from historical accumulation to real-time prediction—is where biology crosses into cognition.

And that crossing belongs to the next chapter.

20.7 — Evolution Without Heroes, Villains, or Goals

The final illusion that must be removed from evolution is narrative itself. Stories demand protagonists and antagonists, victories and failures, ascent and downfall. Evolution has been burdened with all of these—often unconsciously—because humans are storytellers before they are analysts.

UED refuses the story.

Evolution has no heroes. It has no villains. It has no goals.

There is no virtue in survival and no shame in extinction. A lineage that persists is not rewarded, and one that disappears is not judged. Both outcomes are structurally equivalent responses to consequence pressure under different constraints.

Predation is not evil. Cooperation is not good. Parasitism is not immoral. Altruism is not noble.

They are strategies, nothing more.

Each strategy represents a different way of negotiating delta: where costs are paid, how risks are distributed, how consequence is exported or absorbed. In some environments, cooperation lowers individual cost and persists. In others, it is exploited and collapses. In some contexts, aggression pays. In others, it accelerates extinction.

Evolution does not care which strategy feels right. It only records which strategies remain affordable long enough to leave traces.

This reframing is uncomfortable because it removes moral anchoring. It tells us that nature is not a teacher of ethics, nor a guarantor of justice. Reading lessons into evolutionary outcomes is a human projection, not a biological fact.

UED does not deny ethics. It simply refuses to source them from evolution.

By stripping evolution of heroes and villains, UED also dissolves the illusion of inevitability. Nothing in evolutionary history had to happen. Every outcome was contingent, fragile, and reversible. Intelligence was not destined. Humans were not prewritten. Our existence is not a culmination—it is a local persistence event.

This perspective restores humility without nihilism.

If evolution has no goals, then meaning cannot be inherited from biology. It must be constructed elsewhere. If nature does not reward goodness, then ethics must be chosen, not justified by survival. If survival itself is neutral, then responsibility emerges not from destiny, but from awareness.

This is where the chapter completes its arc.

Evolution explains how configurations persist. It does not explain why they should.

And that gap is not a flaw. It is an opening.

Some systems eventually develop mechanisms that do not merely endure consequence but attempt to anticipate it. They compress history into memory, sensation into signal, and future risk into present action.

That move—from blind accumulation to active prediction—is not evolutionary destiny. It is another expensive strategy that only a few lineages can afford.

Those systems are called nervous systems.

Evolution remembers the past without caring. Nervous systems attempt to borrow from the future.

That transition—from consequence recorded to consequence predicted—is where the book must now go next.

From Consequence Remembered to Consequence Anticipated

With evolution stripped of progress, purpose, and moral weight, what remains is something stark and powerful: memory without awareness. Evolution remembers what worked only by leaving it behind. It has no foresight, no preference, and no urgency. Its ledger updates slowly, written in generations, not moments.

This slowness is not a flaw. It is the cost of blindness.

Evolution can only correct by erasing. It discovers unaffordable strategies by letting them disappear. Every adjustment is paid for in extinction, wasted variation, and lost lineages. The process is robust—but brutally inefficient.

And yet, under certain conditions, a radical shift occurs.

Some systems become expensive enough, fragile enough, and fast enough that waiting for generational correction becomes untenable. The cost of failure within a single lifetime grows too high. For these systems, relying on evolutionary memory alone is no longer affordable.

So a new layer appears.

Instead of waiting for consequence to accumulate across generations, these systems begin to sense consequence before it completes. They detect misalignment early. They stabilize pathways in advance. They trade metabolic cost for prediction.

This is not a break from evolution. It is evolution discovering a faster accounting method.

Where evolution accumulates consequence historically, nervous systems attempt to compress consequence temporally. They turn external delta into internal signal. They convert past outcomes into learned pathways. They reduce future cost by acting before resolution becomes unavoidable.

This is the decisive transition.

Evolution filters after the fact. Nervous systems intervene before the fact.

Nothing mystical enters here. No new principles are added. The same UED machinery applies—delta, constraint, consequence—but now operating at millisecond scales rather than geological ones.

With this shift, life stops being merely persistent and becomes anticipatory.

The next chapter does not introduce consciousness, meaning, or mind yet. It begins with something simpler and more fundamental: how living systems detect delta, stabilize pathways, and convert pain and pleasure into structural guidance.

This is where biology learns to predict.

Evolution remembers what survived. Nervous systems attempt to avoid what would not.

That is the bridge the book now crosses.

Evolution remembers the past. Nervous systems attempt to borrow from the future.

Chapter 21 — Nervous Systems

Prediction as Delta Compression

When Waiting Became Too Expensive

For most of life's history, waiting was acceptable. A configuration failed, and the cost was paid quietly—by disappearance. Variation tried again. Time absorbed the loss. Nothing noticed.

This works when systems are cheap.

Simple organisms can afford to fail repeatedly. Their bodies are small, their energy budgets minimal, their commitments shallow. If a strategy collapses, evolution removes it across generations, and nothing within the system needs to understand what happened. Consequence arrives late, but the delay is tolerable.

At some point, that changes.

As organisms grow larger, slower to reproduce, and more internally complex, the cost of failure shifts from being evolutionary to being personal. A single mistake can now erase years of accumulated structure. Waiting for selection becomes ruinously expensive. A system that only learns by disappearing is no longer viable.

This is the pressure point where nervous systems emerge.

They are not an upgrade in intelligence. They are an upgrade in economics.

A nervous system is nature's way of paying for consequence early, in small installments, instead of all at once through extinction. It does not eliminate error. It reduces the price of error by detecting misalignment before resolution completes.

This reframing matters. Nervous systems are often treated as the beginning of mind, awareness, or experience. In UED, they begin as something far more modest and far more profound: a mechanism to compress future cost into present signal.

Prediction appears here not as foresight, but as debt management.

Rather than waiting for delta to resolve catastrophically, the system internalizes traces of past consequence. It turns external gradients into internal urgency. It trades energy for speed. The goal is not accuracy or truth—it is survival efficiency under time pressure.

This is why nervous systems are not evenly distributed across life. They appear only where the math works. Where energy intake can support signaling overhead. Where the environment changes fast enough to reward early detection. Where bodies are costly enough that waiting for evolutionary correction becomes unacceptable.

Seen this way, the nervous system is not a leap toward consciousness. It is a concession to reality.

Life does not become aware because it wants to understand the world. It becomes predictive because waiting became too expensive.

Everything that follows—learning, memory, pain, pleasure, even the illusion of control—grows out of this single economic shift. Once consequence is anticipated rather than endured, experience begins to take shape.

But nothing mystical has entered yet.

At this stage, the nervous system is still doing one job only: detecting delta early enough to act before resolution becomes irreversible.

Understanding comes later.

21.1 — Sensing: Delta Detection, Not Information Gathering

The most persistent misunderstanding about nervous systems is the belief that they exist to *represent the world*. We speak as if eyes capture reality, ears record sound, and nerves transmit information about "what is out there." This language is intuitive—and deeply misleading.

In UED, sensing is not about knowing. It is about detecting misalignment.

A sensory receptor does not ask, *What is this object?* It asks, *Where does difference exceed tolerance?*

Every sensory modality—light, pressure, temperature, chemical concentration—operates on the same principle. A receptor responds only when a gradient crosses a threshold. Uniformity produces silence. Change produces signal. The nervous system does not receive a picture of the world; it receives alerts of deviation.

This is why sensing is fundamentally selective. Vast portions of reality pass through the body unnoticed because they do not generate actionable delta. The nervous system ignores what is aligned and reacts to what is not. This selectivity is not a flaw—it is the entire point.

Sensing, then, is not passive reception. It is active filtering.

A receptor is tuned to a narrow band of difference. Outside that band, nothing exists as far as the system is concerned. This is true for infrared light invisible to human eyes, ultrasonic frequencies inaudible to human ears, and chemical gradients imperceptible to organisms lacking the relevant receptors. Absence of sensation is not absence of reality—it is absence of delta relevance.

Seen this way, all senses are structurally identical. Vision is not special. Hearing is not special. Smell, touch, proprioception—all of them detect gradients of different kinds, but they perform the same operation: flagging where equilibrium is disturbed enough to matter.

This also explains why sensory signals do not arrive labeled with meaning. A spike in a neuron does not carry truth, objecthood, or interpretation. It carries urgency. It says only: *Something has changed enough that waiting may be costly.*

Meaning, recognition, and categorization are layered later, if at all. At the level of sensing, there is no world-model, no representation, no interpretation. There is only delta crossing a boundary.

This reframing also clarifies a subtle point: sensing is always partial and delayed. By the time a gradient is detected, it has already formed. By the time a signal propagates, the situation may have shifted further. Nervous systems do not eliminate delay—they manage it.

Accuracy is secondary. Timing is everything.

A crude but fast signal often outperforms a precise but slow one. This is why reflexes exist. This is why organisms jump at shadows. False positives are cheaper than missed deltas when resolution can be lethal.

From a UED perspective, sensing is therefore the first act of prediction—not because it knows what will happen, but because it intercepts consequence early enough to intervene.

The nervous system does not gather information to understand the world. It detects delta to survive it.

Once this is clear, the next step becomes inevitable. If signals flag deviation, then systems that experience the *same* deviation repeatedly can reduce cost further by stabilizing responses.

That stabilization is called learning—and it is not what we usually think it is.

21.2 — Signal Before Meaning

Once sensing is understood as delta detection, a second illusion must be removed: the idea that nervous signals carry meaning. We often assume that a signal *represents* something—an object, a threat, a sound, a memory. This assumption quietly imports cognition into systems that operate long before cognition exists.

In UED, a signal has no meaning. It has priority.

A nervous signal is not a message about the world; it is an instruction to redistribute resources now. It says nothing about *what* caused the delta. It communicates only that a threshold has been crossed and that delay may be costly.

This is why early nervous systems operate without interpretation. Reflex arcs do not ask what is happening. They route energy immediately. Withdrawal from heat, contraction under pressure, orientation toward chemical gradients—these responses precede any form of understanding. Meaning would slow them down.

Signals exist to move the system, not to explain the situation.

This distinction matters because it dissolves the false hierarchy between "lower" and "higher" brains. Even in complex organisms, the vast majority of nervous activity never becomes conscious, never acquires narrative, and never enters awareness. It remains what it has always been: fast delta-routing.

Meaning appears only when systems begin to reflect on signals rather than act on them directly. That reflection is optional, expensive, and late. The nervous system works perfectly well without it.

This also explains why nervous signals are standardized and abstract. A neuron fires or it does not. It does not encode color, fear, or danger. Those qualities are interpretations layered by later circuitry. At the signal level, everything is reduced to timing, frequency, and intensity—parameters optimized for speed and reliability, not richness.

UED calls this semantic austerity. Nervous systems strip the world down to what is actionable. Everything else is noise.

This austerity also clarifies why signals can be misinterpreted so easily. Because they do not carry meaning, context must supply it. The same signal intensity can correspond to different external conditions depending on the system's history, state, and expectations. Confusion arises not because signals are ambiguous, but because meaning is being inferred where none exists intrinsically.

Crucially, this reframing removes the idea that nervous systems aim at truth. They aim at viability. A signal that is "wrong" but timely can save the system. A signal that is "accurate" but late can destroy it.

This is why nervous systems tolerate illusions, biases, and false alarms. They are not errors in the system. They are the cost of operating under time pressure with incomplete information.

Signal comes before meaning because survival comes before understanding.

Once this is accepted, learning itself must be reframed. Learning is not about storing meanings or representations. It is about stabilizing routes that have proven affordable in the past.

The nervous system does not remember facts. It remembers which pathways paid off.

That shift takes us to the next section.

21.3 — Learning: Pathway Stabilization Under Repeated Delta

If signals carry no meaning and sensing merely flags misalignment, then learning cannot be the acquisition of knowledge in any representational sense. UED replaces that familiar picture with a simpler, more structural one.

Learning is pathway stabilization.

When a nervous system encounters a delta and routes a response, consequence follows. If that routing reduces cost—by avoiding damage, restoring alignment, or maintaining viability—the system retains a trace of that pathway. If it increases cost, the pathway weakens or disappears. Nothing needs to be understood. The ledger updates automatically.

This is why repetition matters. A single successful response proves nothing. Noise, luck, or coincidence can always masquerade as strategy. Repeated exposure to similar deltas, however, filters chance out of the picture. Pathways that consistently reduce consequence are reinforced. Others decay.

Learning, then, is not accumulation of experience. It is commitment to structure.

Each reinforced pathway narrows future behavior. This narrowing is often misunderstood as improvement. In UED, it is better described as irreversibility. Once a pathway stabilizes, alternative responses become harder to access. The system becomes faster, more efficient—and less flexible.

This tradeoff is unavoidable. Flexibility is expensive. Stabilization saves time and energy but reduces optionality. Learning therefore always carries a hidden cost: it mortgages the future to the past.

This perspective clarifies why habits form without intention and why unlearning is difficult. The system is not clinging to behavior out of stubbornness. It is honoring a previously affordable resolution. Changing it requires paying the cost again, often repeatedly, until a new pathway proves cheaper.

Memory, in this framework, is not stored experience. It is frozen consequence. The nervous system does not remember what happened; it remembers what worked well enough to survive repetition. Emotional intensity, reward, and punishment merely accelerate this freezing by amplifying consequence signals.

This also explains why learning can mislead. If an early environment rewards a pathway that later becomes maladaptive, the system will continue to deploy it until the accumulated cost becomes undeniable. Learning is conservative by design. It assumes continuity until forced to revise.

UED therefore dissolves the myth that learning leads toward truth. Learning leads toward economy under familiar conditions. When conditions shift, learning becomes liability until new stabilization occurs.

Seen this way, learning is not wisdom. It is compression.

The nervous system compresses history into structure so that it can act faster next time. That compression reduces the time between delta detection and response. It also embeds the past into the present, shaping future perception and action without awareness.

This compression is powerful, but it has limits. Stabilized pathways can anticipate only what resembles what has already occurred. Novelty, ambiguity, and sudden shifts expose the fragility of learning-based prediction.

To manage that fragility, nervous systems go one step further. They do not only stabilize past routes. They attempt to act before delta fully manifests.

That move—from stabilized response to early intervention—is what we call prediction.

21.4 — Prediction: Compressing Time Into Structure

Prediction is often portrayed as a sophisticated mental feat—a sign of intelligence, foresight, or planning. In UED, prediction is none of these. It is a structural shortcut born out of necessity.

Prediction exists because time is expensive.

Once learning stabilizes pathways, the nervous system gains a new option. Instead of waiting for delta to cross a sensory threshold and then reacting, it can activate pathways before full misalignment occurs. This does not require imagining the future. It requires only one assumption: that the near future will resemble the immediate past closely enough to justify early action.

Prediction, therefore, is not about seeing ahead. It is about acting early.

Structurally, prediction is the compression of temporal sequence into spatial arrangement. The nervous system embeds expected consequence into its wiring. Certain patterns of input now trigger responses sooner, with less evidence, and sometimes even in the absence of direct sensory confirmation.

This early activation is cheaper than waiting. If prediction is correct, the system avoids high-cost resolution later. If prediction is wrong, the cost is usually smaller than the cost of inaction. This asymmetry explains why nervous systems favor anticipation over accuracy.

Prediction trades certainty for speed.

This trade explains many otherwise puzzling behaviors. Organisms flinch before contact. They prepare muscles before movement completes. They experience fear before danger is confirmed. These are not cognitive errors. They are economic optimizations under uncertainty.

UED clarifies a critical distinction here. Feedback corrects after consequence appears. Prediction intervenes before consequence completes. Feedforward pathways are therefore more valuable than feedback pathways in fast-changing or high-risk environments, even though they are more error-prone.

This also explains why prediction fails under novelty. Prediction depends entirely on historical compression. When conditions deviate too far from prior patterns, early activation becomes misaligned. The system continues to pay in advance for consequences that no longer arrive, or fails to anticipate ones that do.

From a UED perspective, surprise is not mysterious. It is the moment when compressed history no longer maps onto present delta.

Prediction also introduces a subtle but profound side effect. By acting early, the system alters the very conditions it is predicting. This creates loops—self-reinforcing or self-defeating—where anticipation shapes outcome. Nervous systems do not merely respond to the world; they begin to co-create their own consequence landscape.

At this stage, still nothing mental has been added. No beliefs, no intentions, no goals. Prediction is simply delta management under time pressure, encoded into structure.

But once a system begins acting before certainty, it must also decide how early is too early. Acting too soon wastes energy. Acting too late risks damage. This balance introduces thresholds—points at which accumulated signal suddenly triggers action.

Those thresholds are not arbitrary. They define when prediction turns into commitment.

This is where timing becomes discrete, not continuous. This is where ATT—Alignment Threshold Time—quietly enters nervous function.

Before naming it formally, one more piece must be understood: how nervous systems mark success and failure of prediction without concepts, language, or judgment.

That marking appears as pain and pleasure.

21.5 — Pain and Pleasure: Consequence Signaling, Not Reward

Pain and pleasure are usually treated as psychological experiences—one to be avoided, the other to be pursued. This framing immediately moralizes them. Pain becomes bad. Pleasure becomes good. From there, motivation, desire, reward, punishment, and ethics are smuggled in.

UED cuts beneath all of that.

Pain and pleasure are not values. They are signals.

More precisely, they are consequence signals that report how prediction and action are interacting with delta. They do not evaluate behavior. They mark cost.

Pain signals that delta is escalating faster than the system's capacity to manage it. It is not a lesson. It is not a warning issued by nature. It is a structural alarm indicating that resolution is becoming expensive or irreversible. Tissue damage, extreme stress, suffocation, metabolic failure—all generate pain because they represent delta that is no longer being negotiated gently.

Pleasure signals the opposite. It marks successful alignment or efficient delta reduction. When an action restores balance, reduces threat, or stabilizes a pathway that lowers future cost, pleasure appears—not as a reward, but as a confirmation that the chosen pathway is economically favorable.

Seen this way, pain and pleasure sit after prediction, not before it. They do not motivate action in advance. They update the ledger after action has occurred, reinforcing or weakening the pathways that led there.

This reframing explains several long-standing confusions.

First, pain is not always bad for survival, and pleasure is not always good. Systems can learn to tolerate pain if the long-term delta budget improves, and they can pursue pleasure even when it leads to catastrophic cost. Addiction, self-harm, and compulsive behavior are not moral failures—they are signaling distortions, where pleasure no longer correlates with alignment.

Second, pain and pleasure are not proportional to truth. A false prediction can feel good if it temporarily reduces stress, and a correct prediction can feel painful if it demands immediate cost. The nervous system tracks affordability, not accuracy.

Third, chronic pain and chronic pleasure are both signs of malfunction. Persistent pain indicates unresolved delta that the system cannot escape. Persistent pleasure indicates over-reinforcement of a narrow pathway that crowds out adaptive flexibility. In both cases, signaling has lost its calibrating role.

UED therefore rejects the idea of reward and punishment at the nervous-system level. There is no judge. There is no intent to teach. There is only feedback—sometimes clean, sometimes corrupted.

Pain says: *this pathway is too expensive.* Pleasure says: *this pathway currently pays.*

Nothing more.

This also explains why pain and pleasure are unreliable guides to long-term well-being. They evolved to optimize short-horizon delta management, not lifetime flourishing or ethical behavior. Expecting them to guide meaning or morality is a category error.

Their true function is narrower and sharper. They accelerate learning by amplifying consequence signals. They make some outcomes unforgettable and others repeatable. They bias stabilization without requiring cognition.

Once this is understood, a deeper pattern becomes visible.

Pain and pleasure operate through thresholds. Below a certain level, delta is tolerated silently. Above it, signaling escalates suddenly. Action is forced. Attention collapses onto the signal. Delay becomes impossible.

These thresholds are not psychological artifacts. They are structural boundaries that govern when prediction must give way to immediate resolution.

That boundary—the moment when accumulated delta triggers decisive action—is not continuous. It is threshold-gated.

And that brings us directly to the final structural role of nervous systems: how they act as local time compressors by triggering alignment at critical moments.

That role is governed by Alignment Threshold Time.

21.6 — Nervous Systems as Local Time Machines

By the time sensing, learning, prediction, and consequence signaling are in place, the nervous system is doing something extraordinary—without ever intending to.

It is reshaping time.

Not by bending it, not by escaping it, but by compressing it.

In earlier chapters, time was reframed as resolution duration: how long delta remains unresolved before equilibrium asserts itself. Nervous systems intervene directly in this process. They do not wait for resolution to unfold at natural pace. They attempt to force resolution early, at carefully chosen moments, by acting when thresholds are crossed.

This is where nervous systems become local time machines.

They borrow against the future by paying costs now. They pull potential consequence forward into present action. Instead of letting delta accumulate silently until catastrophic resolution, they trigger alignment while there is still room to maneuver.

This operation is not continuous. It is threshold-based.

Below a certain level of delta, the system waits. Above it, action is unavoidable. Muscles contract. Attention narrows. Behavior locks in. These transitions feel sudden because they are sudden. The system is not gradually deciding. It is crossing a structural boundary.

This is Alignment Threshold Time (ATT) operating inside biology.

ATT is the moment when accumulated signals force resolution—not because the system wants to act, but because delay has become unaffordable. Freeze responses, panic, fight-or-flight activation, sudden insight, fatigue collapse—all follow the same rule. Delta accumulates quietly until a threshold is reached, and then time seems to "speed up" or "slow down" subjectively because resolution is being forced.

From the inside, this feels dramatic. From the outside, it is mechanical.

Nervous systems therefore do not experience time uniformly. They experience windows of tolerance separated by compulsory action. Between thresholds, prediction and learning attempt to manage delta cheaply. At thresholds, the system commits decisively, often irreversibly.

This explains why nervous systems privilege speed over accuracy. Near thresholds, the cost of delay outweighs the cost of error. Acting fast—even wrongly—is cheaper than waiting. This is why fear responses override reasoning, why pain collapses attention, why reflexes bypass deliberation.

Time compression is also why nervous systems break under overload. If deltas arrive faster than thresholds can reset, the system loses its ability to stage resolution. Stress accumulates. Signals saturate. Everything begins to feel urgent, even when nothing is. Anxiety is not excess imagination—it is ATT miscalibration.

Seen through UED, nervous systems are not fragile because they are psychological. They are fragile because they operate close to threshold limits by design. They are optimized to survive under time pressure, not to remain calm under abundance.

This reframing dissolves another illusion: that brains perceive time. They do not. They allocate resolution opportunities. Subjective time expands and contracts because threshold spacing changes under load.

Once this is understood, the final piece falls into place.

If nervous systems compress time by triggering action at thresholds, and if prediction stabilizes pathways that succeed at this compression, then something subtle emerges. Successful prediction begins to feel like control. Fast resolution feels like agency. Early action feels like authorship.

But this feeling is an aftereffect, not a mechanism.

That illusion—of control arising from successful delta compression—is the last thing this chapter must prepare the reader to confront.

And that belongs to the next section.

21.7 — Where Control Illusions Begin

Once a nervous system can predict, act early, and trigger resolution at thresholds, something new appears—not in structure, but in experience.

It feels like control.

This feeling is persuasive because it correlates with success. When early action reduces cost, when prediction prevents damage, when pathways fire quickly and consequences remain manageable, the system experiences continuity and stability. From the inside, this stability is misread as authorship.

But structurally, nothing new has been added.

The nervous system did not gain agency. It gained timing advantage.

Control is not a mechanism. It is a sensation produced by successful delta compression. When prediction aligns well with unfolding consequence, the system experiences reduced friction. Actions appear smooth. Outcomes appear responsive. This smoothness is interpreted as "I am in charge."

When prediction fails, the illusion collapses instantly. Surprise, panic, helplessness, and confusion rush in. The system feels out of control not because control was lost, but because the illusion depended on compression working.

This explains a deep asymmetry in experience. Control is felt only when things go well. Failure never feels like shared causality or structural inevitability—it feels personal. The nervous system retroactively claims success and disowns failure, even though both arise from the same mechanisms.

UED makes this explicit: agency is not a driver; it is a byproduct.

The nervous system routes signals, stabilizes pathways, and triggers thresholds. None of this requires a self. None of it requires intention. Yet when these operations are fast and effective, the system generates a narrative center to explain coherence. That narrative is not false—it is secondary.

This is where intelligence begins to be mistaken for wisdom.

Prediction that works repeatedly gives rise to confidence. Confidence gives rise to overcommitment. Overcommitment hardens pathways. Hardened pathways resist revision even when conditions change. What once felt like control becomes rigidity.

This is not a psychological flaw. It is a structural risk inherent in compression-based systems.

The illusion of control is strongest precisely where prediction has been most successful. This is why expertise blinds, why habits dominate, why intelligence can become fragile under novelty. The system trusts its compressed history more than unfolding reality.

UED therefore draws a sharp boundary here.

Control is not mastery over nature. It is temporary alignment under stable constraints.

Once constraints shift, control evaporates. The nervous system does not lose power—it loses relevance. Its predictions no longer map, its thresholds misfire, its signals amplify unnecessarily.

This section does not yet dismantle the self, free will, or consciousness. It only places a marker.

What feels like "me deciding" begins here—not because a self exists, but because successful compression produces coherence. That coherence demands an explanation, and the mind will later provide one.

For now, the reader needs only one realization.

Nervous systems do not create agency. They create the conditions under which agency feels real.

Understanding that prepares the ground for the next transition. Once prediction begins observing its own success and failure, experience turns inward. Signals stop being merely acted upon and start being noticed.

That turn—from action to observation—is where mind, awareness, and consciousness enter the discussion.

But they do not enter as mysteries.

They enter as structural consequences of everything laid down so far.

From Prediction to Experience

At this point, the nervous system has done everything it was ever structurally required to do. It detects delta, stabilizes pathways, anticipates consequence, compresses time, and triggers resolution at thresholds. Survival becomes faster. Failure becomes rarer. Life extends its delay against equilibrium.

Nothing else is needed for persistence.

And yet, something else appears.

Not as a mechanism. Not as an addition. But as an emergent side-effect.

When prediction becomes reliable enough, when pathway activation becomes fast enough, when threshold crossings become routine, the system no longer operates in pure urgency. It gains small pockets of slack—moments where action is not immediately forced.

In those pockets, signals are no longer just routed. They are noticed.

This noticing does not improve survival directly. It does not speed response. It does not manage delta more efficiently. It simply arises when the system has enough stability to register its own operations without immediately acting on them.

This is the critical turn.

Experience appears not because the system wants to reflect, but because reflection becomes affordable. The system begins to sense not only external delta, but internal signaling itself. Prediction starts observing prediction. Pathways begin to monitor their own activation.

Still, nothing metaphysical has happened.

There is no observer added. There is no inner entity installed. There is no jump to meaning or self.

There is only a new layer of consequence: signals now affect not only behavior, but other signals.

This recursive interaction creates what we later call awareness. But at this stage, it is raw, unstructured, and contentless. It is not about understanding the world. It is about registering the fact that signaling is occurring.

From the outside, this looks like consciousness beginning. From the inside, it feels like experience emerging. Structurally, it is simply delta interacting with its own compression machinery.

This transition matters because it marks the boundary between biology and mind.

Everything up to this point could operate without experience. Everything beyond this point cannot avoid it. Once signals can be noticed, not just acted upon, a new class of phenomena becomes unavoidable: attention, confusion, emotion, identity, suffering, insight.

These are not new forces. They are second-order consequences.

The next part of the book does not introduce them as mysteries to be solved or illusions to be dismissed. It introduces them as inevitable outcomes of the structures already in place.

Prediction gave life speed. Experience gives life depth.

What that depth does to systems—how it helps, how it harms, and how it misleads—will define the remainder of the journey.

This is where UED crosses from life into mind.

When a system begins predicting consequence, experience appears—but not yet understanding.

Part VII — Mind, Awareness, And Human Dynamics

This part completes the arc without breaking the rules already installed.

Up to now, UED has shown how systems manage delta across physics, time, energy, life, and nervous prediction—without invoking mystery. Here, the same discipline is applied to the most emotionally charged territory of all: the human inner world. The aim is not to psychologize physics or spiritualize biology, but to structurally demystify mind without flattening experience.

Mind is treated as what it actually does: it amplifies, circulates, and sometimes traps delta. Desire, fear, attention, and thought are not flaws or virtues; they are energetic gradients competing for resolution. When unmanaged, they loop. When misunderstood, they escalate. When clearly seen, they settle on their own. Calm is not imposed—it appears when delta volatility drops.

Awareness enters not as intelligence or morality, but as resolution clarity—the ability to distinguish signal from noise before action hardens into consequence. This is why intelligence without awareness accelerates chaos: it increases leverage without improving alignment. The "self" is reframed not as an entity, but as a priority allocator that frequently misassigns urgency.

Emotion is brought back into the body where it belongs. It is not irrational disturbance but field feedback—real-time reporting of how predicted futures are colliding with physiological limits. Anxiety reflects future-delta overload; depression reflects gradient collapse. Neither is moral, neither is personal failure, and neither can be solved by instruction alone.

Finally, morality, dharma, and ethics are stripped of ideology and restored to function. They are not belief systems or identity badges; they are social delta stabilizers—mechanisms evolved to prevent destructive interference when many self-referencing delta engines coexist. Morality is difficult not because humans are weak, but because self-generated gradients constantly distort shared alignment.

By the end of this part, psychology no longer feels vague, subjective, or exceptional. It becomes structurally continuous with physics, yet remains humane, lived, and emotionally honest. Human suffering is not dismissed—but it is no longer mysterious. And human responsibility is grounded, not preached.

This is where UED proves its final claim: it can explain the human condition without reducing it, and clarify it without freezing it into formulas.

Chapter 22 — The Mind as a Delta Engine

Why the Mind Never Stays Still

The most common complaint about the mind is that it will not stop. Thoughts arise uninvited. Attention drifts. Desires pull. Fears intrude. Even in moments meant for rest, the inner movement continues. This persistence is often treated as a defect—as a lack of discipline, focus, or maturity.

UED asks for a different starting point.

The mind does not stay still because it is not designed to.

Stillness is not its function. Resolution is not its job. The mind exists to circulate delta—to keep gradients in motion until they are resolved elsewhere. Expecting it to remain quiet is like expecting a river to stop flowing because the landscape feels noisy.

This misunderstanding creates unnecessary conflict. People attempt to silence the mind through force, instruction, or ideology. When these efforts fail—as they almost always do—the failure is personalized. The individual concludes that something is wrong with them, when in fact the system is operating exactly as intended.

Mental activity is not evidence of weakness. It is evidence of unresolved gradients.

Desire pulls the future into the present. Fear drags imagined cost forward. Attention follows the steepest gradient available, not the one most socially approved. These movements are not moral choices. They are structural responses to perceived imbalance.

The mistake lies in assuming that the mind should behave like a judge or a ruler. It is neither. It is an engine—one that converts delta into motion.

When gradients are strong, motion is intense. When gradients settle, motion slows on its own.

This is why attempts at mental control so often backfire. Suppression does not remove delta; it merely blocks its outlet. The pressure remains and finds another path—often more disruptive than the original one.

Understanding this changes the posture of engagement completely.

Instead of asking, "How do I stop my mind?" the more accurate question becomes, "What delta is being circulated, and why has it not resolved?"

This shift removes blame without removing responsibility. The mind is not the enemy, but it is not the solution either. It is a messenger, a mover, a distributor. Asking it to resolve what belongs to life, body, or environment is a category error.

This opening reframes the entire chapter. The goal is not to tame the mind, purify it, or transcend it. The goal is to understand its mechanics well enough that unnecessary friction disappears.

Once the mind is seen as a delta engine, its restlessness stops being personal. It becomes readable.

And what can be read clearly no longer needs to be fought.

22.1 — Desire, Fear, and Attention as Delta Forces

Desire and fear are usually treated as psychological traits or moral tendencies. One is encouraged, the other is condemned. Attention is treated as a skill to be trained, disciplined, or controlled. This framing makes all three feel personal, intentional, and blameworthy.

UED removes that burden.

Desire, fear, and attention are not character traits. They are delta forces.

Desire is future-oriented delta. It arises wherever there is a perceived gap between the present state and a preferred state. That gap does not need to be real or achievable. It only needs to be *felt* as unresolved. The mind responds by circulating energy toward that imagined resolution. Desire is not greed, ambition, or weakness—it is gradient awareness pulling action forward.

Fear is the same mechanism, inverted. It is projected cost amplification. When a future state is perceived as threatening, the delta is magnified by urgency. Fear does not invent danger; it exaggerates potential consequence to accelerate response. This exaggeration is often inaccurate, but structurally useful under uncertainty.

Both desire and fear operate without permission. They are not chosen. They arise automatically wherever prediction detects imbalance. Moral judgment arrives later and is largely irrelevant to their operation.

Attention is the allocator that decides which delta gets processed next. It does not follow intention reliably because intention is a weak gradient compared to unresolved threat or promise. Attention flows toward the steepest gradient available—toward what feels most urgent, incomplete, or unstable.

This explains why attention drifts during tasks that are conceptually important but energetically flat. Importance does not move attention. Delta does.

When multiple gradients compete, the mind experiences noise. Thoughts collide. Focus fragments. The system oscillates between unresolved deltas, attempting to reduce pressure without a clear outlet. This oscillation is often misinterpreted as lack of willpower. In reality, it is gradient competition.

Understanding these forces removes a common confusion. Desire and fear are not opposites. They are siblings. Both arise from the same predictive machinery. Both aim to resolve perceived imbalance. One pulls toward a future; the other pushes away from it. Attention simply follows whichever gradient promises the fastest reduction in uncertainty.

This is why suppressing desire or fear rarely works. Suppression adds resistance but does not flatten the gradient. The force remains and often strengthens. The mind continues to circulate delta until something resolves.

Seen structurally, the mind is not malfunctioning when it fixates, worries, or longs. It is doing what engines do—responding to pressure differentials.

The next confusion arises when this circulation becomes repetitive, dense, and seemingly endless. People call it thinking. More precisely, it is motion without resolution.

To understand that, thought itself must be reframed—not as content, but as movement.

22.2 — Thought Is Motion, Not Content

Thought is usually treated as something *about* something. We ask whether a thought is correct or incorrect, useful or useless, positive or negative. This immediately pulls attention toward content, meaning, and interpretation.

UED asks us to look elsewhere.

Thought is not primarily content. Thought is motion.

A thought is the mind's way of moving delta along available pathways. It is circulation, not storage. When a gradient is present—desire, fear, uncertainty, incompletion—the mind does not sit with it. It moves. That movement appears as images, words, memories, plans, and arguments. These are not the engine; they are the exhaust.

This is why two people can think very different thoughts under the same pressure. The content varies because pathways differ, but the motion is identical. The mind is attempting to reduce gradient by moving energy through whatever routes are most accessible.

Seen this way, the problem is rarely *what* one is thinking. The problem is how fast and how densely delta is circulating.

High-volatility minds produce many thoughts quickly. Low-volatility minds produce fewer. The difference is not intelligence, education, or character. It is gradient intensity and pathway resistance. When gradients are steep and outlets are blocked, thought accelerates.

Repetition is the clearest sign of this. Repeated thoughts do not indicate obsession with content; they indicate unresolved motion. The mind keeps sending delta through the same loop because no exit has been found. Changing the words, reframing the story, or arguing with the thought does nothing because none of those actions alter the underlying gradient.

This explains why "thinking it through" so often fails. Adding more motion to an already saturated loop increases heat, not resolution. The system becomes louder, not clearer. What feels like problem-solving is often just delta recycling.

It also explains why attempts to stop thought directly are ineffective. You cannot stop motion by commanding it to halt while the gradient remains. Suppression increases resistance, which steepens the gradient and accelerates circulation elsewhere.

UED therefore reframes a fundamental misunderstanding: clarity does not come from better thoughts. It comes from lower motion.

When gradients flatten, thought slows naturally. When outlets open, loops dissolve without effort. Silence appears not because the mind has been trained, but because it has nothing urgent left to move.

This reframing removes identification with thought. If thought is motion, then "my thoughts" are not me any more than blood flow is me. They are processes responding to pressure. Observing them becomes possible only when we stop demanding that they carry meaning.

Once thought is understood as motion, the most troubling mental phenomenon becomes easier to name.

Overthinking is not excess thought. It is motion trapped in a closed loop.

That is the next structure to examine.

22.3 — Overthinking as a Closed Delta Loop

Overthinking is one of the most misunderstood human experiences. It is described as excess intelligence, lack of discipline, anxiety, obsession, or weakness. Entire industries are built around "managing" it. Almost all of these descriptions miss the structure completely.

Overthinking is not too much thinking. It is delta with nowhere to go.

In UED terms, overthinking occurs when a gradient is detected, prediction is activated, but no resolution pathway is available or trusted. The mind continues to circulate delta because its job—to reduce imbalance—has not been completed. Each pass through the loop slightly alters content but leaves the underlying gradient untouched.

This is why overthinking feels exhausting and unproductive at the same time. Energy is being spent, but nothing is changing. The system is active without progress.

Closed loops form when:

- action is perceived as risky,
- consequences feel irreversible,
- responsibility is unclear,
- or resolution is socially or emotionally constrained.

In these conditions, the mind cannot discharge delta outward. It cannot act, decide, or release. So it turns inward and loops.

Importantly, the loop is not chosen. No one decides to overthink. The loop is the least costly option available when all exits feel blocked. Thinking becomes a substitute for action, not because it helps, but because it is safer than committing.

This also explains why advice like "just stop thinking" or "distract yourself" fails. Distraction merely reroutes the loop temporarily. The gradient remains intact. As soon as distraction fades, circulation resumes.

Overthinking is self-amplifying. Each loop pass reinforces the pathway, increasing its accessibility and speed. Over time, the system learns the loop. What began as situational becomes habitual—not because the person enjoys it, but because the nervous system has stabilized the only available route.

This stabilization is often mistaken for personality: "I am an overthinker." UED rejects that label. There is no overthinking type. There is only persistent unresolved delta under constraint.

The critical mistake is attempting to solve overthinking *inside the loop*. Insight rarely arrives there. The loop is too noisy, too recursive, too saturated. Resolution requires changing the conditions that made looping the cheapest option.

That change may involve:

- external action,
- bodily regulation,
- environmental shift,
- social clarification,
- or simply allowing consequence to occur instead of postponing it.

Notice what is absent from this list: mental force.

Force increases resistance. Resistance steepens gradients. The loop tightens.

Overthinking relaxes only when delta finds a legitimate exit—when the system believes that resolution, even if imperfect, is survivable. The moment action becomes affordable, the loop loses purpose and dissolves on its own.

This is why calm often appears suddenly, without effort, once a decision is made—even a difficult one. The gradient collapses. Motion stops because it is no longer needed.

Understanding overthinking structurally does something crucial. It removes shame.

Nothing is wrong with the mind when it loops. It is signaling that resolution has been deferred too long.

The next confusion arises when people attempt to fix this by tightening control—by discipline, suppression, or self-command. That attempt introduces a new gradient: resistance.

And resistance creates friction.

That friction is what keeps many minds permanently overheated.

22.4 — Control Attempts and Mental Friction

When overthinking becomes unbearable, the most natural response is control. People try to discipline the mind, suppress unwanted thoughts, impose routines, or command calm through effort. At first glance, this seems reasonable. If motion is the problem, control should fix it.

Structurally, it does the opposite.

Control introduces resistance, and resistance increases friction. Friction does not remove delta; it converts it into heat. In mental terms, that heat appears as tension, irritability, fatigue, and a sense of constant inner struggle.

The key misunderstanding is assuming that the mind is disobedient. It is not. It is responsive. When force is applied, the system reacts by increasing gradient elsewhere. Suppressed thoughts return with greater intensity not because they are rebellious, but because the underlying delta has not been resolved—only blocked.

This is why discipline-based approaches often work briefly and then fail dramatically. Short-term control can redirect delta temporarily, but it does not flatten gradients. Over time, the pressure accumulates behind the barrier. When it breaks through, the release is stronger and more chaotic than before.

Mental friction also explains why effort feels exhausting even when nothing outward changes. Energy is being spent resisting motion rather than allowing resolution. The system runs hotter while going nowhere.

UED makes a sharp distinction here. Effort can redirect delta, but it cannot dissolve it. Control can postpone consequence, but it cannot cancel it. Every attempt to dominate the mind creates a second-order problem layered on top of the original gradient.

This is where many philosophies and productivity systems go wrong. They confuse suppression with stability. Suppression narrows pathways forcibly. Stability emerges only when gradients naturally flatten.

Another subtle cost of control is rigidity. When control becomes habitual, the system learns to maintain tension as a baseline state. Calm begins to feel unfamiliar or unsafe. The mind equates vigilance with survival, and relaxation triggers anxiety. This is not psychological weakness; it is a predictable consequence of long-term friction.

UED reframes mental discipline not as strength, but as temporary scaffolding. It can be useful in emergencies, but it is not a solution. Relying on it continuously guarantees overheating.

The alternative is not passivity. It is gradient literacy—the ability to recognize where resistance is amplifying cost and where allowing resolution would actually reduce it.

This distinction sets up a crucial redefinition.

Calm is not achieved by winning against the mind. It appears when there is nothing left to fight.

Understanding that requires a structural definition of calm itself—one that does not depend on effort, control, or virtue.

That is the next section.

22.5 — Calm as Low Delta Volatility

Calm is often portrayed as an achievement. Something to be cultivated, practiced, imposed, or earned through discipline. This framing quietly assumes that calm is an *action*—a behavior the mind can perform if trained well enough.

UED rejects that assumption completely.

Calm is not something you do. Calm is what remains when delta volatility drops.

Volatility matters more than magnitude. A large but steady gradient can be lived with. A small but rapidly fluctuating gradient creates unrest. What people experience as mental disturbance is rarely the presence of desire or fear alone; it is the rapid switching and competition between gradients that exhausts the system.

Calm appears when gradients stop colliding.

This can happen in several ways:

- a decision collapses competing futures into one,
- an action releases deferred consequence,
- a bodily state stabilizes signaling,
- or an environment removes persistent triggers.

Notice what is common across all of these: delta resolves or settles, even if imperfectly. The mind does not need to be convinced. It slows because there is less pressure to circulate.

This reframing dissolves a crucial myth. Calm is not the absence of thought. Thought can continue under calm. What disappears is urgency. Motion loses heat. Loops lose their grip.

This is why calm often follows acceptance—not because acceptance is virtuous, but because it ends internal resistance. Resistance itself is a gradient. When resistance stops, one entire layer of delta vanishes immediately.

UED also distinguishes calm from sedation. Sedation flattens sensation artificially, suppressing signals without resolving underlying gradients.

Calm is different. It is responsive, alert, and available. It arises from alignment, not numbing.

This distinction matters because many practices mistake dullness for peace. True calm does not reduce sensitivity; it reduces noise. Signals remain, but they no longer compete destructively.

Another misconception is that calm must be maintained. In reality, calm is transient by nature. New gradients will arise. Life will reintroduce imbalance. Expecting permanent calm is itself a gradient that generates disturbance.

Calm is not a state to hold. It is a byproduct of alignment in the present moment.

Once this is understood, a subtle danger becomes visible. Intelligence—when applied without awareness—can reduce volatility briefly while increasing long-term instability. Clever strategies, mental reframing, and cognitive tricks may smooth immediate gradients but harden pathways that later resist change.

This is why some of the most intelligent minds suffer the most. They generate motion faster than it can be clarified.

That observation sets the stage for the final section of this chapter: why intelligence alone is not enough, and why awareness becomes unavoidable once the mind's mechanics are understood.

22.6 — Intelligence Without Awareness: A Dangerous Engine

Once the mind is understood as a delta engine, a sobering pattern becomes impossible to ignore: intelligence amplifies whatever dynamics are already present. It does not correct them by default. It accelerates them.

Intelligence increases speed, range, and combinatorial power. It can project futures further ahead, compare more options, simulate more outcomes, and generate more internal motion per unit time. This is useful only when the system can distinguish which motion matters.

Without that distinction, intelligence becomes a multiplier for chaos.

An intelligent mind without awareness does not resolve delta more effectively; it circulates it faster. Desire becomes elaborate. Fear becomes sophisticated. Overthinking gains arguments, justifications, and narratives that reinforce the loop. What was once simple pressure becomes a self-sustaining storm.

This is why intelligence often correlates with suffering when awareness is weak. The system has more leverage but no governor. It can delay action indefinitely by endlessly refining prediction. It can justify avoidance convincingly. It can generate reasons to resist resolution while believing it is being responsible.

UED frames this cleanly: intelligence increases throughput, not clarity.

Throughput without clarity raises volatility. It sharpens gradients instead of flattening them. It makes internal friction more efficient. The mind becomes excellent at producing motion and terrible at allowing resolution.

This also explains a common paradox. Highly intelligent individuals often recognize their patterns clearly yet remain trapped by them. Recognition alone does not reduce delta. It can even increase it by adding a meta-layer of frustration: *I see the loop, so why can't I stop it?*

Because seeing is not the same as prioritizing.

Awareness is not intelligence's partner. It is its constraint.

Without awareness, intelligence behaves like an engine without brakes—powerful, responsive, and prone to overheating. It rewards speed over direction. It confuses cleverness with progress. It treats motion itself as evidence of control.

UED therefore draws a firm boundary. Intelligence belongs to the mind. Awareness does not. Awareness does not think faster, calculate better, or plan deeper. It performs a different function entirely: it separates signal from noise before action hardens into consequence.

Until that separation exists, intelligence remains dangerous—not morally, but structurally. It amplifies unresolved delta and delays commitment under the illusion of sophistication.

This is the final insight required to complete the chapter.

The mind can move delta. It cannot see which delta should move.

That capacity—to see without accelerating—is not mental. It is awareness.

And once awareness enters the system, the engine no longer overheats. It becomes readable.

That transition—from motion to clarity—is where the next chapter must begin.

From Mental Motion to Resolution Clarity

By the end of this chapter, the mind stands revealed for what it actually is: a powerful, tireless engine built to circulate delta, not to judge it, end it, or transcend it. Desire, fear, attention, thought, and even overthinking are no longer personal defects or virtues. They are mechanical consequences of gradients moving through structure.

Nothing here needs to be fixed. Nothing here needs to be silenced. What needs to change is where resolution is attempted.

As long as the mind is asked to do what it cannot—see clearly without amplifying motion—it will overheat. As long as intelligence is trusted to resolve what it can only accelerate, instability will persist. The struggle many people experience is not against life, emotion, or circumstance; it is against a misassignment of roles.

This chapter completes that correction. The mind moves. It does not clarify.

Once this is understood, a natural pause appears. Motion continues, but identification weakens. Thought still flows, but urgency softens. The engine keeps running, yet something steps back—not to control it, not to stop it, but to see it without adding force.

That seeing is not another mental act. It does not belong to thought, analysis, or intelligence.

It belongs to awareness. Awareness does not compete with the mind. It does not replace it. It does not argue with it. Awareness simply reveals which gradients matter and which are noise, before commitment hardens into consequence.

Only then does resolution become possible without friction. The next chapter begins exactly there—not with self-improvement, not with discipline, not with higher intelligence—but with a quieter, more decisive shift: clarity without motion.

That shift is awareness. *When the engine is seen clearly, it no longer overheats.*

Chapter 23 — Awareness as Resolution Clarity

Why Seeing Changes Everything Without Doing Anything

Most human effort is spent trying to *do* something about inner life. Think better. Decide faster. Control emotion. Fix habits. Improve discipline. Even when people turn toward reflection, the question remains action-oriented: *What should I do now?*

Awareness enters as a quiet disruption to that entire approach.

Awareness does not act. It does not intervene. It does not correct.

And yet, when awareness is present, everything changes.

This is deeply counterintuitive because we are conditioned to believe that change requires effort. We equate improvement with activity and clarity with control. Awareness violates this assumption completely. It alters outcomes without adding motion, without introducing resistance, and without competing with the mind.

That is its defining feature.

Awareness is not another layer of thinking. It is not better intelligence. It does not generate strategies, explanations, or solutions. In fact, the moment awareness tries to *do* something, it collapses back into the mind and loses its function.

What awareness does instead is simpler—and more radical.

It reveals what is already happening before reaction hardens into consequence.

This revelation alone reorganizes the system. When signal and noise are clearly distinguished, false urgency dissolves. When urgency dissolves, motion slows. When motion slows, resolution becomes possible without force.

No instruction is required.

This is why awareness has been so difficult to define across traditions. It does not behave like a faculty. It does not accumulate. It cannot be strengthened through repetition the way skills can. It does not belong to time in the same way the mind does. It arrives, or it does not—and when it arrives, struggle weakens instantly.

The mind cannot do this because the mind *is* motion. It amplifies delta by design. Asking it to clarify priorities is like asking a storm to rank its own winds.

Awareness stands outside that process—not spatially, not metaphysically, but functionally.

It does not suppress desire or fear. It does not silence thought. It simply sees them as gradients rather than commands. And when gradients are seen clearly, they lose their absolute authority.

This is the central correction of the chapter.

Awareness does not fix life. It prevents misalignment from escalating.

Seeing changes everything not because it adds power, but because it removes false leverage. What no longer feels urgent no longer drives action. What is no longer misidentified as signal no longer consumes energy.

From here onward, the discussion must stay precise. Awareness will not be elevated into spirituality, reduced into cognition, or romanticized as virtue. It will be treated exactly as it functions: resolution clarity—the capacity to recognize what matters before the system commits.

Once this capacity is installed, the chaos produced by intelligence alone becomes understandable—and avoidable.

The next step is to define that primary function clearly: how awareness separates signal from noise, and why the mind cannot do this job no matter how intelligent it becomes.

23.1 — Signal vs Noise: The Primary Function of Awareness

At the most fundamental level, awareness performs one operation—and it performs it without effort.

It separates signal from noise.

Signal, in UED terms, is delta that has real consequence if ignored. It is imbalance that will propagate, escalate, or collapse into irreversible cost unless addressed. Noise, by contrast, is delta that circulates without meaningful consequence relevance. It may feel intense, repetitive, or emotionally charged, but it does not require action to preserve alignment.

The mind cannot reliably make this distinction.

The mind responds to intensity, not importance. It follows steep gradients, not meaningful ones. It amplifies whatever is loud, urgent, or emotionally charged because those features evolved to demand quick routing. In complex human environments, however, intensity is often manufactured—by memory, imagination, social pressure, or internal narrative—rather than by actual consequence.

This is where awareness enters.

Awareness does not react to intensity. It notices relevance.

It sees that some thoughts, fears, and desires feel urgent but carry no real cost if left alone. It also sees that some quiet, easily ignored signals—fatigue, unease, misalignment—carry deep consequence if neglected. This inversion is critical. Without awareness, systems chase noise and neglect signal.

This separation alone reduces mental motion dramatically. When noise is no longer treated as signal, it loses its claim on attention. It may still arise, but it no longer demands engagement. The mind does not need to fight it. It simply stops feeding it.

Awareness accomplishes this without suppression because it does not oppose noise. Opposition would generate a new gradient. Awareness simply withholds endorsement. What is seen clearly no longer needs to be circulated compulsively.

This also explains why awareness feels stabilizing even in difficult circumstances. Clarity does not eliminate real problems, but it prevents false ones from multiplying. The system can allocate energy to what actually matters instead of dissipating it across imagined threats and symbolic conflicts.

Importantly, awareness does not decide what to do. It decides what not to do first—what not to escalate, what not to prioritize, what not to dramatize. That negative space is where relief appears.

This function cannot be trained into the mind through analysis. Analysis itself is motion. Awareness precedes analysis. It sees before explanation begins.

Once this distinction is installed, intelligence can finally operate safely. Speed and power no longer amplify chaos because they are guided by correct prioritization.

Without this distinction, intelligence remains dangerous no matter how refined it becomes.

This leads directly to the next clarification: why intelligence, when ungoverned by awareness, accelerates disorder instead of resolving it.

23.2 — Why Intelligence Without Awareness Amplifies Chaos

Intelligence is commonly assumed to be self-correcting. We believe that if someone thinks deeply enough, analyzes carefully enough, or gathers sufficient information, clarity will eventually emerge. This belief is comforting—and structurally false.

Intelligence does not reduce chaos by itself. It multiplies whatever it touches.

Intelligence increases range, speed, and combinatorial depth. It can hold more variables, imagine more futures, and generate more internal motion per unit time. This is powerful only if the system already knows which deltas matter. Without that discrimination, intelligence accelerates noise faster than it can resolve signal.

This is why highly intelligent systems—human or artificial—often fail in subtle, catastrophic ways. They optimize locally, overfit narratives, and build elegant solutions to the wrong problems. They are not stupid. They are misaligned.

Without awareness, intelligence mistakes intensity for importance. It treats urgency as relevance and complexity as depth. It generates explanations for feelings, stories for impulses, and strategies for anxieties—each adding another layer of motion to an already saturated system.

Overanalysis is the clearest example. The system believes it is being responsible by thinking more, but it is actually feeding the loop. Each new model adds credibility to noise, each scenario adds urgency to imagined futures, and each refinement delays commitment further. The result feels like diligence while producing paralysis.

UED frames this precisely: intelligence increases throughput, not filtering.

Filtering is awareness's job.

When awareness is absent, intelligence cannot tell when to stop. It keeps optimizing because it has no criterion for sufficiency. More thought always seems safer than action, even when delay itself is the risk.

This is also why intelligent minds are more vulnerable to anxiety and burnout. They can project future consequences far beyond actionable range. They simulate costs that will never arrive. They rehearse losses that do not exist. Intelligence gives these projections detail and plausibility; awareness is what tells the system which projections are irrelevant.

Without awareness, intelligence becomes a liability.

It builds internal worlds faster than reality can correct them. It sharpens fear into conviction and desire into obsession. It creates coherence where none exists and then defends that coherence aggressively.

This does not mean intelligence is flawed. It means intelligence is ungoverned.

Awareness acts as the governor not by slowing intelligence mechanically, but by constraining what intelligence is allowed to amplify. When awareness is present, intelligence becomes precise instead of prolific. It serves alignment instead of acceleration.

The difference is subtle but decisive.

With awareness, intelligence asks: *Does this matter?* Without awareness, intelligence asks only: *Can this be thought further?*

The first leads to resolution. The second leads to endless motion.

This is why clarity often feels like a sudden stop rather than a gradual improvement. When awareness arrives, the system does not become smarter. It becomes selective. Whole lines of thought lose relevance instantly—not because they were disproven, but because they were never signal to begin with.

Understanding this prepares the ground for a final correction in this chapter.

If awareness filters signal from noise, and intelligence amplifies what remains, then something else must be deciding what gets treated as "me"—what claims ownership, urgency, and identity.

That mechanism is usually called the self.

And it is far less solid than it feels.

23.3 — Awareness Has No Agenda

One of the most persistent confusions around awareness is the belief that it *wants* something—that it aims for peace, truth, happiness, morality, or enlightenment. This assumption quietly turns awareness into another agent, another strategy, another improvement project.

UED removes that projection.

Awareness has no agenda.

It does not seek calm. It does not oppose desire. It does not correct fear. It does not prefer outcomes.

The moment awareness is given a goal, it stops being awareness and becomes another form of intelligence—another motion trying to produce a result. That motion immediately adds delta, re-entering the very loop awareness dissolves.

Awareness functions precisely because it is non-participatory.

It does not compete with gradients; it reveals them. It does not suppress mental activity; it exposes its structure. Desire can continue. Fear can continue. Thought can continue. What changes is not their presence, but their authority.

This neutrality is often misunderstood as detachment or indifference. Structurally, it is neither. Awareness does not withdraw from life. It simply refuses to add urgency where none is structurally required.

Because awareness has no preference, it does not resist what appears. Resistance would be an agenda. It would create a counter-gradient. By not resisting, awareness allows gradients to be seen clearly—and gradients that are clearly seen often collapse on their own.

This is the paradox many people encounter but cannot explain: when something is fully seen without judgment, it loses its grip. Not because it was wrong, but because it was over-prioritized.

Awareness does not remove desire; it removes false centrality.

Awareness does not erase fear; it removes imagined inevitability.

This is why struggle dissolves without effort in the presence of awareness. Struggle requires opposition. Awareness offers none. Without opposition, the extra layer of delta created by resistance vanishes instantly.

Another crucial implication follows.

Because awareness has no agenda, it cannot be cultivated through ambition. Trying to "be more aware" is already a misstep. That attempt recruits the mind, introduces evaluation, and re-creates hierarchy: *more aware vs less aware*. Awareness itself does not recognize such scales.

Awareness either clarifies or it does not. When it clarifies, things fall into proportion. When it does not, intelligence resumes its default behavior.

This is also why awareness feels ordinary rather than exalted. It lacks drama because drama is motion. Awareness is quiet not because it is passive, but because it does not amplify.

In UED terms, awareness is zero-delta observation.

It neither injects energy nor extracts it. It allows existing energy to reveal its own structure. That is why it feels relieving without being numbing, stabilizing without being controlling.

Understanding this neutrality is essential before moving further. Otherwise, awareness is easily mistaken for morality, discipline, or transcendence—all of which distort its function.

Awareness is not here to make life better. It is here to make priorities accurate.

Once that accuracy is present, the system reorganizes itself naturally. And when reorganization occurs without force, something that feels deeply personal begins to loosen its grip.

That something is the sense of self as the unquestioned center of urgency.

To understand that, awareness must now be applied to the mechanism that assigns priority in the first place—the self itself.

23.4 — The "Self" as Delta Prioritizer

The self is one of the most emotionally charged concepts in human life. It is defended, cultivated, criticized, exalted, and attacked. Some traditions insist it must be strengthened. Others insist it must be dissolved. Both positions assume the same thing—that the self is a fixed entity with intrinsic authority.

UED takes a simpler and far less dramatic position.

The self is not an entity. It is a prioritization mechanism.

Functionally, the self is the structure that decides which deltas are treated as *mine*, which consequences feel personal, and which urgencies demand immediate response. It does not generate gradients. It selects among them.

This selection is not deliberate. It is learned.

Repeated experiences stabilize certain pathways, and those pathways begin to define what feels central. Over time, the system develops a bias: some deltas are amplified automatically, others are ignored. That bias is experienced as identity.

"I am ambitious." "I am anxious." "I am responsible." "I am sensitive."

These statements do not describe essence. They describe habitual delta prioritization.

The self forms wherever prediction, learning, and consequence signaling repeatedly converge around similar patterns. It is not invented in one moment. It accretes gradually, shaped by environment, survival demands, social feedback, and bodily conditioning. Once stabilized, it feels solid because it is structurally reinforced, not because it is fundamentally real.

This explains why the self feels threatened by awareness.

Awareness exposes prioritization as process rather than truth. It reveals that what feels central is often merely familiar. When awareness sees a delta clearly without granting it special status, the self loses its monopoly over urgency.

This loss is frequently misinterpreted as danger. People report feeling diminished, disoriented, or "not themselves." In reality, nothing essential is being removed. What is dissolving is misplaced priority.

UED therefore rejects both extremes.

The self does not need to be strengthened. That only hardens bias. The self does not need to be destroyed. That creates instability and confusion.

The self needs to be seen accurately.

When seen accurately, it resumes its proper role: a tool for coordinating action under constraint, not an authority over reality. It can still allocate resources, protect boundaries, and maintain continuity—but it no longer distorts scale.

Why does the self prioritize wrongly so often?

Because it evolved under conditions where speed mattered more than accuracy. Early misalignment could be fatal. The system learned to treat familiar gradients as existential. In modern environments, those same priorities persist long after the original constraints have vanished.

The self is therefore conservative by design. It overreacts rather than underreacts. It inflates some deltas and ignores others. This bias is not malicious. It is outdated.

Awareness does not fight this bias. It outgrows it.

When awareness is present, the self is no longer the sole arbiter of importance. It becomes one input among many. Decisions become quieter, cleaner, and less dramatic—not because life becomes trivial, but because proportion is restored.

This redefinition is crucial for the next step.

If suffering persists even when external conditions are manageable, the cause is rarely life itself. It is almost always misplaced priority—noise treated as signal, familiarity treated as necessity, identity treated as destiny.

That structural error—not emotion, not desire, not thought—is the real engine of human suffering.

And that is what the next section must make precise.

23.5 — Misplaced Priority: The Root of Suffering

Once the self is seen as a delta prioritizer, suffering can finally be addressed without moral blame, psychological drama, or spiritual abstraction. What appears as suffering is rarely caused by the presence of pain, difficulty, or challenge alone. Those are unavoidable features of life. Suffering arises when priority is assigned inaccurately.

In UED terms, suffering is not excess delta. It is misallocated urgency.

Some gradients are treated as existential when they are not. Minor threats are amplified into identity crises. Temporary discomfort is interpreted as permanent damage. Social signals, imagined futures, and symbolic losses are elevated to the same level as physical survival. The system reacts appropriately—but to the wrong input.

This is why suffering often persists even when nothing objectively terrible is happening. The environment may be stable, resources sufficient, and relationships intact, yet the inner system remains under siege. The siege is internal, maintained by prioritization that no longer matches consequence reality.

Awareness dissolves suffering not by eliminating gradients, but by correcting scale.

When awareness is present, it becomes obvious that many thoughts, fears, and desires carry emotional intensity without structural consequence. They feel loud but lead nowhere. Once seen as noise, they lose their claim on energy. The system stops feeding them, and they collapse naturally.

At the same time, awareness often reveals quieter signals that were previously ignored: fatigue, misalignment, resentment, bodily strain, ethical discomfort. These signals lack drama, so the self overlooks them. But their consequence is real. Neglecting them accumulates cost over time.

This inversion is critical.

Suffering persists when noise is treated as signal and signal is treated as noise.

The mind cannot correct this inversion because it is driven by intensity. The self cannot correct it because it is invested in familiar priorities. Awareness alone can reorder the field without adding force.

Importantly, this reordering does not require resolution of every problem. Many deltas remain unresolved. Life continues to be imperfect. What changes is that the system stops bleeding energy into false emergencies.

Relief, in this framework, is not the absence of challenge. It is the absence of misplaced importance.

This explains why relief can arrive suddenly, even when circumstances remain unchanged. Nothing external has improved. What has improved is prioritization. The nervous system stops mobilizing resources against threats that never arrive.

It also explains why advice and reassurance often fail. Being told that something "isn't a big deal" does nothing if the self continues to treat it as central. Only awareness—direct, unmediated seeing—can strip false urgency of its authority.

This section also clarifies why suffering feels personal. Priority is assigned internally. When it misfires, the cost is borne privately, even if the cause is abstract or imagined. That privacy makes suffering feel isolating and real, even when it is structurally unnecessary.

UED does not trivialize suffering. It explains it.

Suffering is not weakness. It is not failure. It is not punishment.

It is the natural outcome of a system responding correctly to incorrect priorities.

Once this is seen, the question shifts again.

If awareness can reorder priority without effort, does that mean action is no longer required? Does clarity imply withdrawal, passivity, or disengagement from life?

That misunderstanding must be corrected immediately.

Awareness does not eliminate action. It changes when and how action occurs.

That clarification belongs to the final section of this chapter.

23.6 — Awareness and Action: Timing, Not Withdrawal

A common fear arises once awareness is understood as non-interfering clarity: *If I stop pushing, will I stop acting?* This fear assumes that action requires pressure, urgency, or struggle—that without internal force, life will stall.

UED shows the opposite.

Awareness does not remove action. It restores timing.

Most dysfunctional action is not caused by laziness or passivity; it is caused by premature commitment or delayed commitment. Systems act too early on noise, or too late on signal. Both errors are costly. Awareness corrects neither by adding effort, but by ensuring that action occurs at the right threshold.

When awareness is present, false urgencies fall away. Action is no longer triggered by intensity alone. At the same time, real signals become impossible to ignore. Awareness does not create indecision—it removes distraction.

This is why awareness often leads to faster, cleaner decisions rather than hesitation. The system is no longer negotiating with competing gradients. Once signal is clear, action follows naturally, without drama, justification, or internal resistance.

UED reframes decisiveness here.

Decisiveness is not force. It is absence of internal contradiction.

When noise is filtered out, there is nothing to debate. The system moves because movement is aligned, not because it is commanded. This kind of action feels quiet, even when it is firm. It does not require self-coercion. It does not generate regret loops.

This also explains why awareness does not imply withdrawal from life. Withdrawal happens when the system is overloaded or misaligned. Awareness reduces overload by preventing unnecessary mobilization. Energy that was previously spent on internal conflict becomes available for engagement.

Awareness therefore improves participation, not detachment.

It also changes the experience of effort. Effort is no longer expended resisting the mind or forcing motivation. It is applied outwardly, where it belongs—into tasks, relationships, and decisions that actually carry consequence.

Another subtle shift occurs here. When awareness governs timing, action stops being reactive. It becomes responsive. The system no longer flinches at every gradient. It waits until resolution is structurally required—and then acts without hesitation.

This is not passivity. It is precision.

Awareness does not make life slower. It makes it less wasteful.

This distinction is critical as the book moves forward. Without it, awareness could be mistaken for disengagement or spiritual bypassing. In UED, awareness is not an escape from consequence. It is the means by which consequence is met without unnecessary amplification.

With this clarification, the chapter's logic is complete.

Awareness clarifies before motion. Intelligence executes after clarification. Action occurs when threshold is genuine.

What remains to be addressed is what happens after clarity—how the body reports alignment and misalignment once priorities are set correctly.

That reporting appears as emotion.

And emotion, when understood structurally, no longer needs to dominate, confuse, or overwhelm.

It simply needs to be read correctly.

That is where the next chapter begins.

From Clarity to Feeling

With awareness in place, the internal architecture finally settles into its correct order. The mind continues to move, intelligence continues to execute, and the self continues to coordinate—but none of them operate blindly anymore. Priority is no longer assigned by intensity alone. Timing is no longer hijacked by noise.

What changes most is not thought, but how consequence is registered.

Once signal is distinguished from noise and action occurs at the right threshold, the body begins to report alignment more cleanly. Sensations sharpen. Emotional responses become more specific. Feedback no longer arrives as overwhelming turbulence; it arrives as information.

This is the crucial transition.

Emotion is not the enemy of awareness. Emotion is what awareness allows to be read correctly.

Before clarity, emotion has to shout. It must escalate to be noticed. Anxiety swells, sadness deepens, anger sharpens—not because the situation demands it, but because prioritization has failed upstream. Emotion compensates for confusion by increasing intensity.

After clarity, that compensation is no longer necessary.

Emotion becomes what it always was structurally meant to be: field feedback. A bodily report of how current conditions align—or fail to align—with capacity, limits, and consequence. It no longer needs to dominate decision-making because decision-making is no longer blind.

This is why awareness often changes emotional life without suppressing it. Feelings still arise, but they move through the system instead of lodging inside identity. They inform rather than command.

The next chapter does not treat emotion as something to regulate, heal, or transcend. It treats emotion as a language of the body-field, reporting the success or strain of alignment in real time.

Once awareness clarifies priority, emotion can finally speak at its natural volume.

And when emotion is heard correctly, it stops needing to overwhelm.

That is the shift that now unfolds.

When clarity arrives first, emotion no longer needs to shout.

Chapter 24 — Emotion as Field Feedback

Why Emotion Was Never Meant to Lead

Emotion has been placed in a role it was never designed to play. In some traditions, it is treated as a weakness to be controlled. In others, it is elevated as a source of truth to be followed. Modern psychology oscillates between these extremes—either regulating emotion or validating it—without ever asking a more fundamental question: what is emotion actually for?

UED offers a precise correction.

Emotion is not a guide. Emotion is not a judge. Emotion is not a command.

Emotion is feedback.

It arises after interaction, after prediction meets reality, after delta collides with capacity. It reports how the system is doing—not what the system should believe, decide, or become. Confusion begins when this report is mistaken for instruction.

This confusion is understandable. Emotions are vivid. They arrive in the body with urgency, color, and intensity. They demand attention. When something speaks loudly from within, it feels authoritative. But loudness is not accuracy, and intensity is not wisdom.

Emotion answers a narrow but vital question: *What is happening to the system right now as a result of current alignment or misalignment?*

It does not answer:

- What is true?
- What is right?
- What should I do?

- Who is to blame?

Those questions belong elsewhere.

When emotion is treated as a leader, it creates chaos. Anxiety dictates avoidance. Anger dictates reaction. Sadness dictates withdrawal. When emotion is treated as an enemy, a different chaos emerges. Suppression distorts signals. Unread feedback accumulates. The system loses its internal diagnostics.

Both errors stem from the same mistake: misunderstanding emotion's role.

Emotion was never meant to sit at the steering wheel. It was meant to sit on the dashboard.

This reframing immediately changes how emotion is approached. The goal is no longer to eliminate it, justify it, or obey it. The goal is to read it correctly. Just as pain in the body does not tell a story but signals strain or injury, emotion in the psyche does not narrate reality—it signals consequence.

This is why emotions are often accurate but misleading. They are accurate about the state of the system and misleading about the meaning of that state. Without clarity, the mind fills that gap with stories, identities, and judgments that intensify suffering.

UED does not ask you to detach from emotion. It asks you to demote it to its proper function.

Once emotion is recognized as field feedback, its power to overwhelm diminishes without being denied. Feelings still arise. They may still be strong. But they are no longer confused with truth, destiny, or selfhood.

This opening reframes the entire chapter. Emotion will not be pathologized, romanticized, or spiritualized. It will be treated as what it has always been beneath the stories: a real-time report from the body-field about how present conditions are being borne.

Once that is understood, even the most difficult emotional states—anxiety, depression, despair—become intelligible without becoming trivial.

They are not enemies. They are messages.

And like all messages, their value depends on whether we know how to read them.

24.1 — Emotion as Consequence Reporting

Emotion does not initiate events. It registers what has already happened.

In UED terms, emotion appears *after* delta interaction—after a system encounters resistance, overload, misalignment, relief, or resolution. This sequencing matters. When emotion is mistaken as a cause, people chase the wrong layer. They try to think differently, suppress feelings, or generate motivation, while the underlying condition remains unchanged.

Emotion is downstream.

It is the body-field's way of saying: *this is what the current configuration is doing to us.*

This is why emotion often arrives suddenly and without explanation. The explanation is not missing—it is simply not cognitive. The body detects consequence faster than language can assemble narrative. By the time the mind asks "why do I feel this way?", the system has already processed interaction at a deeper level.

Seen clearly, emotion is not mysterious at all. It follows the same logic as physical feedback:

- Pain follows tissue strain.
- Heat follows friction.
- Fatigue follows sustained effort.
- Emotion follows consequence accumulation.

Nothing more. Nothing less.

This also explains why emotional responses can feel disproportionate. The body does not report *events*; it reports load. A small external trigger can release a large emotional response if unresolved delta has accumulated quietly over time. Conversely, a dramatic event may produce little emotional reaction if the system is already aligned or prepared.

Emotion measures systemic impact, not narrative importance.

Another critical correction emerges here: emotions are not personal in the way we think. They feel personal because they occur inside the body, but structurally they are impersonal responses to conditions. Two people can experience identical emotions for entirely different reasons, and one person can experience different emotions in the same situation at different times—depending on capacity, history, and current alignment.

This is why emotional labeling often fails. Naming an emotion does not explain it. Saying "this is anger" or "this is sadness" is like labeling a warning light without opening the hood. The label may help communication, but it does not reveal cause.

UED replaces emotional storytelling with emotional diagnostics.

The key question shifts from:

- *Why do I feel this?* to
- *What condition is being reported right now?*

When this shift happens, emotions stop feeling random or hostile. Even difficult emotions regain dignity. They are no longer signs of weakness or flaws in character. They are indicators that something is happening in the system that requires attention—either action, rest, boundary change, or realignment.

Importantly, this does not mean every emotion requires action. Some emotions simply require acknowledgment. Once the system knows it has been heard, the signal fades naturally. Others persist because the condition persists. Ignoring them only increases volume.

Emotion, in this framework, is honest but unsophisticated. It does not interpret. It does not plan. It does not optimize. It reports.

And because it reports from the body-field, it reports in sensations—tightness, heaviness, agitation, warmth, collapse, expansion. The story comes later. Often, the story is wrong.

Understanding emotion as consequence reporting is the first step in restoring emotional health without control, suppression, or indulgence.

Once emotion is seen this way, the next question becomes unavoidable:

How does the body measure consequence so quickly and so reliably?

That requires looking at the body itself—not as a container for emotion, but as a measurement instrument.

That is where the next section begins.

24.2 — The Body as a Measurement Instrument

Emotion is not generated in the mind and then felt in the body. It is measured in the body and then interpreted by the mind.

This reversal corrects a deep misunderstanding in how emotions are explained and treated. The body is not a passive receiver of psychological states; it is an active sensing system, continuously measuring alignment, strain, safety, capacity, and overload across multiple layers at once.

In UED, the body is understood as a high-bandwidth measurement instrument operating below language and faster than cognition.

Neural signals, hormonal shifts, muscle tone, breathing patterns, posture, and visceral sensation together form a distributed sensing network. This network does not care about stories, identity, or social meaning. It cares about consequence—what current conditions are doing to viability and stability.

This is why emotional signals often feel physical before they feel mental. Tightness in the chest, heaviness in the limbs, agitation in the gut, fog in the head—these are not metaphors. They are measurements. The body is registering gradients, loads, and mismatches in real time.

Because the body measures before the mind explains, emotion often arrives without a clear narrative. The mind then scrambles to attach a story that justifies the sensation. This story may be useful—or it may be completely wrong. The accuracy of the story does not change the validity of the sensation.

Emotion is true as a signal, not as an explanation.

Another important correction follows from this. Emotional sensitivity is not fragility. It is often measurement precision. Systems that sense finely report early. Systems that are numbed or overloaded report late or not at all. When people say someone is "too sensitive," they are often misreading a system that detects misalignment earlier than others.

The cost of this sensitivity is only high when signals are misinterpreted or ignored. When read correctly, early detection prevents accumulation and collapse.

This also explains why repeated emotional patterns form baselines. The body adapts to sustained conditions. Chronic stress, unresolved conflict, prolonged insecurity, or ongoing misalignment recalibrate what the system considers "normal." Emotional tone flattens or sharpens accordingly. What once registered as alarming becomes background noise, and what once felt calm may feel unfamiliar.

This adaptation is not a failure. It is calibration.

However, calibration can drift away from healthy ranges when conditions remain suboptimal for too long. In such cases, emotional feedback becomes distorted—not because the body is malfunctioning, but because it has adapted to survive within constraint.

This is why emotional healing cannot rely on thought alone. Thinking addresses interpretation. The body requires changed conditions—rest, safety, resolution, or realignment—to recalibrate.

UED emphasizes this because it dissolves blame. Emotional states are not moral verdicts or psychological defects. They are readings taken by an instrument under specific conditions. Change the conditions, and the readings change.

Understanding the body as a measurement instrument also explains why emotion cannot be permanently bypassed. Even highly intellectual or spiritually inclined individuals eventually encounter emotional feedback. The body continues to measure whether the system is aligned, regardless of belief.

Emotion, therefore, is not something to transcend. It is something to trust as data, while refusing to confuse it with instruction.

Once this is clear, a critical question arises.

If emotion reports consequence accurately, why do some emotional states become chronic, overwhelming, or disabling—especially anxiety and depression?

The answer lies not in emotion itself, but in what kind of delta the body is measuring.

That distinction begins with anxiety.

24.3 — Anxiety: Future-Delta Overload

Anxiety is often described as fear without an object, worry without cause, or imagination running wild. These descriptions capture the experience but miss the structure. In UED, anxiety has a clean and precise definition:

Anxiety is the body measuring anticipated delta that exceeds perceived capacity.

It is not fear of what *will* happen. It is overload happening now, caused by projection.

The nervous system does not distinguish sharply between present and vividly anticipated conditions. When future scenarios are simulated with sufficient intensity, the body responds as if those deltas are already active. Muscle tone changes, breathing shortens, hormones mobilize. The system prepares to manage imbalance before imbalance actually arrives.

This is adaptive—up to a point.

Prediction is one of the nervous system's most powerful tools. It allows preparation, avoidance, planning, and learning. But when prediction multiplies faster than resolution capacity, anxiety emerges. The system is asked to hold too many unresolved possibilities at once.

This is why anxiety feels like urgency without direction. There is mobilization, but no clear pathway for discharge. The system is on alert, but cannot act, because the threat is not yet real, specific, or singular.

In UED terms, anxiety is future-delta congestion.

Several features of anxiety follow directly from this framing:

- Restlessness arises because energy is mobilized without a target.
- Racing thoughts occur because the mind searches for resolution pathways that do not yet exist.
- Somatic tension appears because the body is preparing for action that never arrives.
- Reassurance fails because reassurance addresses narrative, not delta load.

This also explains why anxiety is highly sensitive to context. A person may feel calm in one environment and anxious in another without any conscious reason. The difference is not psychological—it is predictive. Certain contexts activate learned future-delta patterns based on history, memory, or symbolic cues.

Anxiety is therefore not irrational. It is mis-timed.

The system is acting correctly in the wrong temporal frame.

Importantly, anxiety is not solved by eliminating uncertainty. Uncertainty is unavoidable. It is solved by reducing the number of simultaneous unresolved futures the system is trying to hold.

Awareness plays a critical role here. When awareness is present, prediction narrows. Instead of running multiple catastrophic or hypothetical scenarios in parallel, attention returns to what is actually required *now*. This does not deny the future. It defers it to when it becomes actionable.

This is why grounding techniques work—not because they distract, but because they collapse projection into present-moment resolution. The system is reminded that current delta is manageable.

Anxiety also explains why modern life amplifies distress. Constant information flow, symbolic threats, abstract risk, and social comparison flood the system with future-deltas that cannot be acted upon. The body is asked to prepare endlessly without release.

Seen this way, anxiety is not weakness. It is a system doing too much prediction with too little closure.

The solution is not to think positively or suppress fear. It is to restore correct timing—allow prediction only where action or preparation is structurally possible.

Once anxiety is understood as overload, another emotional state comes into focus—one that looks opposite on the surface but shares a structural relationship with delta.

That state is depression.

24.4 — Depression: Delta Collapse and Low-Gradient States

Depression is often misunderstood as sadness, lack of motivation, or negative thinking. These descriptions focus on surface symptoms and miss the underlying structure. In UED, depression is not defined by how it feels emotionally, but by what has happened to delta.

Depression is a low-gradient state. It is what the system reports when usable difference has collapsed.

Where anxiety is overload—too many future deltas—depression is the opposite condition. The system no longer detects sufficient gradient to justify movement. Energy is present, but direction is missing. Effort feels pointless not because the person is weak, but because the system cannot see a path where effort meaningfully changes consequence.

This is why depression feels heavy, flat, and slow. There is no urgency. There is no pull. The system is not resisting life—it has stopped seeing *difference worth resolving.*

In UED terms, depression is not excess pain. It is insufficient signal.

Several features follow directly from this framing:

- Loss of motivation occurs because motivation requires gradient. Without perceived difference between action and inaction, movement cannot initiate.
- Emotional numbness arises because feedback amplitude drops when delta collapses. The body reports little because little is registering as consequential.
- Fatigue without effort appears because maintaining existence without gradient still consumes energy, but yields no directional reward.
- Hopelessness is not pessimism—it is the perception that future delta will not meaningfully differ from the present.

This explains why encouragement, advice, or "thinking positively" often fails. These approaches attempt to inject narrative without restoring gradient. The system does not need better stories; it needs conditions that reintroduce difference.

Depression also clarifies why pressure worsens collapse. Telling someone to "try harder" increases demand without increasing gradient. The system experiences this as further evidence that effort does not change outcome. Delta collapses even more.

UED makes an important distinction here between grief, fatigue, and depression. Grief still has gradient—loss is sharply felt. Fatigue still has direction—rest resolves it. Depression is different. It is what remains when repeated effort, disappointment, or misalignment has flattened expectation itself.

This is why depression often follows long periods of unresolved struggle. Delta has been held open for too long without resolution. Eventually, the system conserves energy by reducing signal. Collapse is not failure—it is protective shutdown.

Seen this way, depression is not an enemy to be fought. It is a message that the system can no longer afford to mobilize under current conditions.

Recovery, therefore, is not about forcing activity or restoring emotion artificially. It is about reintroducing small, real gradients—conditions where action produces tangible, immediate difference. This may be physical, relational, environmental, or structural. The scale matters less than the reliability.

Once even minor gradients are trusted again, signal strength begins to return. Emotion reappears. Energy mobilizes. Direction slowly reforms.

Depression, in UED, is not a permanent state and not an identity. It is a phase of low-delta economy.

Understanding this removes shame. Nothing is broken. The system has adapted to survive prolonged misalignment by minimizing expenditure.

With anxiety and depression now seen as opposite distortions of delta—one from overload, the other from collapse—the reader is ready for a critical refinement.

Not all emotions that feel intense are important. Not all emotions that feel dull are insignificant.

The next section addresses how emotional intensity often misleads, and how informational value must be read differently.

24.5 — Emotional Intensity vs Informational Value

One of the most persistent misunderstandings about emotion is the assumption that intensity equals importance. Loud emotions are treated as urgent. Subtle emotions are dismissed as negligible. This inversion quietly drives much of human misalignment.

In UED, emotional intensity is not a measure of truth, depth, or consequence. It is a measure of signal amplification.

Amplification occurs when feedback is not being read or resolved. The system increases volume to gain attention. This is not wisdom; it is compensation.

This explains why anxiety can feel overwhelming even when nothing catastrophic is happening, and why a quiet sense of unease can precede major life failures without being noticed. The system does not shout when it is confident it will be heard. It shouts when it has been ignored.

Emotional intensity therefore measures urgency felt, not impact pending.

This distinction is critical. Many low-grade signals—fatigue, mild resentment, ethical discomfort, bodily strain—carry enormous long-term consequence if neglected. They speak softly because the system expects correction. When correction does not come, they either accumulate silently or erupt later as crisis.

Conversely, many intense emotional states are short-lived and low-consequence. They flare because prediction has overshot or because symbolic meaning has inflated delta artificially. They demand attention, but not necessarily action.

Misreading this dynamic creates two common errors:

First, people overreact to noise. They mobilize energy, justify behavior, or make life-altering decisions based on emotionally intense but structurally trivial signals. This leads to regret, instability, and self-distrust.

Second, people underreact to signal. They tolerate discomfort that quietly erodes health, relationships, and integrity because it does not scream. By the time it does, the cost is already high.

UED reframes emotional literacy as signal discrimination, not emotional control.

The question is no longer:

- *How strong is this feeling?*

But:

- *What kind of delta is this reporting?*
- *Is this immediate overload or slow misalignment?*
- *Will this resolve on its own, or will it accumulate cost if ignored?*

This reframing also explains why social environments distort emotional reading. Social validation amplifies some emotions and suppresses others. Anger may be rewarded. Sadness may be minimized. Anxiety may be normalized. These cultural filters alter emotional volume without changing informational value.

This distortion is especially dangerous at the collective level. Societies often mobilize around intense but shallow emotional signals while ignoring quiet structural decay. The same misreading that harms individuals destabilizes systems.

UED insists on a correction here: volume must be decoupled from authority.

A calm signal may require immediate attention. A loud signal may require none.

Once this is understood, emotional regulation becomes unnecessary. The system does not need to suppress emotion. It needs to interpret it accurately.

When interpretation improves, volume naturally adjusts. Signals that are read stop shouting. Signals that matter but were ignored become clearer—not louder.

This prepares the ground for the next section, where the question becomes practical and unavoidable:

If emotion is feedback, and intensity is unreliable, how should one respond?

Not by control. Not by indulgence.

But by learning when emotion resolves through action, and when it resolves through acknowledgment.

That distinction determines whether emotion remains a guide—or becomes a burden.

24.6 — Emotional Regulation Reframed: Reading Before Responding

Once emotion is understood as feedback and intensity is separated from importance, the entire idea of "emotional regulation" needs to be rewritten. Most regulation strategies are built on a false assumption—that emotion is something to be managed, reduced, or overridden. In UED, regulation is not about control. It is about correct response timing.

Emotion resolves when its report is understood and met at the right layer.

This immediately explains why many popular techniques partially work and then fail. Distraction works briefly because it interrupts attention, not because it resolves the underlying condition. Suppression works temporarily because it dampens signal, not because it corrects alignment. Expression helps sometimes because it releases pressure, but can also entrench patterns if it substitutes narration for resolution.

UED reframes regulation as a three-step structural process:

1. Read the signal accurately
2. Identify the layer it belongs to
3. Respond at that layer—or do nothing

Most emotional suffering comes from skipping the first two steps and jumping straight to reaction.

For example, anxiety does not always require action. Often it requires *narrowing prediction*—bringing attention back to what is actually actionable now. Depression does not always require motivation. Often it requires restoring gradient through rest, environment change, or removing impossible demands. Anger does not always require expression. Often it requires boundary correction or withdrawal from misaligned interaction.

In each case, the emotion resolves not because it was controlled, but because its report was honored structurally.

This also explains why some emotions dissolve simply by being acknowledged. When a signal is finally recognized as valid and accurate, amplification is no longer necessary. The system has achieved its objective—communication. Silence follows naturally.

Other emotions persist because the condition persists. No amount of mindfulness or reframing will eliminate hunger, exhaustion, chronic disrespect, or misalignment. Attempting to "regulate" these emotions without changing conditions is not strength—it is misinterpretation.

UED draws a hard line here:

If an emotion keeps returning, it is not asking to be managed. It is asking for condition change.

Regulation becomes harmful when it delays necessary action. Many people pride themselves on emotional control while slowly eroding their health, boundaries, or integrity. Calmness in such cases is not alignment—it is muted distress.

At the same time, UED also protects against impulsivity. Not every emotion requires immediate response. Some signals are transient fluctuations that resolve naturally when attention shifts or when prediction recalibrates. Acting on every emotional spike creates instability.

The skill, therefore, is not control. It is discernment.

This discernment emerges automatically when awareness is present. Awareness does not suppress emotion; it contextualizes it. It sees whether the signal is early-warning, overload, depletion, or residue. It knows when to wait and when to move.

When regulation is reframed this way, emotional life becomes simpler, not harder. There is less inner debate, less self-judgment, less fear of feeling "too much" or "too little." Emotion becomes part of the system's navigation toolkit, not a problem to solve.

This prepares the reader for the final consolidation of the chapter.

If emotional health is not about positivity, control, or resilience narratives, then what is it actually about?

The answer is neither emotional strength nor emotional absence.

It is signal fidelity.

That is the final correction before the chapter closes.

24.7 — Emotional Health as Signal Fidelity

Emotional health has been misdefined for so long that it now carries expectations it can never fulfill. It is often equated with happiness, calmness, positivity, or emotional stability. These ideals are not only unrealistic—they are structurally incorrect.

In UED, emotional health has a precise and modest definition:

Emotional health is the system's ability to generate, transmit, and resolve emotional signals without distortion.

Nothing more is required.

A healthy system feels emotions clearly, specifically, and temporarily. Signals arise when conditions demand them, and they fade when conditions change. There is no need for suppression, dramatization, or prolonged narration. Emotion does its job and moves on.

Distortion enters when any part of this cycle is disrupted.

Signals become distorted when:

- They are ignored repeatedly
- They are amplified for identity or validation
- They are suppressed for control or performance
- They are misread as meaning rather than measurement

Over time, distortion produces chronic emotional states—persistent anxiety, flatness, irritability, or numbness. These are not "deep emotions." They are feedback loops that no longer correspond cleanly to present conditions.

This is why emotional numbness is not neutrality. It is loss of resolution capacity. The system stops reporting because reporting has failed too many times. Silence here is not peace; it is shutdown.

Similarly, emotional volatility is not aliveness. It is signal instability. The system cannot stabilize gradients long enough to resolve them, so emotion oscillates wildly.

UED reframes emotional maturity accordingly. It is not about being unshaken. It is about being accurately shaken—only when something genuinely requires it, and only for as long as it does.

This reframing removes both pride and shame. Feeling deeply is not virtue. Feeling little is not strength. What matters is whether emotion corresponds to real delta and whether it returns to baseline once addressed.

Emotional health also implies recoverability. Healthy systems rebound. They do not store unnecessary residue. They do not turn every experience into identity. Memory exists, but it does not dominate signal generation.

This has a subtle but powerful consequence for self-understanding. You are not your emotions—not because emotions are unimportant, but because they are reports, not definitions. Identity confusion arises when signal is mistaken for selfhood.

When emotional health is restored as signal fidelity, self-trust returns. The system knows that when something feels wrong, it probably is—and when it stops feeling wrong, it can move on without doubt or guilt.

This closes the emotional loop cleanly.

Emotion has now been fully integrated into UED:

- It is not a flaw
- It is not a truth oracle
- It is not a control problem

It is a field-reporting layer operating within the same delta–equilibrium–consequence architecture as everything else.

With this integration complete, the book is ready to take its final step in this part.

Emotion is individual. But humans do not live alone.

What happens when millions of emotional systems interact inside shared structures?

That question leads directly to morality, dharma, and ethics—not as ideals to obey, but as collective delta stabilizers.

That is where the next chapter begins.

From Inner Signals to Shared Stability

With emotion correctly placed as field feedback, the inner world finally becomes readable without being dramatized. Anxiety, depression, anger, and calm are no longer treated as personal verdicts or psychological anomalies. They are understood as signals arising from interaction between capacity, prediction, and consequence.

At this point, something important becomes clear.

Emotions do not arise in isolation. They are generated inside shared environments.

Every human emotional system operates within layers of family, culture, economy, technology, and history. The body may report alignment or overload individually, but the conditions producing those reports are often collective. A system can be emotionally healthy and still suffer if the surrounding structure is misaligned.

This recognition creates a necessary pivot.

If emotion is feedback, then social systems must also manage feedback. Without shared stabilizers, individual clarity is insufficient. Even well-aligned minds will be strained by environments that reward noise, amplify false urgency, or punish honesty.

This is why ethics, morality, and dharma emerge naturally in any society—not as moral ideals, but as engineering solutions to collective instability. They exist to reduce destructive delta amplification between people, to smooth interaction, and to prevent emotional contagion from becoming chaos.

Seen this way, morality is not about virtue. It is about system survivability. Dharma is not obedience to tradition. It is alignment with consequence in shared fields. Ethics are not rules imposed from above. They are constraints evolved to keep human deltas from tearing each other apart.

The next chapter does not ask what is right or wrong. It asks a more fundamental question: What structures are required to stabilize emotional and behavioral feedback when individual systems interact at scale?

Once emotion is understood, morality stops being mystical, preachy, or ideological.

It becomes necessary. And that necessity now comes into focus.

Once emotion is understood as feedback, the next question is how societies stabilize millions of such signals without collapsing.

That question leads directly to morality, dharma, and ethics—not as ideals, but as structural necessities.

Chapter 25 — Morality, Dharma, Ethics: Social Delta Stabilizers

Why Morality Exists at All

Morality is often treated as an abstract ideal, a cultural artifact, or a religious imposition. In debates, it is defended, attacked, moralized, or dismissed—yet rarely examined for what it actually does. UED begins this chapter by stripping morality of mystique without stripping it of necessity.

Morality exists because humans generate delta faster than they can resolve it individually.

The moment intelligence emerged, behavior became amplified. Choices began to ripple beyond immediate survival. Actions started carrying delayed, distributed, and indirect consequences. Without constraints, this amplification would make coexistence impossible. Groups would fracture under unmanaged gradients of power, desire, fear, and advantage.

Morality did not arise to make humans virtuous. It arose to make human interaction sustainable.

This is why moral systems appear everywhere—across cultures, epochs, and belief systems. The details differ, but the function remains consistent: limit destructive delta exchange, preserve trust, and prevent local advantage from becoming systemic collapse.

UED reframes morality as social physics.

Just as physical systems require constraints to remain stable, social systems require behavioral constraints to prevent runaway amplification. Remove friction from a physical system and it explodes. Remove ethical constraints from a social system and it implodes—sometimes slowly, sometimes violently.

This reframing also explains why morality becomes most visible when it fails. In stable periods, ethical norms operate quietly in the background. When power concentrates, inequality widens, or speed accelerates, moral breakdown surfaces. Outrage follows—not because people suddenly care more about virtue, but because the system is becoming unstable.

Another crucial correction emerges here. Morality is not about judging character. It is about managing interaction. Judgement is a byproduct, not a function. When morality is reduced to judging others, it loses its stabilizing role and becomes another source of conflict.

This is why highly moralistic societies are often deeply unethical in practice. They focus on symbolic compliance while ignoring structural damage. The signal has been replaced by performance.

UED also dissolves the false opposition between morality and intelligence. Intelligence without morality is not freedom—it is amplification without containment. It accelerates consequence beyond recoverability. Morality, when functioning properly, does not suppress intelligence. It channels it.

This opening sets a critical boundary.

Morality is not optional. It is not transcendent. It is not proof of goodness.

It is a constraint layer required wherever intelligent agents interact over time.

With this foundation, the chapter can now examine how ethics, morality, and dharma function—not as ideals to worship, but as mechanisms that keep shared fields from tearing themselves apart.

The next section begins at the most basic level of this stabilization: bonding.

25.1 — Ethics as Bonding Mechanics

At the most basic level, ethics are not about right and wrong; they are about whether bonds hold. Every human system—families, teams, markets, nations—depends on bonds that can carry load without breaking. Ethics emerge wherever those bonds are stressed by competing deltas.

In UED, ethics function as bonding mechanics.

A bond forms when two or more agents coordinate action over time. The moment coordination begins, delta appears: uneven power, unequal information, asymmetrical risk, mismatched reward. Without constraints, these gradients amplify. One side extracts more, trusts less, or defects earlier. The bond weakens, then fails.

Ethical norms exist to prevent this failure.

Trust, honesty, fairness, promise-keeping, restraint—these are not moral ornaments. They are load-bearing rules that allow interaction to persist under strain. They reduce friction in exchange, lower emotional noise, and prevent every interaction from turning into a defensive negotiation.

This is why ethical behavior feels costly in the short term but pays off structurally. It limits immediate extraction in favor of long-term stability. Unethical behavior often feels efficient initially because it exploits gradient, but it quietly destroys the very bonds that made extraction possible.

UED reframes this trade-off cleanly:

- Ethics reduce short-term delta advantage
- Ethics increase long-term consequence resilience

This is not altruism. It is systems engineering.

When ethical norms are present, transaction costs drop. People do not need constant monitoring, enforcement, or threat. Emotional energy is conserved. Coordination becomes lighter. Creativity increases. Bonds can carry more load because less energy is spent guarding against betrayal.

When ethical norms erode, the opposite happens. Suspicion rises. Rules multiply. Surveillance replaces trust. Even honest actors are forced into defensive postures. The system becomes heavy, brittle, and expensive to operate.

Importantly, ethics are not universal rules imposed equally everywhere. They are context-sensitive bonding constraints. What stabilizes a family differs from what stabilizes a market. What stabilizes a small tribe differs from what stabilizes a global supply chain. The function remains the same even when the form changes.

This also explains why ethical breakdown often precedes visible collapse. Bonds fail quietly before systems do. Trust erodes before institutions crack. By the time collapse is obvious, ethical mechanics have already been bypassed for too long.

UED therefore treats ethics not as ideals to debate, but as signals of bond health. When ethical norms are mocked, selectively applied, or publicly performed while privately violated, it is not a moral crisis—it is a structural warning.

Bonds are carrying more delta than they can sustain.

This understanding prepares the ground for the next step. Ethics stabilize bonds, but bonds exist inside larger interaction fields. When stabilization fails, societies introduce stronger constraints—norms, sanctions, and collective judgments.

That escalation is what we usually call morality.

And morality, as the next section shows, is not about virtue either—it is about preventing individual delta from becoming collective damage.

25.2 — Morality as Interaction Stabilization

When ethics quietly stabilize bonds, morality appears when those bonds are under threat. Morality is not the first layer of social order; it is the reinforcement layer. It activates when ethical mechanics alone are insufficient to contain delta exchange between individuals and groups.

In UED, morality functions as interaction stabilization under stress.

As societies grow, interactions multiply and anonymity increases. Not every bond can be personal. Not every action can be negotiated. Delta gradients—power, wealth, influence, access—begin to widen. At this scale, simple trust-based ethics cannot carry the load. Systems require stronger constraints to prevent localized advantage from cascading into systemic damage.

Morality emerges to meet this need.

Moral norms define behavioral boundaries—lines that should not be crossed because crossing them destabilizes the field. These boundaries are not about purity or intention; they are about containment. They exist to limit amplification, to slow down harm propagation, and to protect the many from the few who can exploit gradients most effectively.

This explains why morality is often enforced collectively. Sanctions, shame, praise, exclusion, and punishment are tools societies use to dampen behaviors that threaten stability. These tools are crude, imperfect, and sometimes unjust—but their function is not mystical. It is structural.

Morality acts where ethics fail.

A crucial distinction must be made here between corrective consequence and moral outrage. Corrective consequence aims to restore stability. Moral outrage often aims to assert identity, power, or belonging. When outrage replaces correction, morality loses its stabilizing role and becomes another source of delta amplification.

UED draws this line sharply.

- Correction reduces future harm
- Outrage increases emotional load without resolution

This is why highly moralistic environments can be deeply unstable. They focus on symbolic enforcement rather than structural correction. Behavior is judged, but conditions remain unchanged. Delta continues to accumulate beneath performative compliance.

Another important insight follows: morality is always reactive. It does not prevent all harm. It responds to harm that has already exceeded ethical containment. This reactivity is why moral systems often appear harsh, inconsistent, or emotionally charged. They are operating under pressure.

Morality also tends to lag reality. New forms of power—technological, economic, informational—create new pathways for damage faster than moral frameworks can adapt. The resulting mismatch produces confusion and conflict. Old moral rules strain under new gradients.

Seen this way, moral failure is not primarily personal. It is often systemic. Individuals operate inside structures that reward destabilizing behavior and punish restraint. When morality collapses under such conditions, blaming individuals misses the deeper problem.

UED reframes morality accordingly: not as a measure of goodness, but as a stress response of social systems trying to maintain coherence.

This reframing does not excuse harm. It clarifies where responsibility actually lies—both individually and structurally.

With morality now understood as stabilization under pressure, the chapter can move to a concept that has been deeply misunderstood and ritualized: dharma.

Dharma is often treated as duty, loyalty, or tradition. In UED, it has a much sharper meaning—one that aligns directly with consequence rather than conformity.

That correction comes next.

25.3 — Dharma Reframed: Alignment With Consequence

Dharma is one of the most overloaded words in human thought. It has been stretched across religion, culture, duty, loyalty, profession, identity, and morality until it has lost operational clarity. In many contexts, dharma is reduced to obedience—follow your role, your group, your tradition, your leader. UED strips this confusion away and restores dharma to its original structural function.

Dharma is alignment with consequence.

Nothing more. Nothing less.

In UED, dharma does not ask *who* you serve or *what* you believe. It asks a simpler, sharper question: *Do your actions create consequences that the system—human, social, ecological—can actually absorb?*

This immediately separates dharma from loyalty theater. Loyalty prioritizes belonging over consequence. Dharma prioritizes consequence over belonging. When loyalty conflicts with consequence, dharma breaks with loyalty—quietly, often at personal cost.

This is why dharma is difficult to practice and easy to corrupt.

Dharma is contextual, not absolute. What stabilizes one situation may destabilize another. A soldier's dharma differs from a teacher's, not because of hierarchy, but because the consequence landscape differs. Dharma is always role-relative, not identity-bound.

UED makes a critical distinction here:

- Role is functional participation in a system
- Identity is symbolic attachment to status or story

Dharma belongs to role, not identity.

When dharma is confused with identity, it becomes rigid. People defend harmful actions by invoking labels—nation, religion, profession, ideology. Consequence is outsourced to authority. Alignment collapses, but loyalty is preserved. This is how dharma degenerates into dogma.

True dharma operates in the opposite direction. It is quietly corrective. It adjusts behavior when consequence becomes unpayable—even if tradition approves and authority demands compliance. It does not announce itself. It does not seek recognition. It simply refuses to generate destructive ripple.

This is why dharma often appears invisible in stable systems and conspicuous only during crisis. When systems are aligned, dharma blends into normal functioning. When systems drift, dharma becomes uncomfortable, even disruptive. It exposes misalignment without drama.

UED also clarifies why dharma cannot be universalized into a single rulebook. Consequence landscapes vary across time, scale, and technology. What was dharmic in a low-speed, low-impact society may be catastrophic in a high-speed, high-impact one. Treating ancient prescriptions as eternal commands is not reverence—it is negligence.

Dharma evolves because consequence evolves.

This reframing also dissolves moral arrogance. Dharma is not about being right. It is about being responsible to outcome. One can act with good intention and still violate dharma if consequences accumulate harm. Conversely, one can act against social approval and still uphold dharma by preventing collapse.

In UED, dharma is not sacred behavior. It is consequence-aware action.

With dharma restored to alignment rather than allegiance, the chapter must now confront an uncomfortable truth: if ethics and dharma are structurally necessary, why do humans fail so consistently to uphold them?

The answer is not wickedness. It is interference.

That interference—internal and external—is what makes morality hard.

25.4 — Why Morality Is Hard

If ethics stabilize bonds and dharma aligns action with consequence, a natural question follows: why do humans fail at morality so persistently, even when they understand it? The answer is not ignorance and not inherent evil. In UED, the answer is simpler and more uncomfortable.

Morality is hard because the self interferes with alignment.

Every human operates as a delta engine. Desire seeks gain, fear seeks protection, ego seeks validation, and identity seeks continuity. None of these are flaws. They are survival mechanisms. But when these forces dominate decision-making, they distort consequence perception. The system begins to prioritize immediate relief or advantage over long-term stability.

This is the core tension.

Morality requires acting in ways that do not maximize short-term delta extraction, while the self is optimized precisely to detect and exploit short-term gradients. The conflict is structural, not psychological. Expecting humans to behave morally without acknowledging this conflict is unrealistic.

UED reframes moral failure accordingly.

Humans do not fail morally because they choose harm. They fail because short-term delta masks long-term consequence.

This masking is amplified under pressure. Scarcity, competition, threat, humiliation, and urgency all compress time horizons. When the system feels endangered, it narrows focus. Ethics become optional. Dharma becomes abstract. Survival dominates.

This explains why moral clarity collapses fastest during crises. War, economic stress, political polarization, and technological disruption all intensify gradients. Under these conditions, individuals and groups justify actions they would otherwise condemn. The behavior changes, not because values vanished, but because perceived capacity shrank.

Another source of difficulty is asymmetry. Consequence often arrives delayed, distributed, or externalized. Those who benefit from harmful actions may never directly experience the cost. This decoupling weakens feedback. Without feedback, alignment fails.

UED highlights this mismatch sharply:

- Immediate gain is local and visible
- Long-term harm is diffuse and invisible

Systems that reward visible gain while hiding diffuse cost train moral blindness. Over time, such systems normalize misalignment. Individuals adapt not because they are immoral, but because alignment appears irrational within the incentive structure.

Self-interference also operates at the identity level. People defend actions that protect their role, status, or group even when those actions generate harm. Identity creates blind spots. It reframes consequence as attack. Moral criticism is experienced as threat, triggering further defensiveness.

This is why moral conversations so often fail. They target behavior but ignore identity load. When identity feels endangered, alignment becomes impossible.

UED does not excuse this failure. It explains it so it can be addressed structurally rather than emotionally. Moral exhortation without gradient correction does not work. Shame without capacity change does not work. Ideals without incentive alignment do not work.

Morality becomes possible only when:

- Consequence is visible
- Time horizons are extended
- Capacity is sufficient
- Identity is not under siege

This is why moral behavior flourishes in stable environments and erodes under sustained stress. Stability is not decadence—it is moral infrastructure.

Understanding why morality is hard does not lower standards. It redirects effort. Instead of demanding virtue from overloaded systems, UED demands better alignment conditions.

This prepares the reader for the next escalation.

Even when individuals strive for alignment, morality does not operate in a vacuum. It is distorted by social gradients—power, wealth, status, and asymmetry.

Those gradients determine whose morality is enforced, whose is ignored, and whose is punished.

That distortion must be confronted next.

25.5 — Social Gradients and Moral Distortion

Morality does not operate on a flat surface. It operates on gradients—of power, wealth, status, visibility, and protection. These gradients distort moral application even when moral principles are widely agreed upon. This distortion is not accidental. It is structural.

In UED, moral inconsistency is not primarily hypocrisy. It is gradient asymmetry.

When gradients are steep, morality becomes selective.

Those at the top of a gradient can generate large consequences with minimal immediate cost. Those at the bottom experience amplified consequences for small deviations. The same action produces radically different outcomes depending on position in the gradient. Over time, this difference reshapes moral enforcement.

This is why morality often feels unfair rather than wrong.

Rules exist, but enforcement flows downhill. Moral outrage targets the weak while excusing the powerful. Systems publicly praise ethical ideals while privately rewarding unethical success. This is not because societies are blind. It is because enforcing morality upward is costly, while enforcing it downward is cheap.

UED makes this explicit: Morality follows enforcement feasibility, not moral clarity.

Social gradients also distort perception. When someone benefits from a system, they tend to see it as fair. When someone is harmed by it, they see injustice everywhere. Both perceptions arise from position, not virtue. Without recognizing gradient effects, moral debates devolve into mutual incomprehension.

Another distortion appears as performative morality.

When real correction is impossible—because power is concentrated or consequence is externalized—societies substitute performance for alignment. Moral language intensifies. Symbols multiply. Outrage cycles accelerate. None of this reduces harm. It only redistributes emotional load.

Performative morality is a signal of structural helplessness.

People express moral judgment loudly when they cannot change conditions quietly.

UED also explains why moral rules tend to be harsher for behaviors that threaten visible order and softer for behaviors that threaten invisible foundations. Theft by the poor is punished more severely than exploitation by the powerful. Public disorder attracts more sanction than systemic harm. This is not because societies value order more than justice, but because order violations are easier to detect and punish.

This gradient distortion creates moral cynicism. People begin to believe morality itself is fake, weaponized, or naïve. That cynicism further erodes ethical behavior, creating a feedback loop where misalignment accelerates.

Importantly, UED does not romanticize rebellion here. Ignoring morality because it is unevenly enforced does not restore alignment. It simply adds more delta into an already stressed system. The problem is not morality itself, but moral asymmetry.

Correcting this requires design, not preaching.

Moral stability improves when:

- Power gradients are constrained
- Consequences are made visible
- Incentives align with long-term outcomes
- Accountability scales with impact, not status

Without these conditions, moral language becomes noise. With them, moral behavior becomes practical rather than heroic.

This section completes a critical insight: morality fails not only because individuals struggle, but because systems skew consequence distribution.

That insight now leads to a necessary clarification.

If morality is structural and gradients distort it, how should responsibility be assigned without turning morality into sanctity, guilt, or hatred?

That correction belongs to the next section.

25.6 — Responsibility Without Sanctity

Once morality is understood as structural stabilization rather than virtue, responsibility must be redefined just as carefully. Most moral systems collapse at this point because they confuse accountability with sanctity. One produces correction. The other produces polarization.

UED draws a firm boundary.

Responsibility is not about purity. It is about participation in consequence chains.

To be responsible does not mean to be morally superior, intention-pure, or emotionally correct. It means that one's actions—direct or indirect—have entered a system and altered its trajectory. Once participation occurs, responsibility exists, regardless of motive.

This is why intention alone cannot absolve impact. A well-intended action can destabilize a system just as effectively as a malicious one. UED does not dismiss intention, but it places it upstream of consequence, not above it. Consequence is what systems must absorb. Responsibility exists wherever absorption is required.

This framing immediately removes moral sanctity.

Sanctity claims exemption. Responsibility accepts accounting.

When individuals or institutions declare themselves morally pure, they place themselves outside correction. They transform responsibility into identity and accountability into attack. This is how moral systems become brittle. They protect symbols rather than stabilize outcomes.

UED insists on a different posture: responsibility without holiness.

This posture allows correction without hatred. It enables accountability without humiliation. It focuses on repair rather than punishment theater. The goal is not to prove wrongness, but to restore alignment.

This distinction matters deeply in complex systems. Modern actions often involve distributed agency—supply chains, algorithms, policies, institutions. No single actor "causes" the outcome, yet outcomes occur. If responsibility is tied to intention or sanctity, it disappears. If it is tied to participation, it remains traceable.

UED therefore reframes accountability as proportional participation.

- Greater impact → greater responsibility
- Greater leverage → greater obligation
- Greater insulation from consequence → greater corrective burden

This is not moral heroism. It is structural fairness.

Importantly, responsibility in UED is not retroactive condemnation. It is forward-looking correction. The question is not *who is bad*, but *what must change to prevent recurrence*. Hatred adds delta. Correction reduces it.

This is why systems obsessed with blame remain unstable. They satisfy emotional demand but fail structurally. Systems focused on correction appear less dramatic, but they endure.

Responsibility without sanctity also protects individuals. It allows people to acknowledge harm without identity collapse. It allows learning without shame paralysis. It preserves dignity while insisting on change.

This section completes a crucial internal alignment: morality can now function without moralism.

What remains is to confront the modern reality.

The world today is high-speed, high-impact, and tightly coupled. Delta propagates faster than ever before. Consequences travel globally. Old moral frameworks strain under this load.

The final section of this chapter addresses that reality directly—ethics not as tradition, but as design constraint in a high-delta world.

25.7 — Ethics in a High-Delta World

The modern world has crossed a threshold that older moral frameworks were never designed to handle. Speed has increased. Scale has exploded. Distance has collapsed. Actions that once affected dozens now affect millions. Decisions made in abstraction propagate through supply chains, information networks, ecological systems, and psychological fields almost instantly.

In UED terms, we are living in a high-delta world.

Delta is no longer local. It is amplified by technology, leverage, and connectivity. A small decision can trigger massive redistribution of consequence. This amplification does not wait for intention, reflection, or ethical maturity. It operates automatically.

This is why ethics can no longer remain symbolic or tradition-bound.

In a high-delta environment, ethics must function as design constraints, not cultural reminders.

Older moral systems assumed:

- Slow consequence
- Local impact
- Clear actors
- Direct feedback

None of these assumptions hold anymore.

Today, consequences are delayed, distributed, and often invisible to those who initiate them. Feedback loops are broken. Harm is abstracted. Responsibility is diluted. Under these conditions, appealing to virtue or conscience is insufficient. The system itself must be structured to prevent runaway damage.

UED makes this explicit: Ethics must move upstream—into architecture.

This means ethics can no longer be treated as personal character alone. They must be embedded in:

- Incentive structures
- Decision pipelines
- Technological affordances
- Institutional checks
- Information flow constraints

When ethics are left to individual goodwill in high-delta systems, failure is guaranteed—not because people are bad, but because amplification overwhelms restraint.

This reframing also explains why moral outrage dominates modern discourse while moral correction stagnates. People sense instability, but lack leverage to alter system design. Emotion substitutes for action. Ethics becomes performative because architecture remains untouched.

UED rejects this substitution.

In a high-delta world:

- Good intentions without constraint cause harm
- Intelligence without ethics accelerates collapse
- Freedom without consequence awareness destabilizes systems

Ethics, therefore, must scale with impact. The greater the reach of an action, the tighter the ethical constraints must be—not as punishment, but as protection. This applies equally to governments, corporations, technologies, and individuals with disproportionate influence.

This section also clarifies why ethical debates feel increasingly polarized. High delta increases stakes. When stakes rise, disagreement intensifies. Without shared consequence accounting, groups retreat into ideology. Ethics becomes identity warfare instead of stabilization.

UED offers a different path.

Ethics are not about agreement. They are about survivability under amplification.

The ethical question of the future is not "What should we believe?" It is "What must be constrained so consequence remains absorbable?"

This completes the arc of the chapter.

Morality is no longer mystical. Dharma is no longer traditional. Ethics are no longer optional.

They are structural requirements in a world where delta outruns intuition.

With this realization, Part VII reaches its deliverable:

Psychology has become physics-like— not cold, not mechanical, but precise, readable, and accountable.

The book is now ready to move beyond human dynamics into its final responsibility: how these principles must guide system design, governance, and the future of intelligence itself.

That transition begins next.

From Human Conduct to System Architecture

At this point, morality has been fully removed from the realm of sermons and returned to the realm of structure. Ethics are no longer personal badges. Dharma is no longer inherited obligation. Right and wrong are no longer emotional judgments. They are mechanisms—quiet, necessary, and unforgiving when ignored.

This shift matters because individual clarity is no longer enough.

A single well-aligned mind cannot stabilize a misaligned system. Emotional intelligence, ethical intention, and personal awareness collapse under architectures that reward extraction, speed, and insulation from consequence. In such environments, even good actors become carriers of harm.

UED therefore makes a decisive move here.

If morality is stabilization, then design becomes ethics.

The question is no longer: *How should people behave?*

It becomes: *How must systems be structured so harmful behavior is difficult, unprofitable, or impossible?*

This is the boundary where responsibility shifts from psychology to architecture. From intention to constraint. From education to design.

It is also where the future will be decided.

As intelligence scales—human, artificial, collective—delta amplification will outpace moral learning. No amount of awareness training will compensate for systems that magnify consequence faster than feedback can arrive.

The remaining parts of this book address that reality directly.

They ask:

- How UED informs system design
- How governance must change in a consequence-coupled world
- How intelligence itself must be constrained to remain aligned

This is not speculation. It is inevitability.

Once delta is understood, neutrality is impossible. Once consequence is visible, design becomes ethical duty.

With human dynamics now fully integrated into UED, the book turns outward—to power, technology, governance, and the architectures that will either stabilize the future or accelerate collapse.

That turn begins now.

Once morality is understood as stabilization, the final question is no longer how humans should behave—but how systems must be designed to survive consequence.

This sets the stage for the final arc of UED, where individual insight becomes architectural responsibility.

Deliverable of Part VII

Psychology Becomes Physics-Like Without Becoming Cold

By the end of this part, something fundamental has shifted in how the inner world is understood. Psychology is no longer treated as a soft science, a narrative exercise, or a moral battleground. At the same time, it is not reduced to mechanism, chemistry, or computation. It becomes structural—readable, predictable, and accountable—without losing its human texture.

In UED, psychology becomes physics-like in one crucial sense: it obeys constraints, gradients, and consequences.

Thoughts, emotions, desires, fears, awareness, and ethics are no longer mysterious forces or personal flaws. They are behaviors of a system operating under delta, equilibrium pressure, and consequence—exactly like physical systems, biological systems, and social systems.

But it does not become cold, because UED never strips away experience.

Pain is still pain. Anxiety is still overwhelming. Depression is still heavy. Ethical conflict is still difficult.

What changes is not the *felt reality*, but the interpretive layer placed on top of it.

Suffering is no longer framed as weakness or failure. Emotion is no longer framed as irrationality. Morality is no longer framed as virtue signaling.

They are framed as system feedback.

This reframing produces a quiet but profound relief. People stop fighting their own experience. They stop blaming themselves for structural effects. They stop seeking salvation through discipline, positivity, or suppression. Instead, they learn to read signals, correct misalignment, and design conditions that reduce unnecessary load.

This is the opposite of detachment.

It is deep engagement without confusion.

Psychology becomes precise enough to explain why overthinking loops form, why calm feels effortless when it arrives, why awareness clarifies without force, why emotion overwhelms or collapses, and why morality strains under pressure. At the same time, it preserves compassion—not as sentiment, but as structural understanding.

UED achieves something rare here:

- It removes mysticism without removing meaning
- It removes moralism without removing responsibility
- It removes pathology without denying suffering

The human is not reduced to a machine. The machine is not elevated to a human.

Both are understood as systems operating under the same universal dynamics, at different scales and with different costs.

With this deliverable complete, psychology no longer floats separately from physics, biology, or society. It takes its place as a natural chapter in Universal Energy Dynamics.

What remains is the final escalation.

If physics, life, mind, and morality all obey the same underlying rules, then the most dangerous mistakes ahead will not be psychological.

They will be architectural.

That is where the book now turns.

Part VIII — Society, Power, And Civilizations

This part scales Universal Energy Dynamics from individual minds and living systems into the collective arena—groups, institutions, markets, and nations—without changing a single rule of the framework. Nothing new is added; only the *scale of coupling increases*. What appears as politics, power struggles, economic volatility, or geopolitical conflict is revealed as the same delta–equilibrium–consequence mechanics operating across shared fields with delayed feedback and narrative interference.

This part dissolves the illusion that societies are driven primarily by ideas, ideologies, or intentions. Instead, it shows that groups behave as *shared fields*, cultures function as *stabilized alignment protocols*, and institutions exist to *regulate deltas that individuals alone cannot manage*. Power is stripped of mystique and redefined precisely—as the ability to shape gradients and constraints—while leadership failures are traced to forced intent that ignores field criteria.

Markets and media are reframed as consequence-processing systems, not moral judges or truth engines. Volatility is treated as informational misalignment rather than chaos. Finally, geopolitics is exposed as multi-scale delta warfare, where tariffs, technology, energy, and narratives operate as engineered constraints rather than isolated policy choices.

By the end of this part, society no longer feels exceptional or unpredictable. It becomes legible. Civilizations rise, stabilize, oscillate, and fracture for the same structural reasons that bodies fall ill or minds destabilize—mismanaged deltas under delayed consequences.

UED now supports geopolitics, leadership, and civilization analysis with the same physics-grade rigor used for matter, life, and mind—no ideology required, no moral overlay imposed.

Chapter 26 — Groups as Shared Fields

Why Groups Behave "Strangely"

One of the oldest confusions in social thinking is the belief that groups behave as the sum of their members. We assume that if individuals are intelligent, ethical, or calm, the group they form should naturally reflect those qualities. Experience repeatedly contradicts this assumption. Intelligent people panic in crowds. Rational individuals amplify rumors. Ethical persons participate in systems they privately disagree with. The error is not moral or psychological—it is structural.

The moment individuals couple, a new entity forms that did not exist before: a shared field. This field is not an agreement, an ideology, or a collective mind. It is a dynamic consequence space created by shared attention, shared exposure, and shared risk. Once this field emerges, behavior is no longer governed primarily by individual reasoning, but by field-level deltas that no single participant controls.

Groups feel "strange" because we keep evaluating them with the wrong unit of analysis. We ask why people acted against their own judgment, when judgment was no longer the dominant variable. In a shared field, alignment pressure replaces reasoning, resonance replaces deliberation, and contagion outpaces reflection. Nothing pathological is required. No irrationality needs to be assumed. The system is simply operating under a different resolution regime.

This is why moral explanations fail so consistently at scale. They address individual intent, while group behavior is driven by coupling strength, feedback delay, and constraint geometry. A crowd does not "decide" to panic; panic is a low-energy resolution pathway once certain thresholds are crossed. Likewise, harmony in a group is not proof of virtue—it often signals stabilized alignment under favorable constraints.

Understanding groups as shared fields removes both blame and mystification. It explains why control strategies feel necessary yet fail, why culture becomes more powerful than instruction, and why institutions emerge even when nobody explicitly wants them. From this point onward, the question is no longer *"Why did people do this?"* but *"What field did they enter, and what deltas were active there?"*

This shift is essential. Without it, society appears chaotic, power appears arbitrary, and collective failure feels inexplicable. With it, group behavior becomes legible—not predictable in detail, but understandable in structure. And that understanding is the only stable starting point for examining culture, institutions, and power itself.

26.1 — From Individuals to Shared Fields

A group does not begin when people agree to act together. It begins the moment consequences become shared. Shared space, shared attention, shared risk, or shared reward is enough to couple individuals into a single operational field. No ideology is required. No intent is necessary. The field forms automatically as soon as individual actions start influencing collective outcomes.

This is the first structural break from individual behavior. An individual operates with mostly local deltas—personal needs, perceptions, fears, and goals. A shared field introduces non-local deltas: pressures that arise from coordination, visibility, comparison, and expectation. These deltas do not belong to any one person, yet every person inside the field feels them. Once this happens, behavior is no longer governed by private reasoning alone, but by how actions resonate within the shared space.

Importantly, a shared field does not require uniformity. People can disagree, resist, or remain silent, and the field still exists. Coupling is not consensus; it is exposure. Even passive presence contributes to field strength. This is why groups can drift in directions that most participants privately oppose. The field resolves its own deltas using the lowest-energy pathways available, not the most rational or ethical ones.

As coupling increases, independence decreases—not by force, but by consequence. Actions that were harmless at the individual level now carry amplified effects. Small signals become louder. Minor deviations attract disproportionate response. This amplification is often misread as emotional overreaction or loss of character, when it is simply a shift in resolution scale.

The critical mistake is assuming that individual competence transfers upward. It does not. Groups do not inherit intelligence, morality, or awareness from their members. They inherit structure. What matters is how deltas are distributed, how feedback loops close, and how quickly consequences return. A group of thoughtful individuals can generate reckless outcomes if the shared field rewards speed, conformity, or visibility over coherence.

From a UED perspective, the transition from individual to group is not psychological—it is energetic and structural. The unit of behavior changes. Once inside a shared field, people are no longer just actors; they become nodes in a consequence network. Their choices are filtered, amplified, or damped by the field they occupy.

Recognizing this shift is essential. Without it, we keep blaming people for behaviors that are better explained by field dynamics. With it, groups stop appearing mysterious or irrational. They become what they are: resolution systems operating under shared constraints, seeking equilibrium in ways that individuals alone never would.

26.2 — Coupling, Resonance, and Contagion

Once a shared field forms, behavior begins to propagate through mechanisms that do not resemble individual decision-making. The primary drivers are coupling strength, resonance, and contagion—not persuasion, instruction, or logic. These mechanisms explain why groups accelerate in certain directions with surprising speed, often bypassing reflection entirely.

Coupling determines how tightly individuals are linked to the shared field. Strong coupling arises when attention is synchronized, stakes are visible, and feedback is immediate. In such conditions, actions taken by one node rapidly influence others. Weak coupling, by contrast, allows individuals to absorb disturbances locally without transmitting them. The same people behave very differently under these two regimes, not because they change internally, but because the field's transmission efficiency changes.

Resonance occurs when a signal aligns with existing deltas in the field. A message, emotion, or narrative does not spread because it is true or well-argued; it spreads because it fits the field's unresolved tension. Fear resonates in a field already carrying uncertainty. Outrage resonates where legitimacy is fragile. Hope resonates where stagnation dominates. Resonance amplifies signals that match the field's topology and suppresses those that do not, regardless of their factual merit.

This is why reasoning often fails in heated group contexts. Rational arguments frequently target individual cognition, while resonance operates at the field level. A resonant signal requires less energy to propagate than a corrective one, because it follows existing gradients rather than opposing them. In UED terms, resonance is a low-resistance pathway for delta redistribution.

Contagion is the visible outcome of coupling and resonance combined. It is not a loss of self-control or a breakdown of morality. It is the natural spread of a field-aligned state across connected nodes. Emotional contagion, narrative contagion, and behavioral mimicry all follow the same structure: once a threshold is crossed, propagation becomes self-sustaining until the delta is diffused or constrained.

Crucially, contagion does not require belief. People often participate in spreading fear, outrage, or enthusiasm while internally disagreeing with it. The field does not ask for consent; it transmits states through exposure and feedback. This is why attempts to counter contagion with appeals to "think independently" routinely fail. Independence is not removed—it is simply overpowered by shared consequence dynamics.

Understanding coupling, resonance, and contagion reframes group volatility. What appears as mass hysteria or collective blindness is better understood as efficient field-level resolution under high coupling. The danger is not that groups feel too much, but that they resolve deltas too quickly and along poorly constrained pathways.

This insight prepares the ground for the next step: how societies attempt to stabilize these volatile dynamics over time. Culture and institutions emerge not from ideology or wisdom, but from repeated efforts to dampen resonance, regulate contagion, and keep shared fields within survivable bounds.

26.3 — Culture as a Stabilized Alignment Protocol

Culture is often treated as something expressive—values, beliefs, traditions, or identity. In UED, culture is understood more precisely as a stabilized alignment protocol: a set of repeated patterns that historically reduced destructive deltas in a shared field. Culture does not arise to explain the world; it arises to make group-level resolution cheaper and more predictable.

When a group repeatedly encounters similar pressures—scarcity, conflict, coordination problems—it experiments, often unconsciously, with behaviors that lower volatility. The patterns that succeed are retained, ritualized, and transmitted. Over time, they harden into norms, customs, taboos, and expectations. What survives is not what is most rational or moral, but what kept the field coherent under stress.

This is why culture carries a strong bias toward stability rather than truth. A practice can be outdated, inefficient, or even harmful in a new context and still persist if it once served as a reliable stabilizer. Culture resists change not because people are ignorant, but because abrupt cultural shifts reintroduce unresolved deltas at scale. Stability is energetically cheaper than continuous re-alignment.

Rituals, in particular, function as low-energy synchronization tools. They align attention, timing, and behavior without requiring explanation. From a field perspective, rituals dampen noise and prevent divergence. Their meaning is secondary to their effect. When rituals lose relevance but are maintained rigidly, they stop stabilizing and begin accumulating suppressed delta—a common source of cultural fracture.

Cultural conflict emerges when multiple alignment protocols compete within the same shared field. Each protocol pulls toward a different equilibrium configuration, and individuals caught between them experience persistent tension. These conflicts are often moralized, but structurally they are equilibrium competition problems, not value failures.

Importantly, culture operates below explicit awareness. People do not "choose" culture moment to moment; they inherit and enact it as the default resolution pathway. This is why cultural change feels threatening even when it promises improvement. It destabilizes familiar alignment patterns before new ones have proven their reliability.

Seen through UED, culture is neither sacred nor arbitrary. It is a functional memory of what once worked to keep shared fields intact. Understanding culture this way allows critique without contempt and reform without naive disruption. It also sets the stage for understanding why, beyond a certain scale, culture alone is insufficient—necessitating the rise of formal institutions as explicit delta regulators.

26.4 — Institutions as Delta Regulators

As shared fields grow in size, density, and duration, culture alone becomes insufficient to manage collective deltas. Cultural norms rely on slow transmission, implicit compliance, and informal correction. Beyond a certain scale, these mechanisms fail—not because people become worse, but because delta magnitude and propagation speed exceed cultural bandwidth. Institutions emerge precisely at this threshold.

From a UED perspective, institutions are not moral authorities or expressions of collective wisdom. They are explicit delta regulators—structures designed to constrain, buffer, redirect, or delay consequences that would otherwise destabilize the shared field. Law, money, education systems, bureaucracies, courts, and governance frameworks all perform variations of this same function: they shape gradients and impose friction where unchecked resolution would be destructive.

An institution converts volatile, high-energy interactions into slower, rule-bound processes. Legal systems delay retaliation. Financial systems standardize value exchange. Educational systems compress generational learning. Bureaucracy absorbs emotional and narrative noise by replacing personal judgment with procedure. These mechanisms feel impersonal by design. Personalization reintroduces variability; impersonality stabilizes fields.

This is why institutions often feel oppressive even when they are functioning correctly. Regulation is experienced locally as constraint, while its benefit appears globally as stability. When individuals judge institutions solely by personal inconvenience, they miss the structural role those constraints play in preventing field-level collapse.

Institutional failure does not usually come from malice or incompetence alone. It occurs when institutions regulate the wrong deltas or apply outdated constraints to new field conditions. When the environment changes faster than institutional adaptation, regulation lags, deltas accumulate, and pressure builds beneath the surface. What eventually appears as corruption, rigidity, or rebellion is often delayed consequence finally breaching its containment.

Importantly, institutions do not eliminate deltas; they redistribute them across time, space, and population. This redistribution is their strength and their vulnerability. If redistribution becomes too uneven, trust erodes. If it becomes too rigid, adaptation stalls. If it becomes too opaque, narratives replace understanding, and legitimacy collapses.

Seen structurally, institutions are neither villains nor saviors. They are tools—necessary but blunt—for managing shared-field dynamics beyond the reach of culture alone. When they align with real field pressures, societies stabilize quietly. When they diverge, instability is inevitable, regardless of intent.

This completes the progression: individuals form shared fields, shared fields stabilize through culture, and culture scales into institutions. From here, a new asymmetry becomes unavoidable—the unequal ability to shape constraints themselves. That asymmetry is what we call power, and it is the subject of the next chapter.

26.5 — When Shared Fields Become Unstable

Shared fields do not collapse because people suddenly change their nature. They become unstable when delta accumulation outruns regulation. Instability is not an anomaly; it is a signal that the field's existing alignment mechanisms—culture and institutions—are no longer sufficient for the pressures being applied.

One common source of instability is over-coupling. When attention becomes too synchronized and feedback too immediate, the field loses its ability to absorb disturbances locally. Every fluctuation spreads everywhere. Minor signals are amplified into crises, and corrective responses arrive faster than understanding. In such conditions, volatility feels constant, even when underlying conditions are relatively stable. The field is not breaking; it is oscillating due to excessive transmission efficiency.

The opposite failure mode is under-coupling. When feedback is delayed, fragmented, or suppressed, deltas do not disappear—they accumulate. Tension builds invisibly, shielded by apparent calm. When release finally occurs, it appears sudden and disproportionate, leading observers to speak of "unexpected" revolutions, crashes, or social ruptures. In reality, the instability was long present; only its visibility was delayed.

Another destabilizing factor is delta misclassification. Fields often treat structural deltas as moral problems, or narrative deltas as material threats. When the wrong regulation is applied—moral pressure instead of structural adjustment, enforcement instead of redesign—instability intensifies. The system expends energy without resolving the actual source of imbalance, increasing frustration and noise.

Suppressed deltas are particularly dangerous. Cultural taboos, institutional inertia, or reputational fear can prevent necessary adjustments. Suppression creates the illusion of stability while raising the energetic cost of eventual resolution. When release occurs, it bypasses gradual pathways and seeks the fastest available outlet, often appearing chaotic or violent.

Importantly, instability rarely announces itself clearly. Early signals are subtle: rising sensitivity to language, rapid shifts in narrative dominance, declining tolerance for ambiguity, and increased reliance on symbolic actions. These are not causes of instability; they are symptoms of a field approaching its regulation limits.

From a UED standpoint, instability is not failure—it is feedback. It indicates that equilibrium pull has shifted, constraints are misaligned, or consequence channels are blocked. The tragedy is not that fields become unstable, but that societies consistently respond too late or at the wrong level of analysis.

Recognizing instability as a structural condition rather than a moral breakdown changes the response entirely. It shifts attention from blame to redesign, from control to re-alignment. This distinction becomes crucial in the next step, where attempts to "manage" instability give rise to power concentration—and with it, the illusion that shared fields can be commanded rather than understood.

26.6 — The Illusion of Control in Group Management

When shared fields become unstable, the instinctive response is to assert control. Leaders issue commands, institutions tighten rules, and narratives harden into slogans. This reaction feels necessary because instability is experienced as loss of agency. Yet in shared fields, control is often not only ineffective—it is destabilizing.

The core illusion is the belief that intent can override field dynamics. Commands operate at the level of instruction, while instability arises from misaligned deltas, delayed feedback, and structural constraints. When authority attempts to force outcomes without addressing these underlying conditions, it adds energy to the system without providing resolution. The result is compliance on the surface and increased volatility beneath.

Top-down control fails because shared fields do not respond to orders; they respond to consequence geometry. Individuals may obey temporarily, but the field redistributes the imposed delta elsewhere—into resistance, evasion, narrative distortion, or quiet withdrawal. What looks like restored order is often just postponed instability.

This is why strong enforcement frequently coincides with fragility. The appearance of stability masks a narrowing of allowable resolution pathways. As constraints tighten, the system loses flexibility. When pressure exceeds tolerance, failure becomes abrupt and severe. From the inside, this feels like betrayal or sudden collapse; structurally, it is the predictable outcome of over-constrained dynamics.

Management language often worsens the problem by confusing coordination with control. Coordination aligns incentives and feedback so that individuals naturally move in stabilizing directions. Control, by contrast, attempts to override local adaptation. In complex shared fields, adaptation is not noise—it is the mechanism through which stability is maintained. Suppressing it removes the system's capacity to self-correct.

The appeal of control lies in its visibility. Commands are clear, decisive, and narratively satisfying. Alignment work is quiet, slow, and rarely dramatic. This asymmetry leads systems to overinvest in authority and underinvest in structural understanding. Leaders are rewarded for appearing decisive, even when their actions increase long-term instability.

UED reframes leadership and governance as field navigation, not enforcement. Effective leaders do not impose outcomes; they adjust constraints, reopen blocked consequence channels, and reduce unnecessary coupling. Their success often goes unnoticed because stability produces little spectacle.

Recognizing the illusion of control is not an argument against leadership—it is an argument for a different kind of competence. One that respects the limits of intent and works with field dynamics rather than against them. This realization naturally raises the next question: if control fails, why does power concentrate at all? The answer lies in unequal access to gradients and constraints—the subject of the chapter that follows.

From Groups to Power

Once groups are understood as shared fields, a quiet but unavoidable asymmetry becomes visible. Not everyone within a field has equal access to influence its gradients, constraints, or channels of consequence. Some positions allow amplification, others absorption. Some can delay consequences, others must bear them immediately. This uneven access does not arise from intention or virtue; it emerges structurally as fields scale and specialize.

Power, in this light, is not authority granted from above nor dominance seized from below. It is the differential capacity to shape how deltas move within a shared field. The ability to set rules, control resources, frame narratives, or regulate feedback is simply the ability to redirect resolution pathways. Whether exercised gently or violently, power always operates through the same mechanics.

This is why power persists even when leaders change, ideologies collapse, or institutions are rebranded. The field remains, and the gradients reassert themselves. Attempts to eliminate power without redesigning constraints merely relocate it. Attempts to moralize power without understanding its structure leave the field unchanged.

At the group level, power first appears as coordination advantage. At larger scales, it hardens into control over infrastructure, attention, and consequence timing. What begins as facilitation gradually becomes leverage. What begins as regulation can become domination if unchecked by feedback and alignment.

Recognizing this transition is critical. Without it, power appears mysterious or corrupt by nature. With it, power becomes legible—as a function of field architecture rather than personal character. This understanding does not justify power, but it removes the illusion that it can be wished away.

The next chapter takes this insight to its logical conclusion. It examines power not as a political concept, but as a structural one—showing how control of gradients and constraints determines stability, failure, and transformation across societies.

Chapter 27 — Power as Delta Control

Why Power Feels Personal but Isn't

Power is almost always described in human terms—ambition, authority, charisma, dominance, corruption. We attach it to faces, titles, and intentions, as if power flows outward from personality. This framing feels intuitive because power is experienced through people. Yet it consistently fails to explain outcomes. Competent leaders collapse systems. Weak figures preside over stability. Moral intent produces disaster, while cynical maneuvering sometimes holds structures together longer than expected. The contradiction is not ethical; it is structural.

Power feels personal because it is exercised locally, but it operates globally. A decision made by one actor propagates through a field of constraints, incentives, delays, and feedback loops that far exceed individual control. When outcomes diverge from intent, we attribute failure to character or success to genius. In reality, both are often irrelevant. The field responds not to who acts, but to what gradients are altered and which constraints are touched.

This confusion persists because intent is visible and structure is not. We can hear speeches, read policies, and observe commands, but we rarely see how consequence timing shifts, how feedback channels narrow, or how pressure redistributes across a system. As a result, power is judged by appearance—decisiveness, confidence, control—rather than by alignment with field criteria.

UED strips away this illusion. It treats power not as a trait but as a positional capability within a shared field. Power exists wherever someone can delay consequences, redirect flows, impose constraints, or amplify signals. Whether this capability is used benevolently or destructively is secondary to the fact that it reshapes resolution pathways for everyone else.

This is why replacing leaders rarely fixes systems. The gradients remain. The constraints persist. New actors inherit the same structural leverage and face the same equilibrium pressures. If they act against field criteria, failure follows regardless of intention. If they act with it, stability emerges even without brilliance.

Understanding power this way is uncomfortable. It removes the comfort of heroes and villains and replaces it with accountability to structure. But it is also clarifying. Once power is seen as delta control rather than personal dominance, its successes, failures, and limits become legible. From that point onward, the question is no longer *"Who is in charge?"* but *"Which gradients are being controlled, and at what cost?"*

27.1 — What Power Actually Is in UED

In Universal Energy Dynamics, power is defined with no psychological or moral qualifiers. Power is simply the capacity to control deltas by shaping gradients, constraints, and the timing of consequences within a shared field. Wherever this capacity exists, power exists—regardless of titles, legitimacy, or intention.

This definition immediately dissolves several common confusions. Power is not force; it does not require coercion. Power is not popularity; it does not require consent. Power is not intelligence; it does not require understanding. All of these may accompany power, but none of them define it. What defines power is access to the structural levers that determine how resolution occurs.

Gradients determine direction. They create pressure that moves behavior, resources, attention, or risk. Whoever can steepen, flatten, or redirect gradients can guide outcomes without issuing commands. Constraints determine pathways. They limit what can happen, delay when it can happen, and decide where accumulated delta must go. Whoever can impose or remove constraints determines which resolutions are possible and which are forbidden.

Timing completes the picture. The ability to delay consequences is one of the most decisive forms of power. Immediate consequences enforce adaptation. Delayed consequences permit accumulation, narrative management, and redistribution. Actors who can postpone consequences often appear successful long after misalignment has begun. Those who cannot absorb immediate cost for the system's adjustment, regardless of correctness.

This is why power persists beyond individuals. Leaders change, governments fall, ideologies rotate, but the same structural positions continue to shape outcomes. Control over infrastructure, law, capital, technology, and information constitutes power because these elements define gradients and constraints at scale. Occupants of these positions may vary in competence or ethics, but the field-level effect remains consistent.

From a UED standpoint, power is neither good nor bad. It is a field property, emerging wherever asymmetry in delta control exists. Attempts to moralize power obscure this reality and often worsen outcomes by encouraging symbolic action over structural correction. Effective intervention requires identifying where gradients are being manipulated and how constraints are enforced—not judging who wields them.

Seen this way, power is unavoidable in complex systems. The only meaningful questions are how power is distributed, how transparently it operates, and how quickly feedback returns. When these conditions are ignored, power concentrates silently, deltas accumulate invisibly, and systems drift toward instability long before failure becomes visible.

This reframing sets the stage for the next step: understanding how asymmetry in gradient and constraint access creates inequality, and why even well-intentioned leadership repeatedly fails when it acts against field criteria rather than with them.

27.2 — Gradients, Constraints, and Asymmetry

Power does not arise because someone desires control. It arises because gradients and constraints are never evenly distributed in a shared field. The moment a system grows beyond trivial scale, asymmetry becomes structural. Some positions naturally sit closer to sources of flow, leverage points, or bottlenecks, while others are exposed primarily to the consequences of those flows. This uneven access is the foundation of power.

Gradients are directional pressures. They pull behavior, resources, attention, and risk toward certain outcomes. Economic incentives, legal obligations, social approval, and fear all function as gradients. Power increases with the ability to create, amplify, or suppress these gradients. A steep gradient compels movement without coercion; a flattened gradient neutralizes motivation without resistance. Those who control gradients rarely need to command. The field does the work for them.

Constraints define the boundaries of possible resolution. They determine what actions are permitted, delayed, or prohibited. Laws, technical standards, infrastructure design, financial rules, and information access all function as constraints. Control over constraints allows power to shape not just outcomes, but the set of outcomes that can even be imagined. When constraints are invisible or normalized, power becomes difficult to detect and harder to challenge.

Asymmetry emerges when some actors can manipulate gradients and constraints while others must adapt to them. This asymmetry is often mistaken for merit or failure. Success is attributed to competence; hardship to personal weakness. Structurally, however, the difference lies in consequence timing. Those with power can externalize costs, defer adjustment, and redistribute pressure. Those without power absorb immediate consequences, forcing rapid adaptation or collapse.

This is why inequality persists even under reformist intent. Policies that address surface outcomes without redistributing gradient control leave the underlying asymmetry intact. The field reconfigures around the same leverage points, and disparities reappear under new labels. Power is conserved structurally, not politically.

Importantly, asymmetry is not inherently destructive. Systems require differentiated roles to function. Problems arise when asymmetry becomes rigid, opaque, or decoupled from feedback. When those who shape gradients no longer experience the consequences of their adjustments, misalignment accelerates. Stability becomes fragile, and correction becomes costly.

Understanding gradients and constraints clarifies why power is so resilient and why it resists moral appeals. Power responds to field-level incentives, not ethical argument. To alter power dynamics, one must redesign gradients and constraints—not merely replace actors or change narratives.

This realization leads directly to the central failure mode of leadership: the attempt to impose intent without respecting field criteria. When leaders mistake authority for gradient control, they act against the very structures that determine success. That failure is the focus of the next section.

27.3 — Forced Intent vs Field Criteria

Leadership is commonly framed as the ability to impose vision, direction, or will. In UED, this framing is precisely where failure begins. Intent does not move systems; gradients do. When leaders attempt to drive outcomes through intent alone—however noble or intelligent—they collide with field criteria that do not negotiate.

Field criteria are the non-negotiable conditions under which a system can resolve deltas: capacity limits, feedback delays, coupling strength, resource constraints, and consequence pathways. These criteria define what is possible at a given scale and time. Intent that aligns with them feels effective and decisive. Intent that violates them produces resistance, distortion, or delayed collapse.

Forced intent occurs when leadership treats authority as a substitute for alignment. Targets are set without adjusting gradients. Rules are tightened without redesigning constraints. Timelines are imposed without accounting for resolution duration. Early compliance is mistaken for success, while underlying deltas quietly accumulate. The system appears responsive until it abruptly isn't.

This mismatch explains why ambitious reforms often fail despite widespread agreement on their goals. The problem is not opposition, ignorance, or sabotage. It is that the proposed change demands resolution pathways the field cannot support. When pressure exceeds capacity, the system adapts by rerouting delta through informal channels—corruption, evasion, burnout, or narrative distortion—rather than by achieving the intended outcome.

Leaders are especially vulnerable to this trap because intent is rewarded symbolically. Decisive action, bold announcements, and visible enforcement create the impression of control. Field alignment, by contrast, produces quiet progress and delayed recognition. The incentive structure favors forcing rather than listening, commanding rather than adjusting.

UED reframes effective leadership as criteria-sensitive intervention. Leaders succeed not by asserting will, but by reading the field accurately: identifying where gradients already point, which constraints are binding, and how much misalignment the system can absorb. They adjust leverage points instead of issuing demands, and they accept slower progress in exchange for sustainable resolution.

Failure, in this context, is not a moral flaw. It is a structural error—treating intent as causal rather than directional. Systems eventually correct this error through consequence, often harshly. When they do, leaders appear to "lose control," but control was never present in the first place.

This distinction sets up the next question: why do leaders continue to fail even after repeated experience? The answer lies in delayed consequences and the false stability they create—producing oscillations that mask misalignment until correction becomes unavoidable.

27.4 — Why Leaders Fail (Even When They Are Competent)

Leadership failure is often attributed to incompetence, corruption, or lack of character. These explanations are emotionally satisfying but structurally weak. In complex shared fields, competence at one scale can produce failure at another. Leaders fail not because they lack skill, but because early success masks growing misalignment.

Many leaders rise by mastering localized deltas. They learn how to motivate teams, optimize processes, manage incentives, and deliver results within limited scope. These skills work because feedback is fast and constraints are visible. Success reinforces confidence, and confidence gradually converts into assumption: what worked before will work again at larger scale.

This is the leadership trap. As scale increases, feedback delays lengthen, coupling intensifies, and consequences diffuse. The leader's actions still produce immediate signals—compliance, applause, performance metrics—but these signals no longer reflect true alignment. They indicate responsiveness, not resolution. The system appears stable while misalignment accumulates beneath the surface.

Competent leaders are especially vulnerable because they trust their judgment. They interpret resistance as inertia, caution as fear, and complexity as inefficiency. In doing so, they suppress feedback that contradicts intent. Advisors soften warnings. Data is filtered. Narratives are refined to sustain confidence. The leader experiences continuity, while the field absorbs stress.

This pattern explains why collapses often appear sudden and personal. A leader who seemed effective yesterday is blamed for failure today. In reality, the failure was structural and cumulative. The system crossed a threshold where delayed consequences could no longer be contained. At that point, correction accelerates rapidly, overwhelming any remaining control.

UED clarifies that leadership competence must evolve with scale. What matters is not decisiveness, but sensitivity to consequence timing. Leaders who adapt learn to slow down, decentralize adjustment, and reopen feedback channels. Those who do not mistake stability for success and momentum for alignment.

This is why leadership turnover rarely fixes systemic problems. New leaders inherit the same accumulated deltas, often under greater pressure to act decisively. Unless they redesign constraints and gradients, they repeat the cycle with different rhetoric.

Understanding failure this way removes the illusion that better people alone can save misaligned systems. It shifts the focus from personality to structure, from blame to design. That shift is necessary to grasp the next phenomenon—why systems rarely fail cleanly, but instead swing between extremes as delayed consequences are released unevenly.

27.5 — Delayed Consequences and System Oscillation

Complex systems rarely collapse in a straight line. They oscillate. Periods of apparent stability are followed by abrupt reversals, policy swings, market crashes, social polarization, or leadership purges. These oscillations are often treated as ideological failures or human irrationality. In UED, they are understood more cleanly as the release patterns of delayed consequences.

Delayed consequences arise when a system has the capacity to absorb misalignment without immediate correction. Power enables delay. Institutions buffer impact. Narratives smooth perception. Early signals of stress are muted, reframed, or statistically averaged out. The system continues to function, giving the impression that intent is working and stability has been achieved.

But delay does not eliminate delta—it stores it. Misalignment accumulates in less visible forms: debt, resentment, inefficiency, ecological stress, trust erosion, or cognitive fatigue. Because feedback is weak or distorted, corrective action is postponed. Leaders interpret calm as validation, not realizing that calm itself is being artificially maintained.

When accumulated delta finally exceeds containment capacity, release occurs unevenly. Instead of gradual adjustment, the system swings. Tight control gives way to deregulation. Optimism flips into panic. Centralization is followed by fragmentation. These reversals feel dramatic because correction happens faster than adaptation, compressing years of unresolved adjustment into a short time window.

Oscillation is often worsened by overcorrection. When delayed consequences surface, response is driven by urgency and narrative pressure rather than structural diagnosis. New constraints are imposed rapidly, old ones dismantled recklessly, and gradients are reversed instead of recalibrated. This creates a new misalignment in the opposite direction, setting the stage for the next swing.

Narrative noise amplifies this effect. Stories of blame, betrayal, or moral failure dominate discourse, obscuring the structural origin of instability. Attention shifts to personalities and slogans, while the deeper issue—mismanaged consequence timing—remains unaddressed. Each oscillation becomes more emotionally charged and less informative.

From a UED perspective, oscillation is not chaos; it is mis-timed resolution. The system is correcting itself, but too late and too abruptly because earlier pathways were blocked or ignored. Stability would have required smaller, continuous adjustments. Delay transforms manageable correction into disruptive shock.

Recognizing this pattern is crucial for any serious understanding of power. The ability to delay consequences is often mistaken for strength. In reality, it increases the amplitude of eventual correction. Systems that appear resilient may simply be accumulating stress quietly.

This prepares the ground for the final element of the chapter: how narratives interact with delayed consequences—masking misalignment, accelerating oscillation, and giving power a temporary illusion of control that ultimately cannot hold.

27.6 — Narrative Noise as Power Multiplier

Narratives play a unique role in power systems because they do not alter structure directly, yet they profoundly affect how deltas are perceived, tolerated, and postponed. In UED terms, narratives act as attention-gradient manipulators. They do not resolve imbalance; they redirect awareness away from it, buying time for those who can delay consequences.

Power relies on narratives not because stories create reality, but because they mask misalignment. When structural stress builds, direct acknowledgment would force corrective action. Narratives intervene by reframing pressure as progress, sacrifice as virtue, delay as strategy, and instability as external threat. The field remains misaligned, but its symptoms are cognitively neutralized.

This is where narrative becomes a multiplier. By shaping what the system pays attention to, narratives reduce corrective feedback. Signals that would normally trigger adjustment are drowned in explanation, justification, or moral framing. The system continues operating under the illusion of control while delta accumulates invisibly.

Narrative noise is especially effective because it operates faster than structural change. Constraints and gradients take time to redesign; stories can be deployed instantly. A failing policy can be rebranded overnight. A weakening institution can be defended through identity. A delayed consequence can be blamed on enemies, markets, history, or fate. Each narrative intervention stretches the delay window a little further.

However, narrative control has strict limits. It cannot change field criteria. It can only interfere with perception and timing. As a result, systems that rely heavily on narrative amplification experience sharper eventual corrections. When reality reasserts itself, narratives collapse suddenly, often in reversal rather than refinement. Confidence flips to cynicism. Unity fractures into polarization. Trust evaporates rapidly because it was sustained symbolically, not structurally.

Another danger of narrative noise is misclassification. Structural problems are framed as moral failures. Material constraints are treated as ideological sabotage. Complexity is simplified into loyalty tests. This mislabeling directs energy toward conflict rather than correction, further destabilizing the field. Power appears active—speeches are made, positions are defended—but resolution pathways remain blocked.

UED clarifies that narratives are not inherently deceptive. They are unavoidable coordination tools. Problems arise when narrative becomes a substitute for alignment rather than a supplement to it. When stories are used to suppress feedback instead of interpret it, they convert power into delay rather than stability.

Ultimately, narrative noise explains why power systems often seem coherent until the moment they fail. The failure is not sudden; only visibility is. When narrative saturation reaches its limit, accumulated deltas re-enter awareness all at once, overwhelming both leaders and institutions.

This completes the anatomy of power failure: gradients manipulated, constraints misaligned, intent forced, consequences delayed, and narratives used to extend the delay. The final lesson is unavoidable—no amount of narrative control can override field resolution indefinitely. That limit, and what it implies for markets and media, is where the next chapter begins.

27.7 — The Structural Limits of Power

Power often appears boundless while it is functioning smoothly. Institutions hold, narratives circulate, consequences are deferred, and systems continue operating despite visible strain. This creates the illusion that power can indefinitely override constraints. In UED, this illusion is temporary. Power always encounters structural limits, not because of moral failure or opposition, but because fields ultimately resolve deltas according to equilibrium criteria.

The first limit is constraint saturation. Every additional rule, control mechanism, or enforcement layer increases friction. Initially, this stabilizes the field by narrowing volatile pathways. Over time, however, excessive constraint reduces adaptability. The system loses degrees of freedom necessary for local correction. When new deltas arise, there are fewer legitimate channels for resolution, forcing pressure into informal, opaque, or destructive routes.

The second limit is feedback starvation. Power that successfully delays consequences also delays information. Signals weaken, distort, or arrive too late to guide correction. Leaders and institutions begin operating on outdated representations of reality, mistaking silence for stability. By the time accurate feedback breaks through, accumulated delta has already crossed manageable thresholds.

A third limit emerges from asymmetric consequence exposure. When those who control gradients are insulated from the costs of their adjustments, misalignment accelerates. Systems depend on shared exposure to maintain coherence. Once consequence asymmetry becomes too large, trust erodes—not as an ethical judgment, but as a rational response to misaligned risk distribution. At that point, compliance decays even if authority remains intact.

Collapse, in this framework, is not punishment. It is forced resolution. When gradual pathways are blocked, the field selects faster, harsher ones. Power structures that appear strong up to the moment of collapse often fall more abruptly precisely because they postponed correction longest. What looks like sudden failure is simply delayed equilibrium asserting itself.

Importantly, power does not disappear when a system collapses; it reconfigures. New gradients emerge, new constraints form, and new actors occupy leverage points. This is why revolutions rarely eliminate power dynamics—they rearrange them. Without redesigning field architecture, old patterns reappear under new labels.

UED therefore reframes the question of power's limits. The issue is not how to make power moral, permanent, or benevolent, but how to keep it coupled to feedback, adaptable to change, and exposed to consequence. Power aligned with field criteria stabilizes quietly. Power that exceeds them destabilizes inevitably.

This recognition closes the chapter. Power is neither villain nor savior; it is a structural capability with strict operating boundaries. When those boundaries are ignored, the field corrects without negotiation. Understanding this prepares us for the next domain—markets and media—where power becomes distributed, indirect, and seemingly neutral, yet operates through the same delta mechanics at scale.

From Power to Markets and Media

Once the structural limits of power are reached, systems rarely return to direct control. Instead, power diffuses. It moves away from visible authority and reappears in mechanisms that feel neutral, automatic, or impersonal. This shift does not reduce power; it redistributes it through aggregation rather than command.

Markets and media emerge as the preferred carriers of this redistribution. They allow influence without enforcement, coordination without instruction, and consequence without explicit attribution. Gradients are embedded in prices, incentives, visibility, and attention rather than in orders. Constraints are encoded in access, algorithms, liquidity, and legitimacy instead of rules alone. Power becomes harder to locate, even as it becomes more pervasive.

This transition explains why modern systems often feel uncontrollable despite extensive regulation. Authority still exists, but outcomes are increasingly shaped by distributed feedback loops that no single actor fully governs. Leaders respond to markets. Institutions react to narratives. Policy follows volatility rather than setting direction. Power appears diluted, yet its effects are sharper and faster.

The danger of this phase is misinterpretation. Because markets and media appear self-organizing, their outputs are treated as truth signals rather than consequence aggregates. Price movements are mistaken for value judgments. Visibility is confused with importance. Volatility is read as chaos instead of feedback. Structural misalignment is obscured by apparent objectivity.

Understanding this shift is critical. Without it, societies oscillate between overconfidence in authority and blind faith in systems they do not understand. With it, markets and media become legible as extensions of the same field dynamics—amplifying, delaying, and redistributing deltas at unprecedented scale.

The next chapter enters this terrain directly. It examines markets as consequence-processing engines and media as attention-gradient shapers, showing how both inherit the same structural strengths and failure modes as power itself—only faster, subtler, and more difficult to correct once misalignment sets in.

Chapter 28 — Markets, Media, and Narratives

Why Markets and Media Feel Objective

Modern societies place extraordinary trust in markets and media because they appear impersonal. Prices move without emotion. Trends emerge without a visible author. Popularity seems earned rather than imposed. This apparent neutrality creates the belief that markets and media reveal reality rather than shape it. What is aggregated is treated as truth.

This belief is understandable—and deeply misleading.

Markets and media feel objective because they remove visible intent. No single actor appears responsible for outcomes, so outcomes are interpreted as natural. A rising price is read as value. A trending narrative is read as importance. Volatility is read as uncertainty. In each case, aggregation is mistaken for judgment.

In UED terms, markets and media are not evaluators; they are resolution surfaces. They collect dispersed actions, compress consequences, and express them as simplified signals—price, volume, visibility, sentiment. These signals feel authoritative because they summarize vast complexity into a single number or story. But compression always discards structure.

What gets lost is alignment. Markets respond to incentives, not meaning. Media responds to attention gradients, not accuracy. Neither system asks whether the outcome is sustainable, coherent, or beneficial in the long run. They only reflect what resolves fastest under existing constraints.

This is why markets can celebrate destructive activity and punish stabilizing effort. This is why media can amplify distortion while ignoring substance. The systems are not broken; they are functioning exactly as designed—aggregating consequences under high-speed coupling and delayed feedback.

The danger begins when these signals are treated as guides rather than symptoms. Price movements become policy inputs. Narrative dominance becomes legitimacy. Volatility becomes justification for reaction rather than inquiry. Structural misalignment is hidden behind the authority of numbers and headlines.

Understanding this is essential before going further. Without it, markets appear wise and media appears informative, while instability seems accidental. With it, both systems become legible as powerful but limited instruments—capable of coordination, yet blind to the deeper equilibrium they continuously reshape.

This reframing sets the stage for the first task: understanding markets not as arbiters of value, but as consequence aggregators—efficient, fast, and structurally indifferent to what they aggregate.

28.1 — Markets as Consequence Aggregators

Markets are often described as mechanisms of valuation or judgment. In UED, this framing is inaccurate. Markets do not decide what is good, useful, or sustainable. They aggregate consequences—rapidly, impersonally, and without memory of intent. Price is not a verdict; it is a compressed signal representing how deltas are resolving under existing constraints.

Every market action embeds multiple forces at once: expectation, fear, opportunity cost, liquidity access, regulatory friction, and time preference. These forces do not align neatly. The market aggregates them anyway, collapsing diverse pressures into a single output. That output feels authoritative precisely because it hides complexity. What looks like clarity is compression.

This explains why markets respond faster than institutions. Institutions must interpret, debate, and redesign constraints. Markets bypass interpretation. They register shifts in gradient immediately—capital moves toward perceived advantage, away from perceived risk. Speed gives markets coordination power, but it also gives them structural indifference. Markets do not distinguish between productive alignment and extractive gain. Both generate signals. Both are rewarded if they resolve deltas efficiently in the short term.

Markets are especially sensitive to consequence timing. Activities that externalize cost or defer damage often appear attractive because their negative deltas do not return immediately. Environmental depletion, financial leverage, labor erosion, and informational asymmetry can all produce favorable market signals while degrading long-term equilibrium. The market is not lying; it is aggregating what is currently visible.

This is why market "rationality" is conditional. Markets are rational with respect to present constraints, not future stability. When constraints are misdesigned or delayed, market outcomes diverge from societal well-being. Attempts to moralize markets miss this point entirely. The issue is not greed or ignorance; it is what consequences are allowed to surface and when.

Another critical feature of markets as aggregators is asymmetry of participation. Not all actors face the same exposure to consequence. Some can exit quickly. Some can hedge. Some can absorb losses indefinitely. Others cannot. The aggregated signal reflects these asymmetries, amplifying the behavior of those with flexibility and muting the realities of those without it. Price therefore encodes power structure as much as supply and demand.

From a UED perspective, markets are neither trustworthy nor untrustworthy. They are accurate within narrow bounds and misleading beyond them. They excel at coordinating short-term adaptation and fail at signaling long-term alignment. Treating market outputs as guidance rather than data leads to reactive policy, speculative bubbles, and delayed correction.

Understanding markets as consequence aggregators reframes their role. They are sensors, not compasses. They tell us what pressures are resolving fastest, not where equilibrium lies. Mistaking one for the other is how volatility turns into crisis and efficiency into fragility.

This sets up the next question: how speed and delay distort these signals further—and why faster feedback does not necessarily mean better information.

28.2 — Feedback Speed, Delay, and Distortion

Markets are celebrated for speed. Prices adjust in real time, capital reallocates instantly, and information appears to be incorporated continuously. This speed is often equated with accuracy. In UED, speed is understood differently—not as clarity, but as a bias toward short-cycle deltas.

Fast feedback favors what resolves quickly. Signals tied to immediate reward, visible risk, or narrative momentum dominate market behavior, while slow-moving consequences fade into the background. Environmental depletion, institutional erosion, social fatigue, and long-term capacity loss do not generate strong price signals until they reach crisis thresholds. By the time they surface, correction is abrupt and costly.

This creates a structural distortion. Markets are highly responsive to micro-alignment—small changes in expectation, liquidity, or sentiment—but relatively blind to macro-misalignment that unfolds over longer horizons. The result is apparent efficiency paired with deep fragility. Systems look adaptive while quietly exhausting their buffers.

Delay compounds this distortion. Many economic activities are evaluated long before their consequences return. Profits are booked today; costs appear years later. Markets treat this gap as opportunity rather than risk because the aggregation mechanism cannot distinguish between delayed and eliminated consequences. What is postponed appears resolved.

Another layer of distortion arises from leverage and abstraction. Financial instruments allow actors to amplify exposure without proportional consequence. Risk is redistributed across time and population, thinning feedback even further. When correction finally arrives, it is systemic rather than local, triggering cascades instead of adjustments.

High-speed trading and algorithmic participation intensify these effects. As feedback loops shorten, market behavior becomes increasingly reflexive. Signals respond to signals, detaching price movement from underlying productive reality. Volatility increases, not because fundamentals are unstable, but because feedback is circulating faster than real-world change.

This is where the myth of market efficiency becomes dangerous. Efficiency in processing available information does not imply relevance of that information to long-term equilibrium. A system can be perfectly efficient at amplifying the wrong signals. In UED terms, it can resolve deltas rapidly while drifting further from sustainable alignment.

Understanding speed and delay clarifies why policy responses based solely on market signals often misfire. Interventions chase volatility instead of addressing structural lag. Stabilization efforts suppress symptoms while allowing misalignment to deepen.

This prepares the transition to the next domain: media. Where markets aggregate consequences through price, media aggregates attention through visibility. Both operate under similar speed biases, and when coupled, they magnify each other's distortions—turning misalignment into spectacle and volatility into narrative momentum.

28.3 — Media as Attention-Gradient Infrastructure

Media is often discussed as a channel for information or a mirror of public opinion. In UED, this framing misses the operative mechanism. Media functions primarily as attention-gradient infrastructure. It does not just transmit content; it shapes where collective awareness flows, how long it stays, and what it ignores.

Attention is a finite resource. In any shared field, what receives attention gains energetic priority. Media platforms—traditional or digital—compete to capture this resource, and in doing so they engineer gradients that pull awareness toward certain signals and away from others. These gradients are not aligned with truth, depth, or long-term relevance. They are aligned with propagation efficiency.

Signals that spread fastest share common traits: emotional charge, simplicity, novelty, and conflict. They require minimal cognitive effort and generate immediate engagement. Calm, complex, or corrective signals require time and patience; they dissipate before achieving resonance. As a result, the attention field becomes skewed toward high-energy, low-resolution content.

This skewing is not manipulation in the conspiratorial sense. It is structural. Media systems optimize for retention, reach, and responsiveness because those metrics sustain the platform. Over time, this optimization reshapes collective perception. What is visible feels important. What is repeated feels true. What is absent feels irrelevant—even when it is structurally critical.

In UED terms, media infrastructure amplifies certain deltas while suppressing others. It magnifies narrative tension, moral outrage, and symbolic conflict because these align with fast attention gradients. Meanwhile, slow-forming misalignments—capacity erosion, systemic risk, institutional decay—remain under-represented because they do not propagate efficiently.

This creates a dangerous inversion. The field becomes highly responsive to narrative fluctuations while remaining insensitive to structural drift. Public reaction accelerates, but understanding lags. Policy responds to visibility rather than consequence. Decision-making becomes reactive, not because leaders are careless, but because the attention field itself is misaligned.

Media's power lies precisely in its indirectness. It does not command behavior; it conditions salience. By shaping what feels urgent and what fades into background noise, it determines which deltas are perceived as requiring resolution. This is why control over media infrastructure often exceeds the influence of formal authority.

Understanding media as attention-gradient infrastructure removes moral outrage from the analysis. The problem is not bias, misinformation, or bad actors alone. The problem is that attention, once engineered at scale, becomes a structural force—one that can destabilize systems by continuously redirecting awareness away from where alignment is actually breaking.

This sets up the next layer of complexity: narratives. If media shapes where attention flows, narratives shape how that attention interprets what it encounters—either clarifying misalignment or masking it entirely.

28.4 — Narratives as Delta Manipulators

Narratives are often mistaken for explanations. In UED, narratives are better understood as delta manipulators—mechanisms that organize perception, compress complexity, and guide emotional response without resolving underlying imbalance. They do not fix misalignment; they make it *bearable*, *meaningful*, or *deferrable*.

A narrative works by selecting a frame. From infinite possible interpretations, it highlights a few elements and suppresses the rest. This selection immediately reshapes the attention field. Certain deltas become visible and emotionally charged, while others recede into abstraction. What feels like understanding is often just coherent omission.

Narratives are powerful because they reduce cognitive load. Complex systems generate ambiguity, delay, and uncertainty—conditions that are uncomfortable for collective awareness. Narratives replace these with clarity, intention, and causality. Someone is responsible. Something is progressing. Something is threatened. The field relaxes, not because alignment has improved, but because uncertainty has been converted into story.

This is where narrative becomes a tool of delay. Structural problems—capacity limits, feedback distortion, misallocated incentives—are reframed as moral struggles, identity conflicts, or temporary disruptions. Attention is redirected toward symbolic resolution: outrage, loyalty, hope, or blame. Meanwhile, the original delta continues to accumulate unaddressed.

Narratives also regulate tolerance. They determine how much instability a system can endure before demanding correction. A convincing narrative can stretch tolerance far beyond healthy limits, allowing misalignment to persist long after corrective action would have been possible. Conversely, destabilizing narratives can collapse tolerance prematurely, forcing reaction before understanding.

Importantly, narratives do not require deception to function. Even accurate narratives can become harmful if they misclassify the delta. When structural misalignment is framed as individual failure, moral weakness, or ideological threat, energy is misdirected. Effort goes into fighting symptoms rather than redesigning constraints.

Narrative competition further complicates resolution. Multiple stories can coexist within the same field, each pulling attention toward a different equilibrium interpretation. This does not balance the system; it fragments it. Shared fields lose common reference points, making coordinated adjustment increasingly difficult. Polarization is not disagreement—it is divergent delta framing.

In UED, the danger of narratives is not that they are false, but that they become substitutes for alignment. When stories are used to explain away pressure rather than guide structural change, they convert power into postponement. The system feels managed while drifting further from equilibrium.

This understanding is essential before interpreting volatility. Without it, volatility appears random or malicious. With it, volatility can be read properly—as either noise generated by narrative and attention dynamics, or as a genuine signal that underlying alignment has shifted and resolution is unavoidable.

28.5 — Volatility: Noise or Alignment Shift?

Volatility is usually treated as a problem to be controlled or a threat to be neutralized. In UED, volatility is neither good nor bad by default. It is a signal—but one that is frequently misread. The critical distinction is whether volatility represents noise generated by distorted feedback, or a real shift in underlying alignment.

Noise-driven volatility emerges when attention gradients, narrative amplification, and short-cycle incentives dominate perception. In such cases, prices, opinions, and reactions fluctuate rapidly without corresponding changes in productive capacity, resource availability, or structural constraints. The system appears unstable, but the instability is superficial. What is oscillating is interpretation, not equilibrium.

This form of volatility is often intensified by attempts to suppress it. Interventions aimed at calming markets or stabilizing narratives without addressing structural drivers tend to compress fluctuation temporarily while increasing future amplitude. The system is prevented from releasing small deltas, so it accumulates pressure until release becomes unavoidable and disruptive.

By contrast, alignment-shift volatility signals a genuine change in field conditions. Resource limits are reached. Energy flows are rerouted. Technological or geopolitical constraints alter production and access. In these cases, volatility is not excess movement—it is reconfiguration. The system is searching for a new equilibrium because the old one is no longer viable.

The challenge lies in distinguishing between these two regimes. Noise volatility is fast, repetitive, and reversible. Alignment-shift volatility is directional, persistent, and resistant to narrative smoothing. One fluctuates around a center; the other relocates the center itself.

Markets and media frequently blur this distinction. Noise is treated as crisis, triggering overreaction. Structural shifts are dismissed as temporary, delaying adaptation. This inversion leads to policy error, misallocation of resources, and strategic blindness. Systems spend energy calming the wrong signals and ignore the ones that matter.

UED reframes volatility as diagnostic rather than disruptive. The question is not *how much* volatility exists, but *what delta it reflects*. Is the system reacting to perception loops, or adjusting to real constraint changes? Suppressing the former is sensible. Suppressing the latter is destructive.

When volatility is correctly interpreted, it becomes an early warning rather than a shock. It signals where alignment has already shifted and where redesign is required. When misread, it becomes a catalyst for oscillation, panic, and narrative excess—accelerating instability rather than resolving it.

This distinction completes the chapter's core logic. Markets aggregate consequences. Media shapes attention. Narratives frame tolerance. Volatility is the visible outcome of how well—or how poorly—these elements remain aligned with underlying field realities.

28.6 — The Illusion of Market Rationality

Markets are often defended as rational because they aggregate vast amounts of information. When outcomes appear extreme or contradictory, the explanation offered is simple: *the market knows something we don't.* In UED, this belief is a category error. Aggregation is not understanding, and responsiveness is not intelligence.

Market behavior emerges from countless local decisions responding to gradients—price, risk, liquidity, visibility—not from any shared comprehension of the system as a whole. No participant sees the full field. Each reacts to partial signals filtered through position, exposure, and time horizon. The resulting coordination looks purposeful, but it is coherence without awareness.

This is why intelligence at the individual level does not cancel field dynamics. Highly educated, well-informed actors still follow gradients when misalignment builds. When incentives pull strongly in one direction, resisting them requires absorbing immediate cost while uncertain benefit lies in the future. Most actors cannot afford that delay, regardless of insight. Rational self-preservation becomes collective drift.

The idea of "smart money" reinforces this illusion. Certain actors appear consistently ahead of trends, giving the impression of superior foresight. Structurally, what they often possess is better positioning—earlier access to information, faster exit routes, or insulation from downside. Their apparent intelligence is frequently an artifact of consequence asymmetry rather than understanding.

Herding behavior is similarly misdiagnosed. It is not stupidity or panic; it is resonance under uncertainty. When feedback is noisy and future outcomes unclear, aligning with visible movement reduces individual risk, even if it increases systemic risk. From the inside, herding feels rational. From the outside, it appears reckless.

This illusion becomes most dangerous when markets are treated as decision-makers rather than sensors. Policies defer to "market reaction" as if it were judgment. Leaders interpret price stability as validation and volatility as warning, without examining what the signals actually reflect.

Structural misalignment persists because the system is listening to its own echo.

UED reframes market rationality as conditional and local. Markets are effective at coordinating short-term adaptation under stable constraints. They are ineffective at anticipating long-term equilibrium shifts or preserving systemic health. Expecting them to do so is not realism—it is abdication.

Recognizing this illusion is not an argument against markets. It is an argument against overloading them with meaning they cannot carry. Markets tell us where pressure is moving *now*. They do not tell us whether the direction is survivable.

This insight prepares the final step: understanding what happens when market signals and media-driven attention loops begin feeding each other—turning localized misalignment into systemic instability through speed, amplification, and mutual distortion.

28.7 — When Markets and Media Destabilize Each Other

Markets and media become most dangerous—not when they fail independently—but when they lock into a shared feedback loop. In this coupled state, price movements influence attention, and attention reshapes price expectations. Neither system is guiding the other; both are amplifying the same unresolved deltas at increasing speed.

This coupling begins innocently. Media reports on market movement. Markets react to anticipated media response. Attention spikes around volatility, and volatility increases because attention has spiked. What started as signal transmission becomes signal generation. The system begins reacting to its own reflection.

In UED terms, this is a classic resonance loop. Markets aggregate consequences into prices. Media aggregates attention into narratives. When these aggregates feed each other without structural grounding, the field loses its external reference. Resolution no longer tracks production capacity, resource constraints, or long-term alignment. It tracks visibility and momentum.

This explains why bubbles inflate faster than fundamentals can justify and why crashes propagate beyond actual damage. Media amplifies expectation. Markets price expectation as reality. Each iteration steepens gradients without changing constraints. Eventually, the system exhausts its ability to postpone consequence, and correction arrives suddenly and nonlinearly.

Speed is the accelerant. Media operates on near-instant attention cycles. Markets increasingly operate on sub-second execution. Structural reality—manufacturing, infrastructure, labor adaptation—moves orders of magnitude slower. The mismatch creates a widening gap between perception and capacity. The longer this gap persists, the harsher the eventual correction.

Another destabilizing factor is narrative synchronization. When a dominant story aligns market interpretation and public attention simultaneously, alternative signals are suppressed. Dissenting data is dismissed as noise. Early warnings are reframed as negativity or sabotage. The field becomes overconfident, not because it is aligned, but because uncertainty has been narratively eliminated.

When correction finally occurs, it feels irrational and external. Confidence collapses into panic. Trust evaporates quickly because it was maintained symbolically rather than structurally. Observers speak of "loss of faith" or "sudden realization," but structurally the system is simply reabsorbing delayed delta.

UED clarifies that destabilization is not caused by speculation, media bias, or human emotion alone. It arises when consequence aggregation and attention aggregation become mutually reinforcing without constraint. In such conditions, even well-designed institutions struggle to intervene, because both perception and price move faster than regulation.

Understanding this coupling completes the picture. Markets and media are powerful coordination tools when loosely coupled and grounded in real feedback. When tightly coupled and driven by narrative momentum, they become systemic amplifiers of misalignment. Volatility then ceases to be informative and becomes transformative—forcing real-world adjustment through shock rather than design.

This realization prepares the transition to the final scale of analysis. When markets and media operate globally, their feedback loops cross borders, override local constraints, and transmit deltas across nations. What emerges is not economics or communication, but geopolitics conducted through distributed field warfare—the subject of the next chapter.

From Aggregation to Geopolitics

When markets and media operate at scale, their influence no longer remains confined to economics or communication. They become global coupling mechanisms, transmitting deltas across borders faster than political systems can absorb or regulate them. Price signals leap jurisdictions. Narratives cross cultures. Volatility in one region reshapes expectations everywhere else. What begins as aggregation becomes pressure projection.

At this stage, traditional distinctions break down. Economic moves become strategic acts. Media narratives become instruments of influence. Supply chains, capital flows, information streams, and technological standards start functioning as levers rather than neutral channels. The battlefield is no longer territory alone—it is gradient control across interconnected fields.

Geopolitical power in this environment is rarely exercised through direct confrontation. It emerges through the ability to shape constraints that others must resolve against. Tariffs alter economic gradients. Technology access defines capability ceilings. Energy routes determine stability thresholds. Narrative dominance influences legitimacy and tolerance. Each lever operates on the same principle: shifting where and how delta must be absorbed.

This is why modern conflict often feels diffuse and persistent rather than decisive. There is no clear beginning or end, only ongoing pressure, adaptation, and counter-adaptation. Stability is not achieved by eliminating conflict, but by managing coupling so that shocks do not cascade uncontrollably across systems.

Understanding this transition is essential before entering the final chapter of this part. Without it, geopolitics appears chaotic, opportunistic, or morally inconsistent. With it, geopolitics becomes legible as multi-scale delta interaction—where nations, markets, technologies, and narratives intersect within a single, continuously resolving field.

The next chapter enters this terrain directly. It reframes geopolitics not as competition between states, but as delta warfare across scales, where the true objective is not victory, but control over stability, timing, and consequence.

Chapter 29 — Geopolitics as Multi-Scale Delta Warfare

Why Modern Conflict Rarely Declares Itself

For most of history, conflict announced its arrival. Armies mobilized, borders were crossed, and war was declared. Today, pressure accumulates without ceremony. Economies strain, technologies fragment, supply chains snap, narratives harden, and instability spreads—often without a single formal declaration. The world feels tense, yet oddly undefined. This ambiguity is not accidental; it is structural.

Modern geopolitics operates in a densely coupled global field where direct confrontation is costly, visible, and difficult to control. Traditional war collapses multiple deltas at once—military, economic, political, and moral—forcing immediate consequence. In a system optimized for delay and deniability, such collapse is inefficient. Pressure applied indirectly travels further, lasts longer, and provokes fewer unified responses.

This is why contemporary conflict rarely declares itself. Instead of invasion, there are tariffs. Instead of occupation, there are technology bans. Instead of blockades, there are regulatory standards. Instead of propaganda ministries, there are narrative ecosystems. Each move shifts gradients and constraints while maintaining plausible normalcy. The field is altered, but the surface remains calm.

From a UED perspective, this is not a shift in ethics but in resolution strategy. States no longer seek decisive victory; they seek advantage in consequence timing. The objective is not to defeat an opponent outright, but to force them to absorb delta faster, longer, or with fewer escape routes. Instability becomes the weapon, and patience becomes the currency.

This form of conflict feels confusing because it violates inherited mental models. There is no clear enemy action to condemn, no singular moment to respond to, no victory to declare. Yet the effects are real: capacity erodes, dependency deepens, trust fractures, and alignment degrades across societies. The battlefield is not territory alone—it is infrastructure, access, standards, and perception.

Understanding this shift is essential before examining specific tools. Without it, modern geopolitics appears chaotic, hypocritical, or opportunistic. With it, patterns emerge. What looks like inconsistency is often coherence at a different scale. What feels like escalation without cause is usually accumulated delta reaching visibility.

This opening reframes geopolitics as continuous multi-scale delta interaction, not episodic war. With that lens in place, the mechanisms of tariffs, technology chokepoints, energy routes, and hybrid conflict become legible—not as anomalies, but as the natural language of power in a tightly coupled world.

29.1 — Geopolitics as a Field, Not a Chessboard

Geopolitics is still commonly imagined as a chessboard—discrete actors making deliberate moves against one another. This metaphor is comforting because it suggests clarity: players, strategies, victories, defeats. In a tightly coupled world, this model no longer holds. Nations do not operate as isolated pieces; they function as interacting fields embedded within larger fields, continuously exchanging pressure, risk, and consequence.

A nation's behavior is constrained not only by its intentions, but by its exposure to global gradients—capital flows, energy dependence, technological access, demographic pressures, ecological limits, and narrative legitimacy. These gradients do not respect borders. They seep through trade, finance, information, and infrastructure, coupling states even when political relations appear distant or hostile. Sovereignty, in this context, is not absolute control; it is managed exposure.

This is why independent policy is largely a myth at scale. Decisions taken domestically immediately encounter external constraints. A tariff shifts global supply chains. A regulation reshapes technological standards.

A currency move ripples through debt markets elsewhere. Each action modifies the field and returns as altered consequence, often through unexpected channels. What looks like choice is frequently selection among constrained options.

Viewing geopolitics as a field clarifies why small actors can exert outsized influence and large actors can appear constrained. Influence depends less on size and more on position within the field—proximity to chokepoints, control of standards, leverage over timing, and ability to absorb delayed consequences. Power accrues to those who can reroute flows rather than confront opponents directly.

Interdependence, often celebrated as stabilizing, also introduces vulnerability. Coupling increases efficiency but reduces insulation. Shocks propagate faster and further. Attempts to weaponize interdependence—through sanctions, supply restrictions, or standards—exploit this vulnerability by forcing asymmetric absorption of delta. The target state does not "lose" immediately; it is forced into prolonged adjustment under unfavorable gradients.

This field-based view also explains why alliances are fluid and contradictions abound. States cooperate in one domain while competing in another because different subsystems are coupled differently. There is no single game being played. There are overlapping resolution problems unfolding simultaneously across economic, technological, security, and narrative layers.

Understanding geopolitics as a field dissolves moral simplifications. Actions are not inherently aggressive or defensive; they are gradient adjustments. Outcomes depend on how pressure redistributes, where constraints bind, and how quickly feedback returns. Stability emerges not from dominance, but from configurations where deltas remain survivable across scales.

With this reframing in place, the instruments of modern geopolitics become intelligible. Tariffs, technology chokepoints, and energy routes are not policy tools in isolation; they are delta levers—mechanisms for reshaping the field without triggering immediate collapse. Those levers are the focus of the next section.

29.2 — Delta Levers: Tariffs, Tech, and Energy

In a field-based view of geopolitics, influence is exercised less through force and more through delta levers—mechanisms that shift pressure, constrain resolution pathways, and redistribute consequences across time and space. Among the most effective of these levers are tariffs, technology chokepoints, and energy routes. They operate quietly, persistently, and asymmetrically, making them ideal instruments in a world where overt conflict is costly.

Tariffs function as economic gradient steepeners. They do not merely raise prices; they alter directionality. By selectively increasing friction at borders, tariffs redirect trade flows, investment decisions, and supply chain architectures. Their power lies in duration rather than intensity. Even modest tariffs, applied consistently, force long-term adaptation. Firms restructure. Costs relocate. Dependencies rewire. The targeted economy absorbs delta continuously, often without a single moment of visible crisis.

Technology chokepoints operate at a deeper layer. Control over critical technologies—semiconductors, software ecosystems, standards, intellectual property—creates capability constraints rather than immediate economic pain. Unlike tariffs, which affect exchange, technology controls affect *possibility*. They define what a system can build, how fast it can adapt, and which futures remain accessible. The impact is slower, but far more structural. Once a capability ceiling is imposed, catching up requires not policy change, but generational effort.

Energy routes represent stability thresholds. Modern economies are energy-conversion systems; disruption at this level cascades into every other domain. Control over extraction, transport, pricing, and currency denomination allows states to shift delta across entire regions. Energy pressure rarely produces instant collapse. Instead, it degrades resilience—raising inflation, constraining industry, and eroding political tolerance. The effect is cumulative and difficult to counter without diversification, which itself takes time.

What makes these levers especially effective is their plausible normalcy. Each can be justified independently—economic protection, national security, market efficiency, regulatory compliance. None require declaration of hostility. Yet their combined effect can rival that of traditional warfare by forcing prolonged misalignment under constrained adaptation.

Indirect pressure also outperforms direct confrontation because it externalizes response cost. The initiating state often absorbs limited immediate consequence, while the target must reorganize internally—politically, economically, and socially. This asymmetry allows pressure to be sustained without escalation, turning patience into strategic advantage.

However, delta levers are not risk-free. Overuse leads to constraint saturation. Supply chains reroute. Alternatives emerge. Allies experience spillover. The field adapts. When multiple levers are pulled simultaneously, unintended coupling can rebound on the initiator, creating inflationary pressure, political backlash, or loss of legitimacy.

From a UED perspective, tariffs, technology chokepoints, and energy routes are not acts of aggression or defense by themselves. They are field interventions. Their effectiveness depends on timing, proportionality, and feedback sensitivity. Used with structural awareness, they can stabilize advantage. Used blindly, they accelerate global misalignment.

This prepares the next step: when such levers are combined deliberately across domains, geopolitics moves from pressure application to constraint engineering—designing the field itself rather than merely pushing against it.

29.3 — Constraint Engineering as Strategy

As geopolitical competition matures, influence shifts from applying pressure to designing the field in which pressure must resolve. This is constraint engineering—the deliberate shaping of limits, standards, dependencies, and pathways so that certain outcomes become easy, others expensive, and some impossible. Unlike direct pressure, constraint engineering does not seek immediate response. It reshapes the environment in which all future responses must occur.

Constraints operate quietly. Standards determine compatibility. Regulations define access. Supply-chain architectures decide resilience.

Legal frameworks set friction. Once embedded, these constraints guide behavior automatically, without constant enforcement. Actors adapt because deviation is costly, not because they are compelled. This makes constraint engineering more durable than coercion and less visible than confrontation.

One of the most powerful forms of constraint engineering is standard-setting. Technical standards, financial norms, and regulatory benchmarks act as invisible borders. They decide whose products integrate smoothly, whose systems interoperate, and whose innovations scale globally. States that dominate standard-setting rarely need to block competitors explicitly; incompatibility does the work.

Dependency design is another critical dimension. When a system relies on external inputs—technology, capital, energy, logistics—it inherits the constraints of its suppliers. Dependency is not inherently harmful; interdependence enables efficiency. The strategic question is who can endure decoupling longer and who bears the greater adjustment cost when constraints tighten. Constraint engineering exploits this asymmetry by hardening exits and narrowing alternatives.

Constraint engineering also explains the shift from dominance to endurance. Traditional power sought control through superiority. Modern power seeks advantage through path dependence. Once systems are locked into particular architectures—financial networks, technological ecosystems, legal regimes—change becomes slow and costly. Influence is exercised not by constant intervention, but by shaping initial conditions.

However, constraint engineering carries long-term risk. Over-engineered fields lose flexibility. Excessive rigidity suppresses innovation and local adaptation. When unexpected deltas arise—technological breakthroughs, demographic shifts, ecological stress—the constrained system struggles to respond. What once provided stability becomes fragility.

UED clarifies that effective constraint engineering requires feedback sensitivity. Constraints must evolve with field conditions. Static limits in dynamic environments guarantee misalignment. The temptation to freeze advantage often leads to eventual reversal, as suppressed deltas seek alternative resolution paths.

Understanding constraint engineering reframes modern strategy. Geopolitical success is less about winning contests and more about designing survivable environments. Those who engineer constraints wisely shape the field without provoking collapse. Those who mistake rigidity for control create systems that appear stable—until they fail suddenly.

This logic naturally extends beyond economics and technology. When constraints are engineered across financial systems, information flows, legal regimes, and security domains simultaneously, conflict becomes hybrid by nature. That expanded battleground is the focus of the next section.

29.4 — Hybrid War Beyond the Military

Hybrid war is often described as a blend of military and non-military tactics. From a UED perspective, this description is incomplete. Hybrid war is better understood as coordinated constraint manipulation across multiple fields, where military force becomes just one optional layer rather than the defining feature.

In hybrid conflict, the objective is not territorial conquest but resolution dominance—forcing an adversary to absorb deltas across economic, informational, legal, technological, and psychological domains simultaneously. Each domain alone may remain below the threshold of open conflict. Together, they overload the system's capacity to adapt coherently.

Financial pressure disrupts capital access and credit stability. Legal actions create uncertainty and delay. Cyber operations distort information flow and trust. Narrative campaigns reshape perception and tolerance. Diplomatic signaling limits alliances. None of these actions require escalation, yet each alters gradients and constraints. The adversary expends energy responding in multiple directions, losing coherence over time.

A defining feature of hybrid war is plausible deniability. Actions are framed as routine governance, market behavior, or defensive necessity. Attribution becomes contested, slowing collective response. This delay is not incidental; it is a strategic asset. While debate continues over intent, deltas continue to accumulate.

Hybrid war also exploits domain mismatch. Military institutions are optimized for kinetic threats, not for legal ambiguity or narrative erosion. Economic systems are not designed to detect strategic manipulation. Media ecosystems amplify noise faster than correction. Hybrid strategies succeed by operating where existing institutions are least prepared to respond.

From a field perspective, hybrid war increases coupling while fragmenting feedback. Systems feel under pressure, but struggle to identify a single source or resolution pathway. This ambiguity leads to reactive measures that often worsen misalignment—tightening controls, suppressing dissent, or escalating rhetoric—without addressing structural vulnerabilities.

Importantly, hybrid war is not inherently aggressive. The same techniques are used defensively to preserve stability under pressure. The distinction lies not in the tools, but in how constraints are applied and whether feedback channels remain open. Overuse or misapplication transforms stabilization into provocation.

UED reframes hybrid war as a consequence of highly coupled global fields rather than a novel doctrine. As systems interconnect, conflict naturally migrates to domains where influence can be exerted without triggering collapse. Military action becomes a last resort, used only when other delta levers fail or when forced resolution is unavoidable.

This understanding sets up the next concern: escalation. When multiple constraint-based actions accumulate, pressure can rebound unpredictably. The same levers that destabilize an adversary can destabilize allies, markets, and the initiator itself. That dynamic—saturation and blowback—is the subject of the next section.

29.5 — Escalation, Saturation, and Blowback

Delta levers feel attractive because they appear controllable. Pressure can be increased gradually, adjusted selectively, and justified procedurally. This creates the impression that escalation is linear and reversible. In reality, delta control saturates. Beyond certain thresholds, additional pressure produces diminishing returns and rising instability—often in directions the initiator does not anticipate.

Escalation begins when multiple levers are applied simultaneously. Tariffs combine with technology bans. Energy pressure overlaps with financial restriction. Narrative campaigns intensify alongside legal action. Each lever is defensible on its own, but together they compress the opponent's resolution space. Adaptation pathways narrow. Flexibility declines. Stress concentrates.

Initially, this compression appears effective. The targeted system slows, fragments, or concedes locally. But as saturation increases, the field seeks alternative routes for resolution. Supply chains reroute through third parties. Technologies are indigenized or substituted. Alliances realign. Informal networks strengthen. What looked like containment becomes reconfiguration.

Blowback occurs when redirected delta returns through coupled fields. Pressure on one economy destabilizes others. Sanctions distort global prices. Financial constraints ripple into allied markets. Narrative campaigns undermine trust beyond their intended audience. Because modern systems are tightly interconnected, no lever acts in isolation. Every intervention perturbs the wider field.

A common error is mistaking early compliance for long-term advantage. Short-term disruption often masks adaptive learning. Systems under sustained pressure reorganize structurally. Once new equilibrium pathways form, reversing pressure does not restore the old order. The initiator faces a changed field—one where leverage has weakened and alternatives have matured.

Escalation also carries reputational cost. Constraint engineering depends on legitimacy. When levers are perceived as excessive or arbitrary, trust erodes. Partners hedge. Neutral actors distance themselves. The field becomes less cooperative and more resistant. Power shifts from influence to enforcement, raising maintenance cost and lowering effectiveness.

UED clarifies that escalation failure is not moral backlash; it is field correction. Systems pushed beyond tolerance do not collapse obediently. They adapt in ways that preserve survival, often at the expense of the initiator's long-term position. The more comprehensive the pressure, the more creative the adaptation.

This does not imply restraint is always optimal. It implies that pressure must be coupled with feedback awareness. Escalation without exit paths converts leverage into instability. Saturation transforms control into fragility.

Recognizing escalation limits leads to a different definition of success. Stability is not achieved by maximum pressure, but by managed coupling—keeping deltas within absorbable bounds while preserving adaptive capacity on all sides. That reframing is essential to understanding the final section of this chapter.

29.6 — Stability as Managed Coupling

Stability in geopolitics is often confused with calm. Periods without open conflict are labeled "peace," and tension is treated as failure. In UED, this interpretation is misleading. Stability is not the absence of pressure; it is the controlled transmission of pressure across coupled systems. Where coupling is unmanaged, calm is temporary. Where coupling is managed, tension can exist without collapse.

Coupling is unavoidable in a globalized field. Trade, technology, finance, energy, and information bind states together regardless of political intent. Attempts to eliminate coupling entirely—through isolation, autarky, or rigid decoupling—introduce new deltas. Efficiency collapses, innovation slows, and internal pressure rises. Total isolation does not produce stability; it merely internalizes instability.

At the opposite extreme, unrestricted coupling creates fragility. When systems are too tightly linked, shocks propagate instantly and globally. Local disturbances become systemic crises. Adaptation time disappears. In such conditions, even small deltas trigger outsized consequences. Absolute integration is as destabilizing as absolute separation.

Managed coupling occupies the narrow but critical middle ground. It accepts interdependence while shaping how stress moves. Redundancy is introduced so failure does not cascade. Buffer zones absorb shock. Time delays are preserved so adaptation can occur before correction becomes violent. Selective decoupling reduces exposure in critical domains without dismantling the entire field.

From a geopolitical perspective, managed coupling requires strategic humility. No actor fully controls the field. Stability depends less on dominance and more on tolerance calibration—knowing how much pressure a system can absorb without triggering runaway adaptation or backlash. This is why restraint, signaling, and phased pressure matter as much as leverage.

Managed coupling also demands feedback openness. Systems must remain sensitive to early signs of saturation and blowback. When warning signals are suppressed—through narrative control, denial, or institutional inertia—pressure accumulates silently. Stability then appears robust until it fails abruptly. True stability feels less dramatic precisely because correction is continuous and distributed.

This reframing explains why some rivalries persist without erupting, while others escalate unpredictably. The difference is not goodwill or ideology; it is field architecture. Where coupling channels are diversified and feedback loops remain active, competition remains survivable. Where coupling is rigid and feedback delayed, escalation becomes inevitable.

UED therefore redefines geopolitical success. It is not the elimination of conflict, nor the imposition of uniform order. It is the maintenance of dynamic equilibrium across interacting systems—allowing pressure to circulate without concentrating to destructive levels. Peace, in this sense, is not a destination. It is a continuously managed condition.

This understanding sets the stage for the final section: why misreading stability as peace leads to strategic blindness, and how periods of apparent calm often conceal the very misalignments that later produce sudden and catastrophic shifts.

29.7 — Misreading Peace, Mismanaging Power

The most persistent error in geopolitics is the belief that absence of visible conflict equals resolution. Calm is treated as success. Silence is taken as consent. Stability is assumed to be durable simply because nothing is breaking—yet. In UED terms, this is a dangerous misreading. Peace is often nothing more than low-visibility conflict under delayed consequence.

When pressure is redistributed rather than resolved, systems can appear stable for extended periods. Trade continues. Markets function. Diplomacy proceeds. Narratives emphasize normalcy. Beneath this surface, however, deltas may still be accumulating—shifted into debt, dependency, resentment, capability erosion, or asymmetric exposure. The field is quiet not because it is aligned, but because correction has been postponed.

This misinterpretation leads to power mismanagement. Leaders mistake calm for validation and double down on existing strategies. Constraint systems harden. Feedback channels narrow. Early warning signals are dismissed as noise or pessimism. Because nothing appears urgent, adaptation is delayed until options become limited.

Historical ruptures often follow this pattern. Long periods of apparent peace are retrospectively described as "missed warnings." In reality, warnings were present, but they did not fit prevailing narratives. Structural tension was reframed as temporary imbalance, manageable risk, or external provocation. By the time instability became undeniable, correction required shock rather than adjustment.

Another failure mode arises from symbolic peace-making. Treaties, summits, and ceasefires create the appearance of resolution without altering underlying gradients or constraints. These actions lower visibility but not pressure. When misalignment persists, such agreements become fragile, breaking under stress because they were never coupled to real field redesign.

UED highlights a crucial distinction: peace that suppresses feedback is unstable; peace that preserves feedback is resilient. Genuine stability tolerates disagreement, competition, and tension because these provide continuous information. Artificial calm, maintained through denial or narrative smoothing, starves the system of corrective signals.

This is why sudden geopolitical shifts feel shocking. The system did not change overnight; visibility did. Accumulated deltas crossed a threshold where suppression was no longer possible. What follows is often described as irrational escalation or unexpected collapse, when in fact it is delayed equilibrium asserting itself.

Recognizing this pattern changes strategic priorities. The goal is not to eliminate tension, but to read it accurately. Power must be managed with awareness of consequence timing, not just leverage capacity. Stability must be assessed structurally, not narratively.

This completes the arc of the chapter. Geopolitics, when viewed through UED, is not a morality play or a contest of wills. It is a continuous process of delta interaction across scales, where survival depends on alignment literacy more than dominance. With this understanding in place, this part reaches its conclusion—and the book turns to its final task: how to engage with these dynamics consciously, without becoming another source of misalignment yourself.

From Civilizations to Consequences

With this chapter, the illusion of geopolitics as a contest of personalities, ideologies, or moral positions finally dissolves. What remains is a far more demanding picture: civilizations are long-duration field experiments, sustained or destroyed not by intent, but by how well they manage deltas across time, scale, and coupling density.

Empires do not fall because leaders become evil or citizens become weak. They decline when pressure is misallocated, when consequences are delayed too long, when constraint systems harden beyond adaptability, and when feedback is mistaken for noise. Civilizations rise when alignment is continuously adjusted—quietly, incrementally, without spectacle. They fall when calm is mistaken for resolution and power is exercised without field awareness.

This reframing unifies everything covered in this part. Groups form shared fields. Power emerges as gradient and constraint control. Markets and media aggregate and amplify consequences. Geopolitics extends these mechanisms across borders and generations. There is no discontinuity between the personal, the social, the economic, and the geopolitical—only scale variation within the same dynamics.

The critical insight is sobering: no civilization escapes consequence. The only variable is how consequence arrives—gradually through adaptive correction, or violently through delayed release. Stability is never permanent. It is maintained, moment by moment, through alignment literacy rather than dominance.

This closes this part not with a warning, but with a responsibility. Understanding these dynamics does not grant immunity. In fact, it raises the stakes. Knowledge without awareness can amplify harm. Insight without humility can accelerate misalignment.

That is why the final part of this book is necessary. Next part does not offer control, optimization, or superiority. It addresses something more difficult and more essential: how to live, decide, lead, and act within these fields without becoming another source of distortion.

From civilizations, the focus now turns inward—not to escape the world, but to engage it without worsening its instability.

Deliverable of Part VIII

A Physics-Grade Foundation for Geopolitics and Leadership

This part completes a critical structural upgrade. It does not merely *support* your geopolitics and leadership books; it reclassifies them. What were previously read as analysis, critique, or perspective are now anchored in a formal, scale-invariant framework that operates with the same internal discipline as physics.

This matters for three reasons.

First, it removes opinion dependency. Geopolitical events, leadership failures, market shocks, and institutional breakdowns no longer need to be explained through ideology, morality, or personality. They are derivable outcomes of delta distribution, constraint design, coupling density, and consequence timing. Readers may disagree emotionally, but the structure remains intact. The analysis stands even when preferences differ.

Second, it establishes cross-scale consistency. The same logic that explains stress in a mind, instability in a group, failure in an institution, or collapse in a civilization now applies without modification. This is the hallmark of a physics-grade framework: no special rules for higher status actors. Leaders, states, and empires are subject to the same dynamics as individuals—only operating at different magnitudes and delays.

Third, it upgrades leadership discourse from prescription to diagnosis. Instead of offering ideals, strategies, or best practices, your leadership work now explains *why* action succeeds or fails regardless of intent. It shows why charisma collapses under delayed feedback, why authority without alignment accelerates instability, and why restraint can outperform dominance. Leadership becomes a problem of field navigation, not virtue signaling.

The practical consequence is decisive. Your geopolitics books stop being read as commentary on events and start functioning as interpretive instruments. Your leadership books stop motivating behavior and start preventing structural error. Readers are not told what to think or do; they are taught how to recognize misalignment before it becomes irreversible.

In short, this part does not add content to your work. It changes its epistemic status. What follows in your writing ecosystem is no longer opinion, ideology, or advice—it is application.

That is what it means to say: *your geopolitics and leadership books gain a formal physics-grade base.*

Interlude — Before Application

Up to this point, Universal Energy Dynamics has been used as a lens—to see more clearly how matter, life, mind, society, power, markets, and civilizations actually behave. The work has been diagnostic, not instructional. It has deliberately resisted the temptation to tell the reader what to do, how to live, or what to believe. That restraint was not caution; it was discipline.

Every framework that survives contact with reality faces the same risk at this stage: misuse. Once a structure explains enough, it attracts certainty. Once it scales across domains, it invites ideology. Once it feels coherent, it tempts authority. Many systems fail here—not because they are wrong, but because they are *applied without humility*.

UED was never meant to replace thinking. It was meant to restore it under pressure. It does not offer answers to all questions. It clarifies which questions are meaningful, which signals matter, and which reactions are likely to worsen misalignment. Used correctly, it sharpens judgment. Used poorly, it becomes another belief system competing for dominance.

This interlude marks a shift in responsibility. Until now, the burden was on the framework to explain itself. From here onward, the burden moves to the reader—to apply without absolutism, to model without overreach, and to act without turning insight into identity.

Next part does not teach doctrine. It teaches restraint with power. Not power over others—but power over interpretation.

The goal is simple and demanding: to use UED as a tool without becoming its servant, to apply it without preaching it, and to let it guide action without freezing into certainty.

Only then does understanding remain aligned with consequence.

Part IX — UED In Practice

This part brings Universal Energy Dynamics down from explanation to use—without turning it into belief, doctrine, or identity. Having shown how UED operates across physics, life, mind, society, power, and geopolitics, this final part addresses the most delicate task: *application without distortion*.

This part does not offer rules, rituals, or universal prescriptions. Instead, it establishes boundaries. It clarifies what UED can predict, what it can only orient, and where certainty becomes misuse. The emphasis shifts from being *right* to being *aligned*—from asserting understanding to maintaining responsiveness to consequence.

UED is presented here as a modeling discipline rather than a philosophy. Readers learn to map constraints, identify deltas, recognize thresholds, and trace consequence channels—while also understanding why most models fail when boundaries are chosen poorly or feedback is ignored. The goal is not control, but legibility.

Finally, UED is translated into a personal operating system—not as self-improvement ideology, but as a way to make decisions under uncertainty without amplifying instability. Stability and growth are reframed as gradient management, and continuity is understood correctly: not as preserving form, but as sustaining process.

The deliverable of this part is deliberately modest and powerful. Readers are not asked to carry a worldview, memorize a system, or adopt a label. They are given enough structure to *apply UED where it matters*, and enough restraint to know when not to.

This is the end of the book, but not the end of the framework. From here onward, UED is no longer something to follow—it is something to use carefully, or not at all.

Chapter 30 — How to Use UED

Without Turning It Into Religion

When Frameworks Become Belief

Every framework that explains enough eventually faces the same danger: it stops being used and starts being believed. The moment a structure feels coherent across domains, it invites certainty. The moment it reduces confusion, it tempts authority. This is not a flaw of the framework—it is a predictable response of the human mind seeking stability.

Most belief systems did not begin as dogma. They began as explanations that worked. Over time, explanation hardened into conclusion, conclusion into identity, and identity into defense. What could once be questioned became something that *must* be protected. The framework stopped responding to reality and began demanding loyalty instead.

Universal Energy Dynamics is especially vulnerable to this drift because it scales cleanly. It applies across physics, life, mind, society, and power without changing its core logic. That coherence is its strength—and its risk. When a framework works at many levels, it becomes easy to mistake orientation for total truth.

UED was never designed to answer everything. It was designed to prevent a specific error: acting without understanding consequence. When it is turned into belief, it produces the very failure it was meant to avoid. Certainty replaces sensitivity. Language replaces observation. Structure becomes slogan.

This chapter exists as a deliberate brake.

It does not defend UED. It limits it. It clarifies where use becomes misuse, where confidence becomes overreach, and where insight turns into distortion. The goal is not to keep UED pure, but to keep it useful under pressure.

A framework that cannot tolerate restraint will eventually demand obedience. UED must do the opposite. It must remain explainable without needing agreement, applicable without requiring adoption, and useful without becoming an identity.

From here onward, the question is no longer *"Is this framework correct?"* It is *"Is this framework being used responsibly?"*

That distinction determines whether UED remains a tool for clarity—or becomes another system that amplifies the very instability it claims to explain.

30.1 — Explain, Don't Preach

The first discipline in using UED correctly is deceptively simple: explain, don't preach. Explanation reveals structure; preaching asserts position. One invites inspection, the other demands alignment. The difference is not stylistic—it is structural.

When UED is used to explain, it maps relationships: where deltas arise, how constraints shape outcomes, why consequences return delayed or distorted. It stops at legibility. It does not require agreement, conversion, or emotional acceptance. The listener is free to test the explanation against experience, reject it, refine it, or ignore it entirely. The framework remains intact regardless of response.

Preaching begins when explanation is replaced by certainty. Language shifts subtly—from "this suggests" to "this proves," from "this tends to happen" to "this always happens." The speaker is no longer describing field dynamics; they are defending a position. At that moment, UED stops functioning as a tool and starts functioning as an identity shield.

This shift is dangerous because it reverses the direction of accountability. In explanation, the framework answers to reality. In preaching, reality is expected to conform to the framework. Any contradiction is dismissed as misunderstanding, resistance, or lack of readiness. Feedback is no longer welcomed; it is filtered.

UED explicitly rejects this posture. It does not claim moral authority, spiritual superiority, or intellectual finality. Its only claim is conditional: *if* certain gradients and constraints exist, *then* certain consequence patterns tend to follow. That conditionality must be preserved in use. The moment it is lost, UED becomes indistinguishable from the belief systems it critiques.

Preaching also signals insecurity. When understanding is stable, it does not need reinforcement. When insight is fresh or partial, it seeks validation through persuasion. This is why early adopters of any framework are most likely to overstate its reach. The urge to convince others often masks unresolved uncertainty within oneself.

Explaining, by contrast, is quieter. It tolerates ambiguity. It allows gaps. It acknowledges limits without apology. It trusts consequence to do the work of correction rather than language.

In practical terms, this discipline shows up in restraint:

- You describe mechanisms, not destinies.
- You offer models, not verdicts.
- You stop when structure is clear, even if conclusions remain uncomfortable.

Using UED responsibly means accepting that explanation is enough. If understanding does not emerge, preaching will not create it—only distortion will.

This discipline is the first safeguard against turning a living framework into a static belief.

30.2 — The Temptation of Absolutism

Absolutism does not begin with arrogance; it begins with relief. When a framework explains enough confusion, it produces a sense of ground under one's feet. Patterns become visible. Noise reduces. Causality feels clearer. The mind, accustomed to uncertainty, instinctively wants to lock this clarity in place. That instinct is where misuse starts.

Absolutism appears when orientation is mistaken for completion. Language shifts subtly but decisively: *this explains everything, this always happens, this is how reality works*. What was once a conditional map becomes an unconditional claim. The framework is no longer used to explore fields; it is used to close questions.

UED is structurally incompatible with absolutism. It deals in gradients, thresholds, feedback delays, and coupling effects—none of which resolve cleanly or universally. Field-based systems do not produce fixed outcomes; they produce tendencies under constraint. Treating those tendencies as guarantees erases the very dynamics UED exists to reveal.

Absolutism is especially seductive because coherence feels like truth. When a model fits many cases, the mind assumes it must fit all cases. Exceptions are dismissed as noise rather than investigated as boundary failures. This is how a framework stops learning. It becomes internally consistent while drifting away from reality.

Another danger of absolutism is premature authority. Once someone believes they *know*, they stop listening. Feedback becomes a threat instead of a correction mechanism. Contradictory evidence is reinterpreted, minimized, or moralized. At that point, the framework is no longer interacting with the field; it is defending itself against it.

UED requires a different posture: structural humility. Every application is provisional. Every model is bounded. Every explanation carries an implicit *until further constraint appears*. This is not weakness; it is how field-based understanding stays alive.

A reliable red flag is emotional attachment to being right. When disagreement feels destabilizing rather than informative, absolutism has already set in. UED, correctly used, should reduce emotional load, not increase it. If certainty hardens identity, the framework is no longer doing its job.

Resisting absolutism means leaving conclusions slightly open—not vague, but conditional. It means allowing consequence to revise understanding. It means accepting that clarity does not require finality.

This restraint protects UED from becoming another closed system. It keeps it responsive, testable, and aligned with the reality it seeks to describe rather than dominate.

30.3 — Vague Spirituality and Conceptual Dilution

A common failure mode of powerful frameworks is not rejection, but soft absorption. The language spreads, the terms circulate, and the structure quietly dissolves. Precision is replaced by comfort. Mechanics blur into mood. This is how frameworks slide into vague spirituality—not through exaggeration, but through dilution.

UED is particularly vulnerable here because it speaks about energy, balance, alignment, and consequence—words already saturated with spiritual overtones. When these terms are detached from their structural meaning and used suggestively rather than analytically, explanation gives way to reassurance. The framework stops describing how systems resolve and starts offering how one should *feel* about reality.

This shift is subtle but damaging. Energy becomes a metaphor for emotion. Alignment becomes agreement or positivity. Balance becomes calmness. Consequence becomes destiny. None of these are what UED means. In UED, energy is not a feeling; it is capacity under constraint. Alignment is not harmony; it is compatibility between gradients and pathways. Balance is not peace; it is temporary equilibrium under active resolution.

Vague spirituality feels attractive because it lowers cognitive effort. It allows one to speak confidently without specifying mechanisms. It expands appeal while reducing accountability. But it also destroys the framework's ability to explain failure. When outcomes contradict expectation, the response becomes interpretive rather than diagnostic: *you weren't aligned enough, the energy wasn't right, the system resisted.* These statements cannot be tested, corrected, or falsified.

Another red flag is substitution. When UED terms are used in place of concrete description—when saying *delta* replaces identifying the actual pressure, or saying *alignment* replaces mapping constraints—the framework has stopped functioning. Language becomes decorative rather than explanatory.

This does not mean UED must become cold or reductionist. It means its language must remain anchored to structure. Metaphor can illustrate, but it cannot replace mechanism. Suggestion can invite, but it cannot substitute for explanation. Once the framework starts offering comfort instead of clarity, it has crossed into territory where belief thrives and understanding fades.

The discipline here is sharp but necessary: If a concept cannot be mapped, constrained, or tested against consequence, it does not belong in UED usage.

Keeping this boundary preserves the framework's integrity. It ensures that UED remains a tool for navigating reality—not a vocabulary for re-enchanting it.

30.4 — Forced Universality and the Loss of Scale

Another common misuse of UED arises not from dilution, but from overextension. Once a framework proves useful across several domains, the temptation is to apply it everywhere, all the time, without boundary checks. This is forced universality—the assumption that because a structure scales, it must apply identically at every scale and in every context.

UED does not support this assumption. It is scale-aware by design. While the same principles—delta, constraint, coupling, consequence—operate across domains, their expression changes with scale, density, and feedback timing. Treating different scales as interchangeable flattens reality and produces confident but shallow explanations.

Forced universality shows up when distinctions are ignored. Psychological stress is described exactly like geopolitical instability. Organizational failure is treated as if it follows the same thresholds as biological collapse. Individual decision-making is modeled with the same tolerance margins as markets or states. The language sounds consistent, but the analysis loses resolution.

This loss of scale awareness is dangerous because it hides limits. Every system has boundaries within which certain models hold. Outside those boundaries, the same variables behave differently. Coupling strength changes. Feedback delays stretch. Local adaptation gives way to systemic inertia. When these shifts are ignored, conclusions feel elegant but fail under pressure.

Another symptom of forced universality is compulsive application. UED is invoked even when it adds no clarity—when simpler explanations suffice or when the system in question is already well understood through direct observation. In such cases, the framework becomes ornamental rather than functional. It signals sophistication instead of providing insight.

Equally important is knowing when not to use UED. Some situations demand immediate action, ethical judgment, or empirical measurement rather than structural modeling. Applying UED in these contexts can slow response or create false distance. Misuse is not always distortion; sometimes it is inappropriate timing.

UED remains effective only when it respects boundaries. Applicability does not mean sameness. Universality does not mean uniformity. The framework clarifies patterns across scales, but it does not erase the need for scale-specific judgment.

Preserving this distinction keeps UED grounded. It prevents the framework from becoming a master explanation that flattens complexity and replaces inquiry. Used with boundary awareness, UED sharpens perception. Used without it, it becomes another way to sound right while seeing less.

This restraint is essential before moving further. Without it, even correct insights turn brittle—unable to adapt when the system changes or when scale quietly shifts beneath the explanation.

30.5 — Prediction vs Orientation

One of the most persistent misuses of any explanatory framework is the demand for prediction. Once a system explains patterns reliably, the next expectation is inevitability: *tell me what will happen*. In UED, this expectation marks the boundary between correct use and overreach.

UED does not predict outcomes in the way deterministic models claim to. It does not forecast exact events, timelines, winners, or forms. What it does provide is orientation—directional clarity about where pressure is accumulating, where stability is weakening, and which resolution pathways are becoming likely or constrained.

This distinction is critical. Prediction assumes closure: fixed variables, stable boundaries, and negligible feedback interference. Field-based systems do not offer this luxury. They evolve as they resolve. Observation alters behavior. Intervention reshapes constraints. Any precise prediction collapses under its own interaction with the field.

UED therefore predicts tendencies, not events. It can indicate that a system under delayed feedback will oscillate, that excessive constraint will trigger adaptation elsewhere, that misaligned gradients will eventually force correction. These are structural certainties. But the *form* of correction—when it occurs, how it manifests, and who absorbs the cost—remains open until resolution happens.

Orientation works differently from prediction. It sharpens judgment without removing uncertainty. It tells you where not to push, what not to ignore, and which signals matter more than surface noise. It narrows the space of plausible futures without claiming authority over any single one.

Misuse begins when orientation is mistaken for foresight. When someone says, *UED predicted this*, the framework has already been distorted. At best, UED explained why such an outcome was possible or likely. Claiming prediction retroactively converts explanation into triumphalism and encourages overconfidence in future application.

Another danger lies in selective validation. When one anticipated pattern appears, confidence spikes. When another does not, it is quietly ignored. Over time, this creates the illusion of predictive power while eroding sensitivity to error. The framework becomes self-confirming rather than self-correcting.

Correct use of UED keeps prediction deliberately weak and orientation deliberately strong. It accepts uncertainty as structural, not as failure. It treats surprise as information, not embarrassment. It allows consequence to refine understanding rather than forcing reality to validate expectation.

This restraint preserves the framework's usefulness. UED remains a guide for navigating complexity, not a crystal ball for controlling it. When users respect this boundary, the framework deepens insight without amplifying risk. When they cross it, they inherit the very instability UED was meant to reveal.

30.6 — The Responsibility of Insight

Insight is not neutral. The moment a framework improves your ability to see structure, it also changes your position within the field. You are no longer just reacting; you are interpreting, anticipating, and influencing. That shift carries responsibility—not moral in the abstract sense, but consequential in the strict UED sense.

Understanding alters exposure. When you see gradients before others do, your actions propagate differently. When you recognize delayed consequences, your tolerance for short-term comfort decreases. When you understand constraint saturation, you become aware of risks that others still dismiss as noise. This asymmetry is not privilege; it is increased consequence load.

Misuse often begins when insight is treated as advantage rather than obligation. The framework becomes a way to feel superior, detached, or protected—*I see what others don't.* This posture quietly disconnects insight from feedback. Instead of reducing instability, it reallocates it. Others absorb the cost of actions informed by understanding they do not share.

UED rejects this separation. Insight does not grant exemption; it tightens accountability. If you act on structural understanding, you must be prepared to absorb more consequence, not less. If your interpretation proves wrong, the cost is yours to carry—not something to be explained away through language.

This is why restraint is an ethical necessity in UED, not a virtue signal. Overconfident application amplifies harm because it shifts pressure onto systems that cannot adapt as quickly. Speaking too early, pushing too hard, or modeling too broadly can destabilize environments that were still correcting locally.

Responsibility also applies inward. Insight increases the temptation to disengage emotionally—*this is just field dynamics.* That detachment can become avoidance. UED does not license indifference. It clarifies causality so that response can be measured rather than impulsive, not absent.

A reliable test of responsible use is this: Does your understanding make you quieter, slower, and more precise—or louder, faster, and more certain?

Correct application narrows action, not expands it. It reduces unnecessary intervention. It respects timing. It allows systems to resolve where possible and intervenes only where misalignment is becoming irreversible.

UED is powerful precisely because it does not protect the user. It exposes them. Once you see structure, you cannot unsee consequence. The framework does not offer moral authority or strategic advantage—it offers clearer visibility into what your actions will disturb.

That visibility is the final discipline of use. Without it, UED becomes another tool for domination disguised as understanding. With it, the framework remains what it was meant to be: a way to engage complex systems without becoming another source of distortion.

From Framework to Modeling

This chapter closes a necessary loop. It establishes that UED is not something to be *followed*, defended, or spread—it is something to be handled. Carefully. Quietly. With awareness of consequence. Once this discipline is in place, the framework can finally be used without turning into identity, authority, or belief.

At this point, the reader should feel a subtle shift. UED no longer sits above reality as an explanation of everything. It settles beside reality as a way of seeing pressure, limits, and movement without rushing to judgment. The urge to declare truths weakens. The need to persuade dissolves. What remains is structural attentiveness.

This is the threshold between understanding and application. Without restraint, application becomes ideology. With restraint, application becomes modeling.

Modeling is where UED belongs. Not as prophecy, not as doctrine, but as a mapping discipline—a way to sketch constraint boundaries, locate deltas, estimate thresholds, and trace consequence channels before action is taken. Modeling accepts uncertainty. It works with partial information. It improves decision quality without promising control.

The next chapter moves deliberately into this terrain. It shows how to build models that respect scale, boundary choice, and feedback—models that clarify risk without overstating certainty. The emphasis shifts from *what UED says* to *what UED helps you see.*

From here on, the framework stops speaking in conclusions and starts working through maps.

Chapter 31 — Modeling with UED

Why Modeling Matters More Than Answers

The most dangerous moment in understanding is not ignorance—it is premature certainty. Answers feel decisive. They close questions, simplify action, and offer psychological relief. But in complex systems, answers age quickly. Reality moves while conclusions stay fixed. This is where failure begins.

Modeling exists to prevent this trap.

In UED, a model is not a miniature version of reality, nor a forecast of what will happen. It is a temporary map of constraints, drawn to reduce blind action under pressure. Its purpose is not to be right, but to be *useful long enough* for consequences to be absorbed without escalation.

Answers collapse uncertainty into claims. Models preserve uncertainty while making it legible. They show where movement is possible, where pressure is accumulating, and where intervention would likely worsen instability. This distinction is essential. Acting on answers assumes control. Acting with models accepts limitation.

Most catastrophic decisions—personal, organizational, or civilizational—were not made without intelligence. They were made without modeling. Leaders reacted to surface signals, narratives, or isolated metrics without mapping constraints, feedback delays, or coupling effects. Action felt necessary, even urgent, but it was structurally blind.

UED insists on a different discipline. Before acting, one asks not *what should be done*, but *what is actually limiting resolution right now*. Before changing direction, one asks not *what outcome is desired*, but *which gradients can move without triggering collapse*. These are modeling questions, not moral or strategic ones.

Modeling also introduces humility. A good model makes you slower. It reduces confidence in sweeping action. It highlights what you do not know as clearly as what you do. If a model makes you eager, certain, or expansive, it is already failing.

This chapter shifts the framework from explanation to application without losing restraint. It teaches how to build models that clarify risk without pretending to remove it, and how to use those models to decide *where not to intervene* as much as where to act.

From here onward, UED is no longer something you cite. It is something you sketch, revise, and sometimes abandon—in conversation with consequence.

31.1 — Models as Constraint Maps, Not Predictions

In UED, a model is not an attempt to simulate reality or anticipate exact outcomes. It is a constraint map—a structured way of seeing what limits resolution, where pressure can move, and which pathways are blocked or fragile. This distinction matters because prediction seeks certainty, while constraint mapping seeks actionable humility.

Traditional models often fail because they aim to forecast results. They assign values, project trends, and extrapolate patterns as if systems remain stable while being observed. Field-based systems do not behave this way. Observation, intervention, and adaptation alter constraints continuously. Any model that promises prediction without accounting for this interaction is structurally incomplete.

Constraint maps operate differently. They begin by identifying boundaries: physical limits, institutional rules, capacity ceilings, psychological tolerances, and temporal delays. These boundaries define what *cannot* happen easily, which is often more informative than what might happen. Knowing where movement is restricted narrows the field of plausible actions without claiming foresight.

In a constraint map, outcomes are secondary. The primary focus is path availability. Which routes allow pressure to dissipate gradually? Which routes amplify stress? Which routes appear open but lead to dead ends?

This orientation transforms decision-making from goal chasing to damage-aware navigation.

Another critical feature of UED modeling is sufficiency, not completeness. A model does not need to include every variable to be useful. It needs to include the right constraints at the relevant scale. Overly detailed models obscure structure. Overly simple models mislead. The discipline lies in selecting boundaries that meaningfully shape resolution.

A reliable indicator of a good constraint map is that it reduces impulsive action. When constraints are visible, many interventions reveal themselves as unnecessary or harmful. The model does not push you toward action; it often tells you to wait, redirect, or do less.

This also explains why UED models are temporary by design. As constraints shift, the map must be redrawn. Holding onto an outdated model is more dangerous than acting without one, because it gives false confidence. Models serve reality, not the other way around.

By treating models as constraint maps rather than predictive engines, UED preserves flexibility. It allows decisions to be revised as feedback arrives. It keeps understanding aligned with consequence rather than frozen in expectation.

With this foundation in place, the next step is to identify where pressure actually resides—because without correctly locating deltas, even the best constraint map points in the wrong direction.

31.2 — Identifying Deltas: Where Pressure Is Accumulating

A model is only as accurate as its identification of deltas. Constraints define where movement is limited, but deltas define where movement is waiting. Misidentify the delta, and every subsequent conclusion—no matter how elegant—will misfire.

In UED, a delta is not noise, disruption, or activity. It is stored pressure—energy that wants to resolve but is being delayed, displaced, or suppressed by constraints. Deltas are often quiet. They accumulate gradually and reveal themselves only through secondary effects. This is why surface-level observation is unreliable; the most visible activity is rarely where the real pressure resides.

A common mistake is to equate urgency with delta. Crises attract attention, but they often represent late-stage release rather than early accumulation. By the time something feels urgent, the delta has already moved or is about to resolve violently. Effective modeling focuses upstream—on conditions that feel stable but are absorbing stress without relief.

Deltas also hide behind functionality. Systems that continue to operate smoothly while degrading internally are particularly dangerous. Efficiency masks exhaustion. Compliance hides misalignment. Growth obscures fragility. In such cases, delta accumulates in reduced resilience, narrowing options rather than visible failure.

Another error is confusing motion with pressure. High activity does not necessarily indicate high delta. Markets can be volatile yet balanced. Organizations can be busy yet aligned. Conversely, stagnant environments often conceal enormous deltas—fear, dependency, deferred maintenance, suppressed dissent—that have nowhere to go.

Identifying deltas requires asking specific, non-dramatic questions:

- Where is effort being applied without progress?
- Where are corrections repeatedly postponed?
- Where does discomfort remain unresolved but tolerated?
- Where does the system rely on delay rather than adaptation?

Deltas also migrate. Pressure displaced from one domain appears in another—economic stress becomes political polarization, personal overload becomes emotional numbness, geopolitical friction becomes supply-chain fragility. Modeling must track these migrations rather than fixating on origin stories.

A useful test of correct delta identification is proportionality. When a small intervention produces large reaction, the delta was already present. When large interventions produce little change, pressure lies elsewhere. Consequence reveals where energy was stored all along.

Once deltas are located, the model gains traction. Constraints explain *why* pressure is stuck; deltas explain *where* it is accumulating. The next step is to understand when accumulation turns into transformation—which requires identifying thresholds.

31.3 — Thresholds: When Gradual Becomes Sudden

Thresholds are where intuition fails. Systems absorb pressure quietly—sometimes for years—until, without warning, behavior changes abruptly. Stability collapses, structures reorganize, and what once seemed manageable becomes irreversible. In UED, these moments are not surprises; they are threshold crossings that were simply misread as continuity.

A threshold is not a breaking point in the dramatic sense. It is a transition condition—a level of accumulated delta beyond which the system can no longer respond linearly. Below the threshold, added pressure produces small, proportional effects. Above it, the same pressure triggers qualitative change. Gradual accumulation suddenly manifests as discontinuity.

This is why systems appear stable right before they are not. The visible variables remain within normal ranges while internal tolerances are being consumed. Capacity erodes quietly. Options narrow. Flexibility declines. Because nothing *breaks*, observers assume nothing *changes*. The threshold is crossed not when damage occurs, but when recovery pathways close.

Modeling thresholds requires abandoning the search for precise numbers. Thresholds are rarely exact values; they are ranges of vulnerability. What matters is not where the line is drawn, but how close the system is to losing adaptive response. Modeling therefore focuses on tolerance margins rather than absolute limits.

Early warning signs of threshold proximity are subtle. Increased sensitivity to small disturbances. Disproportionate reactions. Reduced recovery time. Dependence on control mechanisms rather than self-correction. When a system requires constant intervention to maintain appearance, it is likely operating near a threshold.

False alarms are common, which is why many warnings are ignored. Not every disturbance signals imminent transition. The difference lies in reversibility. If stress recedes and the system returns to baseline without intervention, the threshold has not been crossed. If the system stabilizes only through suppression or delay, tolerance is being consumed.

Another modeling error is assuming thresholds are universal. Thresholds are system-specific. What destabilizes one organization, market, or individual may be irrelevant to another. Copying thresholds across contexts creates false confidence and missed signals. UED modeling demands local sensitivity.

Understanding thresholds changes how action is evaluated. Intervening near a threshold carries far higher risk than intervening early or late. Timing matters more than intensity. Often the most effective action near a threshold is not to push, but to create alternative resolution pathways—reducing delta without forcing release.

With thresholds mapped, the model gains temporal awareness. It no longer asks only *where* pressure exists, but *how close* the system is to transformation. The next step is to understand how pressure moves once it begins to resolve—which requires mapping coupling.

31.4 — Coupling: How Stress Travels

Pressure rarely resolves where it accumulates. In complex systems, delta moves along coupling paths—channels through which stress, constraint, and consequence propagate from one domain to another. Modeling without coupling is like mapping a city without roads: the structures appear accurate, but movement becomes inexplicable.

Coupling describes how tightly systems are linked and how quickly disturbance in one produces response in another. Tight coupling transmits stress rapidly and with little attenuation. Loose coupling slows transmission, allowing dissipation, buffering, or adaptation. Neither is inherently good or bad; their suitability depends on scale, timing, and tolerance.

A common modeling error is assuming coupling follows formal boundaries. It rarely does. Economic stress migrates into social behavior.

Organizational pressure surfaces as cultural rigidity. Technological constraints reshape geopolitics. These cross-domain transfers are not metaphors; they are resolution pathways chosen by the field when direct correction is blocked.

Coupling also explains why well-intentioned interventions backfire. When pressure is relieved in one area without addressing constraints elsewhere, delta reroutes. Decoupling a supply chain might stabilize production while destabilizing labor markets. Regulating one sector can push risk into shadow systems. The model must ask not *what problem was solved*, but *where did the pressure go*.

Another failure mode is over-decoupling. In an attempt to contain risk, systems sever links aggressively. This can localize stress temporarily but often creates pressure reservoirs that grow unseen. When reconnection becomes unavoidable, release is abrupt and damaging. Healthy systems preserve selective coupling—enough to distribute load, not so much that shock cascades uncontrollably.

Modeling coupling requires tracing propagation paths, not assigning blame. The question is not who caused the stress, but how stress is likely to move given existing links and frictions. This shift removes moral distraction and improves predictive orientation without claiming certainty.

Indicators of dangerous coupling include synchronized failures, rapid sentiment shifts, and interventions that must be coordinated across domains to work at all. Indicators of overly loose coupling include isolated optimization, local success paired with systemic fragility, and delayed but massive corrections.

A practical test for coupling awareness is this: if a small change here produces a large change elsewhere, the coupling was tight—even if it was invisible. If repeated large changes here produce little effect elsewhere, the coupling was loose or blocked.

With coupling mapped, the model gains spatial awareness. It can anticipate where stress will surface once movement begins. The remaining challenge is temporal: understanding when consequence returns and who absorbs it. That requires explicit modeling of consequence channels and delay.

31.5 — Consequence Channels and Delay

Most failures are not caused by wrong actions, but by misjudged timing. Consequence rarely returns along the path we expect, and almost never at the moment we are watching. In UED modeling, understanding *how* consequence travels and *when* it returns is as important as identifying deltas or constraints.

A consequence channel is the route through which resolution feedback comes back into the system. Some channels are direct and fast—touch a hot surface, withdraw immediately. Others are indirect and slow—debt, erosion of trust, capability decay, demographic shifts. The more complex the system, the more indirect and delayed the dominant channels tend to be.

Delayed consequence is the primary source of surprise. When feedback is slow, systems misinterpret success. Actions that produce short-term relief appear effective, even when they are accumulating long-term delta elsewhere. Comfort becomes evidence. Stability is inferred. By the time consequence returns, the original action feels distant and unrelated.

Modeling must therefore make delay explicit. If a model cannot answer *when* feedback is likely to appear and *where* it will surface, it is incomplete. Delay compresses accountability. It separates cause from effect so widely that corrective learning fails. This is why intelligent systems repeatedly make the same structural mistakes.

Institutions and narratives often amplify delay. Bureaucracy smooths impact. Media reframes consequence as coincidence. Metrics lag reality. Buffers—financial, political, emotional—absorb stress temporarily, but they also hide depletion. When buffers finally exhaust, correction feels sudden and disproportionate.

Another modeling error is assuming consequence returns symmetrically. It rarely does. Pressure applied in one domain may return in another. Decisions taken by leadership may be absorbed by workers. Policies enacted today may burden future generations. When models assume local consequence, they underestimate harm and overestimate control.

A useful modeling discipline is to trace second- and third-order consequence channels deliberately. Not exhaustively, but honestly. Ask:

- If this works, who pays later?
- If this fails, where will failure appear?
- What buffers are being consumed to maintain appearance?
- Which consequences are invisible because they are delayed?

Delay also explains why restraint often looks ineffective. Choosing not to act rarely produces immediate reward. It preserves optionality and tolerance margins—benefits that only become visible when avoided crises never materialize. Models that ignore delay systematically undervalue restraint.

Correctly modeling consequence channels shifts decision-making from reaction to stewardship. It does not eliminate uncertainty, but it reduces self-deception. When feedback is expected to return late and indirectly, confidence softens, timing slows, and intervention becomes more selective.

With consequence channels and delay mapped, the model becomes temporally aware. It can anticipate not outcomes, but risk horizons. The final challenge is ensuring the model itself is not the problem—which brings us to boundary choice, the most common and most costly modeling failure.

31.6 — Boundary Choice: Where Most Models Fail

The most common reason models fail is not faulty logic or missing data—it is wrong boundary choice. A model can be internally consistent, technically elegant, and completely useless if it excludes the variables that actually matter. In UED, boundary choice is not a technical detail; it is the central modeling decision.

Boundaries determine what is considered inside the system and what is treated as external. They define whose stress counts, which constraints are visible, and which consequences are acknowledged. Every boundary is an assumption, and every assumption carries risk. The danger lies in forgetting that boundaries are provisional, not real.

Overly narrow boundaries are the most frequent error. They focus on what is measurable, controllable, or familiar while excluding slow, indirect, or uncomfortable factors. A financial model ignores social tolerance. A productivity model ignores burnout. A geopolitical model ignores internal cohesion. The system appears optimized—until excluded variables react.

Overly broad boundaries fail differently. They include everything, dilute relevance, and produce analysis paralysis. When too many variables are treated as equally important, structure disappears. Action stalls, not because the model is wrong, but because it has no resolution power. Modeling requires selective exclusion, not maximal inclusion.

Another failure mode is convenience-based boundaries. Models are shaped around available data, institutional jurisdiction, or existing narratives rather than around real coupling. What is easy to model is mistaken for what is important. When reality pushes back, the model is defended instead of revised.

Boundary failure often reveals itself through surprise. When consequence arrives from outside the model, the response is disbelief: *this wasn't part of the system*. In truth, it always was. The boundary merely hid it. UED treats surprise as a diagnostic signal—not of randomness, but of exclusion.

Correct boundary choice is iterative. It evolves with feedback. When reality disagrees with the model, the response is not to refine prediction, but to redraw the boundary. This requires humility. It means letting go of elegant explanations in favor of messier but more accurate maps.

A practical test for boundary health is to ask:

- What have we labeled as external noise?
- Who absorbs consequence without representation?
- Which variables only appear after failure?
- What must stay quiet for this model to hold?

Good models fail gracefully. They reveal their limits early. Bad models persist confidently until collapse forces correction.

Understanding boundary choice is the final safeguard in UED modeling. It ensures that models remain responsive rather than rigid, useful rather than impressive. With this discipline in place, modeling becomes a living practice—one that adapts alongside the systems it seeks to navigate.

The final section completes the chapter by showing how modeling itself must remain adaptive, or it too becomes a constraint.

31.7 — Iterative Modeling and Adaptive Use

A UED model is not something you *finish*. It is something you work with until it stops working—and then revise or discard. Treating a model as a completed object is the final step by which modeling quietly turns into belief.

Iteration is not refinement for precision; it is responsiveness to consequence. As soon as action is taken—or even deferred—the field changes. Constraints shift, deltas migrate, coupling tightens or loosens, delays compress or extend. A model that remains unchanged while reality evolves is no longer a model; it is a fossil.

Most people resist iteration because it feels like error. In UED, iteration is evidence of correctness. A model that never needs revision is almost certainly too abstract, too insulated, or too detached to be interacting with reality at all. Real models attract friction.

Adaptive use begins with a simple rule: The model must answer to consequence, not to intention.

When outcomes diverge from expectation, the task is not to defend assumptions or reinterpret results. It is to ask what the model failed to see:

- Was a delta underestimated?
- Was a coupling path ignored?
- Was a delay longer than assumed?
- Was the boundary drawn too narrowly—or too broadly?

Iteration also requires knowing when to stop modeling. Over-modeling becomes avoidance. When constraints are clear and risk is bounded, continued modeling adds little value and may delay necessary action. UED modeling should reduce activity, not replace decision-making.

Equally important is knowing when to abandon a model entirely. Some systems undergo phase shifts so profound that previous maps become misleading. Holding onto them out of familiarity is more dangerous than starting fresh. Letting go is not failure; it is alignment with a changed field.

Adaptive modeling also changes posture. You stop asking, *What does the model say?* and start asking, *What is the model missing now?* This keeps attention outward, anchored to signals rather than conclusions.

The mark of a mature UED user is not confidence in their models, but ease in discarding them. They are not attached to being right; they are attached to staying responsive.

This completes the modeling discipline. With constraint maps drawn, deltas located, thresholds sensed, coupling traced, consequence channels timed, boundaries respected, and iteration embraced, UED modeling becomes what it was meant to be—not a system of control, but a practice of informed restraint.

The final chapter now turns inward, showing how this same discipline operates at the personal level—where decisions are smaller, consequences are closer, and misuse is easiest to justify.

From Modeling to Living Systems

This chapter completes the shift from explanation to usable discipline. Modeling with UED is no longer about insight for its own sake; it becomes a way of staying aligned while acting under uncertainty. The emphasis is not on mastering complexity, but on avoiding the predictable errors that complexity punishes most harshly—overconfidence, rigidity, and delayed correction.

At this point, UED models should feel lighter, not heavier. They are not something you carry forward intact. They are something you set down, redraw, or abandon as the field changes. Their value lies in the clarity they provide *before* action, and in the humility they preserve *after* consequence returns.

This closing matters because modeling, if left abstract, can still drift into detachment. It can become something done *about* systems rather than *within* them. UED does not allow that separation. The modeler is always part of the field they are modeling. Their choices introduce deltas, alter coupling, and affect consequence channels.

The next and final chapter makes this explicit. It brings UED fully inside the personal domain—not as self-help, optimization, or control, but as an operating logic for everyday decisions. The scale is smaller, the feedback faster, and the temptation to misuse understanding greater.

Where this chapter taught you how to map systems without turning them into belief, the final chapter shows how to act without turning insight into identity.

From maps, the focus now moves to decisions.

Chapter 32 — UED as a Personal Operating System

Why Personal Application Is the Hardest Test

Understanding systems at a distance is comparatively easy. One can analyze markets, institutions, or civilizations without being directly affected by the consequences of interpretation. Personal application is different. Here, you are inside the system you are trying to understand, and every insight immediately collides with emotion, habit, identity, and fear.

This is why the personal domain is the hardest test of Universal Energy Dynamics. There is no abstraction to hide behind. Feedback is faster, discomfort is closer, and the temptation to bend understanding in favor of comfort is constant. At scale, misalignment can be delayed. Personally, it cannot.

UED was never meant to be a self-improvement philosophy or a new way to define oneself. The moment it becomes identity—*I am someone who understands UED*—it stops functioning. Identity demands consistency; living systems require adaptation. The two are incompatible.

Personal application also exposes a subtle trap: the urge to optimize the self. Many frameworks promise efficiency, productivity, or inner mastery. UED does not. It does not offer control over life; it clarifies where control fails and why restraint matters. Used personally, it often reduces ambition before it increases effectiveness. That feels like loss before it feels like stability.

Another difficulty lies in proximity to consequence. When modeling external systems, errors can be observed calmly. When modeling oneself, errors are felt. Discomfort becomes personal, and the mind rushes to reinterpret signals as problems to fix rather than information to read. This is where most misuse begins.

This chapter does not teach techniques, routines, or practices. It offers something more demanding: an operating logic that runs quietly beneath decision-making. It does not replace values, goals, or preferences. It only asks that you see which pressures you are choosing, which gradients you are steepening, and which consequences you are postponing.

If UED can be used here—without becoming belief, excuse, or identity—then it has passed its strongest test.

What follows is not guidance on how to live better. It is a way to live with clearer awareness of consequence, so that fewer decisions need correction later.

32.1 — Decision-Making as Delta Selection

Every decision, whether deliberate or deferred, is a choice about which delta you are willing to carry. This is the most precise way to understand decision-making through UED. You are not choosing outcomes; you are choosing where pressure will accumulate and how long you will tolerate it.

Most decisions are framed as gains or losses, successes or failures. That framing hides the real mechanism. A choice that feels good now often transfers delta forward—into fatigue, dependency, resentment, or fragility. A choice that feels heavy now often reduces future pressure by allowing alignment to occur early. UED does not moralize this distinction; it makes it visible.

Avoiding a decision does not eliminate delta. It merely relocates it. When action is postponed, pressure does not disappear—it embeds itself in narrowing options. What initially feels like freedom becomes constraint. Over time, the system chooses for you, and the delta arrives without consent. This is why indecision often produces harsher consequences than imperfect action.

Seen this way, decisiveness is not about speed or confidence. It is about honest accounting. Before acting, the relevant question is not *what do I want*, but *what pressure am I choosing to absorb*. Short-term discomfort, social friction, loss of comfort, uncertainty, effort—all are deltas. The issue is not whether they exist, but whether they are survivable and resolving.

UED also clarifies why "doing the right thing" often feels wrong initially. Alignment rarely produces immediate relief. Early correction increases awareness of pressure before it dissipates. Misalignment, by contrast, often feels smooth until it fails abruptly. Decision-making based on comfort therefore systematically selects delayed deltas over immediate ones.

Another trap lies in over-optimization. Trying to minimize all deltas at once is impossible. Systems require load to function. Growth, learning, and adaptation all involve carrying pressure temporarily. The discipline is not elimination, but selection—choosing deltas that resolve rather than compound.

A practical sign of a good decision, in UED terms, is this: the pressure becomes clearer, not more confusing. It may feel heavier, but it feels directional.

A poor decision often feels lighter at first and vague later. Pressure diffuses into anxiety, avoidance, or chronic dissatisfaction. Nothing is obviously wrong, yet nothing resolves. That is delta without pathway.

Personal decision-making through UED is therefore not heroic or dramatic. It is quiet, almost unremarkable. You choose the pressure you can live with now so that life does not choose a harsher one for you later.

This is not control. It is responsibility for consequence timing.

With this foundation, the next step is learning how to read internal deltas accurately—without mistaking every discomfort for misalignment or every calm state for stability.

32.2 — Reading Internal Deltas Without Dramatizing Them

Once decision-making is understood as delta selection, the next discipline is learning how to read internal deltas accurately. Most personal misalignment does not come from ignoring signals, but from misinterpreting them—either inflating them into emergencies or dismissing them as weakness.

Internal deltas often announce themselves as discomfort: tension, restlessness, hesitation, fatigue, irritation, or low-level anxiety. The reflex is to treat these states as problems to eliminate. In UED, they are treated as information about pressure distribution. The signal is not asking for immediate action; it is asking to be located.

Dramatization is the most common error. When every discomfort is interpreted as misalignment, life becomes reactive. Small, normal gradients—learning effort, adaptation stress, uncertainty—are mistaken for danger. The system overcorrects, oscillates, and loses continuity. Growth becomes impossible because all pressure is treated as threat.

The opposite error is suppression. When discomfort is normalized indefinitely, internal deltas accumulate silently. Calm becomes a performance. Efficiency hides exhaustion. Emotional flatness is mistaken for stability. In such cases, the absence of signal is not health—it is blocked feedback.

UED reframes the question. Instead of asking *how do I get rid of this feeling*, one asks *what is this pressure trying to resolve*. Some internal deltas are signals of misalignment—values violated, boundaries crossed, capacity exceeded. Others are signals of alignment in progress—new load being integrated, identity shedding, competence forming.

A useful distinction lies in directionality. Healthy deltas feel tense but clear. They point somewhere. Even if uncomfortable, they have orientation. Misaligned deltas feel diffuse. They loop without resolution. They drain without teaching. They persist without narrowing options.

Another indicator is recoverability. After rest or reflection, does the pressure return stronger or weaker? Resolving deltas reduce with honest engagement. Suppressed deltas rebound. Dramatized deltas oscillate.

Reading internal deltas also requires patience. Immediate interpretation often misleads because the system has not yet stabilized enough to reveal structure. Sitting with pressure long enough for pattern to emerge is itself a form of alignment. This is why premature action often worsens confusion.

UED does not ask you to become emotionally detached. It asks you to become structurally attentive. Feelings are not instructions; they are measurements. They tell you where load exists, not what to do about it.

When internal deltas are read correctly, response becomes measured rather than impulsive. Not every signal demands action. Not every calm state indicates health. Clarity emerges not from intensity, but from allowing pressure to disclose its structure.

With this skill in place, personal stability and growth can be reframed—not as opposing goals, but as gradients to be managed rather than chased.

32.3 — Stability vs Growth as Gradient Management

Stability and growth are often treated as opposites. Stability is framed as comfort or stagnation; growth as ambition or expansion. This framing creates unnecessary conflict. In UED, both are understood through a single mechanism: gradient management.

A gradient is the difference in pressure between where you are and where you are trying to move. Growth steepens gradients. Stability smooths them. Neither is inherently good or bad. What matters is whether the gradient remains absorbable over time.

Most personal breakdowns occur not because people avoid growth, but because they steepen gradients faster than their system can integrate. New responsibilities, expectations, identities, or ambitions are added without increasing capacity or recovery pathways. Initially, this feels energizing. Momentum masks overload. Eventually, pressure outruns resolution, and the system collapses into fatigue, anxiety, or disengagement.

Stability, in UED terms, is not staying still. It is maintaining gradients within a range that allows continuous correction. A stable system still changes—it just does not change faster than it can recover. This is why periods of consolidation are not regressions; they are load redistribution phases. They allow previous deltas to resolve before new ones are introduced.

Growth becomes destructive when it is pursued as acceleration rather than integration. Chasing more—more achievement, more clarity, more control—often steepens gradients without creating new pathways. The result is apparent progress paired with declining resilience. Options narrow even as activity increases.

Conversely, excessive stability becomes stagnation when gradients are flattened entirely. When no pressure is allowed, learning halts. Sensitivity dulls. The system becomes brittle because it no longer practices adaptation. Comfort without challenge consumes tolerance silently.

UED reframes the decision not as *should I grow or stabilize*, but as *what gradient can I sustain right now*. This question changes timing. Sometimes growth is appropriate—but only after stability has restored margin. Sometimes stability is necessary—but only to prepare for future steepening.

A practical signal of healthy gradient management is recoverability. After effort, can you return to baseline without collapse? After rest, does capacity increase slightly? If yes, gradients are aligned. If effort leaves lasting depletion or rest fails to restore clarity, gradients are too steep or misdirected.

Another signal is coherence. Growth aligned with capacity feels demanding but integrated. Stability aligned with purpose feels quiet but not empty. Mismanaged gradients feel either frantic or numb.

UED does not advocate balance as a static state. It advocates dynamic calibration—steepening gradients deliberately, smoothing them deliberately, and never confusing intensity with progress.

When stability and growth are managed as gradients rather than identities, personal development becomes less dramatic and more durable. You stop oscillating between burnout and withdrawal. Change continues—not because it is forced, but because the system remains capable of carrying it.

This prepares the ground for a deeper correction: understanding continuity properly—not as preserving who you are, but as allowing process to continue even when form must change.

32.4 — "Perpetual" Correctly Stated: Continuity of Process, Not Form

One of the most persistent personal illusions is the desire for permanence—of identity, role, clarity, or success. We seek states that can be reached and then maintained. In UED, this impulse is a misunderstanding of what continuity actually means. Nothing persists by remaining the same. Continuity is preserved only through change.

"Perpetual" does not mean unchanging. It means self-correcting. Processes endure by allowing their forms to evolve as constraints shift. When form is protected at the expense of process, decay begins—even if everything appears stable on the surface.

This is why personal identities become fragile over time. Roles that once fit begin to constrain. Beliefs that once clarified start to blind. Habits that once stabilized turn rigid. The problem is not change; the problem is refusing it in the name of continuity. What is preserved is not life, but appearance.

UED reframes continuity as unbroken resolution, not fixed structure. A river remains a river not because its shape is preserved, but because flow continues. When flow is blocked to preserve shape, stagnation follows. The same logic applies personally. You remain "you" not by defending who you were, but by allowing the process that made you to keep operating.

This correction is difficult because form feels safe. Process feels uncertain. Letting go of form can feel like loss of self, regression, or failure. In reality, it is often the condition for deeper stability. When outdated structures dissolve voluntarily, collapse is avoided. When they are defended past relevance, collapse becomes corrective.

Another misuse arises when people chase "perpetual growth" as accumulation. Growth that preserves form—more of the same identity, more of the same ambition—eventually saturates. True continuity requires periodic shedding. Capacity is renewed not by stacking layers, but by allowing obsolescence.

A useful internal check is this: Is effort being spent maintaining an identity, or allowing resolution to continue?

Maintenance-heavy living consumes energy without producing adaptation. Process-oriented living consumes energy but returns it as increased resilience.

UED therefore does not ask you to reinvent yourself constantly. It asks you to stop freezing yourself. Let roles end cleanly. Let beliefs update without drama. Let success expire without clinging. These are not losses; they are continuity mechanisms.

When continuity is understood correctly, fear of change reduces. You stop asking *how do I keep this?* and start asking *what must change so this can continue?* That shift alone dissolves much personal conflict.

With this reframing in place, the personal domain widens. Continuity is no longer private—it becomes relational. How your internal state couples with others now matters, because your process does not unfold in isolation.

32.5 — Personal Coupling: How Your State Spreads

No personal state remains personal for long. Humans are inherently coupled systems. Mood, clarity, tension, and instability propagate through interaction even when nothing is spoken. In UED terms, your internal alignment participates in the shared field, whether you intend it or not.

This is why personal instability often produces social friction without clear cause. Irritability spreads. Anxiety amplifies uncertainty. Withdrawal creates gaps others rush to fill. None of this requires intent. Coupling transmits state, not explanation. People respond to pressure before they understand it.

Many individuals underestimate this effect because coupling feels intangible. Yet its consequences are concrete. A calm but misaligned person can destabilize a group more effectively than an openly stressed one, because feedback is blocked. Others sense inconsistency without being able to locate it. Trust erodes quietly. Coordination suffers.

Conversely, a person carrying visible pressure honestly often stabilizes the field. Clarity allows others to orient. Boundaries become legible. Even discomfort, when coherent, reduces collective confusion. This is why emotional suppression is not neutral—it exports uncertainty.

Personal coupling also explains why "working on yourself" in isolation frequently fails. Internal correction that ignores external coupling remains incomplete. If your environment reacts negatively to your internal state, that reaction is not interference—it is feedback. UED treats interpersonal response as part of the system, not as noise to be filtered out.

Another common error is over-responsibility. Recognizing coupling does not mean absorbing everyone else's delta. That leads to exhaustion and blurred boundaries. Healthy coupling requires selective buffering. You allow some pressure to pass through, you dampen some, and you consciously decouple when transmission would cause harm.

Decoupling, however, is not withdrawal. It is temporary insulation to prevent escalation. Done poorly, it becomes avoidance. Done well, it preserves alignment long enough for correction to occur. The distinction lies in intent and duration. Are you creating space for resolution—or hiding from consequence?

UED reframes personal responsibility here. You are not responsible for controlling others' states. You are responsible for not exporting unresolved pressure. This alone reduces collective instability more effectively than any attempt at influence.

A practical check is simple: After interaction, does the field feel clearer or heavier?

If clarity increases—even with disagreement—your coupling is healthy. If heaviness accumulates without articulation, unresolved delta is spreading.

Understanding personal coupling shifts behavior quietly. You speak less reactively. You pause more often. You stop performing calm and start allowing alignment. This does not make you passive; it makes you structurally considerate.

With coupling understood, the final personal discipline becomes clear: knowing when action helps resolution—and when restraint does more than effort ever could.

32.6 — Restraint as the Highest Skill

In a world that rewards action, restraint is easily mistaken for weakness. Yet in UED, restraint is not inaction—it is precision in intervention. It is the ability to sense when a system is already resolving and to avoid becoming the disturbance.

Most personal misalignment does not come from doing too little, but from doing too much at the wrong moment. Intervention applied before structure is clear amplifies confusion. Advice offered before pressure is located creates resistance. Effort applied before capacity is restored accelerates exhaustion. Restraint prevents these errors.

Restraint begins with recognizing that not every delta requires interference. Some pressures resolve through tolerance and time. Growth often requires letting discomfort integrate rather than rushing to remove it. Stability often returns when noise is reduced, not when solutions are added.

This discipline is difficult because restraint offers no immediate reward. Action produces visible effort and social validation. Restraint produces nothing to point to—only the absence of escalation. Its success is invisible, which is why it is undervalued.

UED reframes effectiveness. An effective response does not maximize output; it minimizes unnecessary disturbance. The question shifts from *what should I do* to *would my action help resolution or interrupt it*. Often, the most aligned response is to wait, observe, or reduce engagement until the system clarifies itself.

Restraint also protects against emotional overreach. When insight increases, the temptation to correct others rises. Seeing misalignment does not obligate intervention. In fact, premature correction often exports delta instead of resolving it. Restraint here is not indifference; it is respect for the system's own capacity to self-correct.

Another form of restraint is internal. Not every thought needs expression. Not every feeling needs action. Allowing internal signals to mature before responding improves alignment dramatically. This is why clarity often emerges after silence rather than after analysis.

Importantly, restraint is not permanent. It is situational. When thresholds approach or irreversible harm looms, restraint becomes negligence. The skill lies in sensing the difference. UED does not provide rules; it sharpens sensitivity to timing.

A reliable indicator of appropriate restraint is reduced urgency without reduced awareness. You remain attentive without being compelled. When restraint is correct, tension lowers without dullness. When restraint is avoidance, pressure diffuses into anxiety or guilt.

Restraint is the highest skill because it requires trust—not in outcomes, but in process. It accepts that resolution cannot be forced, only facilitated. Those who master restraint act less, but when they act, they do not need correction later.

With this discipline in place, personal application reaches its final test: living without final conclusions, without closing the system prematurely, and without needing certainty to function.

32.7 — Living Without Final Conclusions

The final discipline of using UED personally is learning to function without closure. The human mind craves final answers because they reduce uncertainty and stabilize identity. But in living systems, final conclusions are rarely correct for long. They freeze understanding while reality continues to move.

UED does not ask you to abandon direction or values. It asks you to abandon the need to close the system. When conclusions harden, sensitivity drops. Feedback is filtered. Correction becomes threatening. The very clarity that once helped now obstructs adaptation.

Living without final conclusions does not mean drifting or indecision. It means holding understanding lightly—firm enough to act, loose enough to revise. Decisions are made, but they are not defended as permanent truths. Beliefs guide, but they are not armored.

This posture changes the emotional relationship to learning. Surprise becomes information rather than embarrassment. Being wrong becomes adjustment rather than failure. Disagreement becomes signal rather than threat. Life becomes navigable again, not because it is predictable, but because you are responsive.

One of the most common personal misuses of insight is narrative fixation. We tell ourselves stories about who we are, what is happening, and where things are going. These stories simplify complexity, but they also narrow perception. UED loosens narrative grip. It invites you to see patterns without demanding a storyline.

Another benefit of open-endedness is reduced internal conflict. When identity is not tied to conclusions, internal contradiction softens. You can hold tension without resolving it immediately. You can act without needing certainty. This reduces oscillation between confidence and doubt.

Practically, this shows up as quieter language. Fewer absolutes. More conditional statements. Less urgency to explain yourself. More willingness to let experience update understanding before explanation solidifies.

Living without final conclusions is not an abdication of responsibility. It is responsibility aligned with reality. You remain accountable for your actions, but you do not pretend to control outcomes you cannot.

This is where UED dissolves into behavior. You stop naming the framework. You stop referencing concepts. You simply notice pressure, respect limits, choose deltas, and adjust when consequence returns.

At that point, UED is no longer something you use. It is something that quietly stops you from misusing yourself.

And that is where this book ends—not with answers, but with the capacity to continue without needing them.

Using UED Without Carrying It

This is where the framework finally disappears.

If Universal Energy Dynamics has been used correctly, it does not remain as a set of concepts, terms, or explanations you carry forward. It recedes into the background, doing its work quietly. You stop *thinking in UED* and start behaving with consequence awareness. The distinction matters.

At this point, there is nothing left to apply consciously. Decisions are made with an instinctive sense of pressure, timing, and tolerance. You hesitate not out of fear, but out of calibration. You act not out of certainty, but out of sufficient clarity. Correction happens early, gently, without drama.

This is why the book ends here.

UED was never meant to become a lens through which you constantly interpret yourself or others. That would be another form of fixation. Its role is more modest and more demanding: to reduce avoidable damage, to prevent premature certainty, and to keep resolution pathways open when life tightens.

When the framework is no longer named, defended, or referenced, it is doing its job. You notice misalignment sooner. You intervene less.

You recover faster. You stop exporting unresolved pressure. You allow systems—internal and external—to resolve where they can, and you step in only where they cannot.

Nothing about this looks impressive from the outside. There is no identity to display, no philosophy to announce, no method to sell. Effectiveness becomes quiet. Stability becomes unremarkable. Growth becomes survivable.

That is the final deliverable of this book.

Not a worldview. Not a doctrine. Not even a framework to remember.

Just a way of moving through pressure without turning understanding into another source of distortion.

From here on, UED does not travel with you as knowledge. It dissolves into judgment, restraint, and timing.

And that is where it belongs.

Deliverable of Part IX

UED Without the Weight of a Library

This part completes the final and most important transition of this book: *from framework to function*. Up to this point, Universal Energy Dynamics has been explained, tested across scales, and carefully protected from becoming belief or ideology. The outcome of this part is deliberately modest and precise.

Readers do not leave with a system to memorize, defend, or continuously reference. They leave with a *working logic* that operates even when the language disappears.

This matters because frameworks often fail at scale not due to lack of depth, but due to excess of it. When understanding requires constant recall, cross-referencing, or conceptual loyalty, it becomes burdensome. It slows action and invites misuse. UED was never meant to be carried as an intellectual load.

This part strips UED down to what survives without explanation:

- The ability to sense where pressure is accumulating
- The habit of choosing deltas consciously rather than reactively
- Awareness of thresholds before collapse forces correction
- Sensitivity to coupling and delayed consequence
- Restraint in intervention when resolution is already underway

With these in place, the reader no longer needs diagrams, terminology, or extended theory. The framework has done its work when it becomes *invisible*.

This is why the metaphor of a "library in the backpack" matters. A library suggests accumulation, reference, and authority. This part does the opposite. It ensures that UED can be used lightly—while thinking, deciding, interacting, and pausing—without turning every moment into analysis.

Readers are not asked to *remember* UED. They are enabled to *notice misalignment sooner*, to avoid overreaction, and to correct course before consequence becomes punitive. This is not mastery; it is *reduced error*.

The deeper implication is restraint. When understanding no longer needs reinforcement, it stops demanding attention. Life regains its natural pace. Decisions become simpler because fewer of them need to be made urgently.

That is the real deliverable of final part.

UED becomes something readers *use without carrying*, apply without naming, and leave behind without losing.

At that point, the framework has fulfilled its purpose—and the book can end without asking anything further of the reader.

After This

Nothing here needs to be carried forward.

What has been described does not continue as an idea, a framework, or a method. It continues only in the same way everything else does—through adjustment, correction, and response to consequence.

There is no stable state to reach after this point. Systems do not arrive anywhere. They either remain responsive or they don't. When responsiveness is present, resolution stays possible. When it is lost, correction returns through force rather than choice.

Universal Energy Dynamics was introduced only to make that difference visible.

If it has done its work, it does not remain active in thought. It does not need to be remembered, referenced, or applied deliberately. It fades into background judgment—into when to act, when not to, and when to let pressure resolve without interference.

Nothing here asks for agreement. Nothing here requires belief. Nothing here needs to be defended.

The framework does not protect against error. It reduces the cost of error by shortening delay. It does not offer certainty. It limits overconfidence. It does not promise stability. It clarifies why stability fails when misread as permanence.

From this point onward, life does what it always does. Gradients shift. Constraints tighten and loosen. Some resolutions arrive quietly. Others arrive late and forcefully. No model prevents this. At best, it keeps attention aligned long enough to avoid unnecessary damage.

If that alignment remains, the framework has already stepped aside.

What follows—decisions, pauses, corrections, restraint—does not belong to this book. It belongs to the same process that was already underway before it was opened.

Nothing ends here. Nothing needs to.

Closing Note

This book was not written to persuade, impress, or complete a theory. It was written to remove a certain kind of confusion—one that survives intelligence, education, and even experience.

For a long time, I noticed a repeating pattern across domains that had nothing obvious in common: physics, psychology, organizations, societies, and personal life. Systems did not fail because people lacked effort or intention. They failed because **pressure was misunderstood**, consequences were delayed, and intervention arrived too early or too forcefully.

Universal Energy Dynamics emerged as a way to speak about that pattern without turning it into belief.

I have deliberately resisted the temptation to make this framework larger than it needs to be. There are no grand claims here, no promises of prediction, no assurances of control. Wherever certainty was tempting, I chose restraint. Wherever explanation could harden into doctrine, I stepped back.

If something in this book feels incomplete, that is intentional. Living systems are incomplete by nature. Any framework that pretends otherwise becomes brittle.

I do not expect agreement from every reader. I do not even expect acceptance. What I do hope is that the language used here helps you notice pressure earlier, intervene less often, and tolerate uncertainty without rushing to resolve it prematurely.

If the book has done its work, you will not feel the urge to quote it, defend it, or apply it everywhere. You will simply find yourself making fewer corrections later.

At that point, this book no longer belongs in your hands.

It belongs where all understanding eventually does—quietly absorbed into judgment, timing, and restraint.

— **Sandeep J. Chavan**

Case Studies

Applying Universal Energy Dynamics

Case Study 1 — The 2008 Global Financial Crisis

When Delta Accumulates Behind Artificial Stability

Financial systems often appear stable precisely when structural pressure is quietly accumulating beneath the surface. The global financial crisis of 2008 offers a clear example of how systems can operate smoothly for years while internal imbalances intensify. From a Universal Energy Dynamics perspective, the crisis was not a sudden failure of markets. It was the delayed resolution of a large quantity of accumulated delta within a tightly coupled financial field.

Constraint

During the years preceding the crisis, financial institutions operated under several powerful structural constraints. Regulatory frameworks allowed complex derivatives to grow faster than oversight mechanisms. Credit rating systems simplified risk into easily tradable categories, and financial incentives encouraged the rapid expansion of mortgage lending. Banks, investment firms, and rating agencies were all bound by institutional constraints that rewarded short-term growth while obscuring long-term exposure.

These constraints did not immediately create instability. In fact, they created the appearance of efficiency and liquidity. However, by limiting the visibility of underlying risk and compressing feedback loops, they restricted the system's ability to resolve pressure gradually.

Delta

Within these constraints, financial pressure began to accumulate. Mortgage loans were increasingly issued to borrowers with weak repayment capacity. These loans were then bundled into complex financial products that circulated widely across global markets.

Each layer of securitization moved risk further from its origin without eliminating it. Instead of resolving imbalance at the local level—through lending discipline or price correction—the system distributed unresolved exposure across institutions, countries, and asset classes.

In UED terms, delta was accumulating quietly inside the financial field. Because immediate consequences were delayed, the system interpreted the absence of visible failure as stability.

Gradient

The gradient driving the system during this period was the enormous incentive to expand credit and generate returns. As housing prices rose, borrowing appeared increasingly safe, reinforcing the gradient toward more lending and more leverage.

Financial institutions competed to accelerate this movement. Capital flowed toward mortgage-backed securities and derivative instruments that promised high yield with apparently manageable risk. The stronger this gradient became, the faster the system moved in that direction, further increasing the quantity of unresolved delta.

ATT — Alignment Threshold Time

For years, the system remained within its tolerance window. Rising asset prices masked the underlying imbalance. As long as borrowers continued making payments and housing prices continued to rise, the system could absorb the growing pressure.

However, every system has a limit to how long unresolved delta can remain contained. In financial markets this threshold is reached when losses begin to propagate faster than institutions can absorb them.

When housing prices stalled and mortgage defaults began to increase, the financial system crossed its Alignment Threshold Time. Gradual correction was no longer possible. The pathways for slow resolution had already been eliminated by excessive coupling and leverage.

Coupling

Modern financial systems are deeply interconnected. Banks lend to each other, institutions share exposure to the same assets, and global markets transmit shocks instantly.

This coupling dramatically amplified the crisis. Once losses began appearing in mortgage-backed securities, uncertainty spread across the entire network. Institutions could no longer reliably determine who held the risk. Credit markets froze as trust between participants collapsed.

What had begun as localized mortgage instability rapidly propagated through investment banks, insurance firms, and international markets. High coupling allowed delta to redistribute through the entire financial field.

Consequence

The visible crisis of 2008 was therefore not the origin of instability—it was the release of accumulated pressure that had been building for years. Major financial institutions collapsed or required government rescue. Global credit markets seized, investment contracted sharply, and economies entered deep recession.

These outcomes were the system's structural response to the resolution of delta. Once the threshold had been crossed, correction occurred rapidly and at enormous cost.

Correction

Following the crisis, governments and financial regulators attempted to restore equilibrium through several corrective mechanisms: capital requirements were increased, derivative markets faced greater oversight, and stress-testing frameworks were introduced to monitor systemic risk.

These reforms aimed to widen the system's tolerance margins and reduce the likelihood of similar delta accumulation in the future. However, UED reminds us that no system can permanently eliminate pressure. The task is not to prevent delta from forming, but to ensure that resolution pathways remain open before thresholds are reached.

Structural Insight

Viewed through the UED lens, the 2008 financial crisis illustrates a recurring pattern in complex systems:

- Constraints suppress local correction
- Delta accumulates behind apparent stability
- Gradients drive further expansion
- Coupling distributes hidden risk
- Thresholds eliminate gradual resolution
- Consequences appear sudden but are structurally delayed

What appears as a market crash is therefore better understood as a late-stage correction within a system that postponed adjustment for too long.

Case Study 2 — The COVID-19 Pandemic

Signal, Noise, and Delayed Consequence in a High-Coupling World

Pandemics are not only biological events. They are systemic events that unfold across medical, social, economic, and informational fields simultaneously. The COVID-19 pandemic demonstrates how complex systems respond when early signals are misread, feedback is delayed, and institutional constraints slow structural adjustment. From the perspective of Universal Energy Dynamics, the global disruption that followed was not solely the result of the virus itself. It was the consequence of unresolved pressure propagating through highly connected systems.

Constraint

Before the outbreak, most national health systems were structured around efficiency rather than resilience. Hospitals optimized for normal patient loads, supply chains minimized inventory to reduce cost, and public health institutions operated within bureaucratic decision frameworks that slowed rapid response.

These constraints were not irrational; they were products of decades of economic optimization. However, they reduced the system's flexibility to absorb sudden shocks. When an unexpected biological event emerged, the system lacked spare capacity to resolve it gradually.

Delta

As the virus began spreading in localized regions, infection pressure started to accumulate. Early cases represented the initial formation of delta within the global health system. At this stage, the imbalance was still relatively small and potentially manageable.

However, because many early signals were uncertain or contradictory, corrective action was delayed. Travel, trade, and everyday social interactions continued largely unchanged. Each day of delay allowed additional transmission chains to form, increasing the unresolved pressure within the system.

The virus itself did not create instability immediately. The instability arose from the growing gap between the rate of transmission and the system's ability to detect, isolate, and contain it.

Gradient

The gradient driving the spread of the virus was simple: human mobility and social interaction. In a globally connected world, people move constantly through dense networks of travel, commerce, and communication.

This connectivity created a powerful gradient for viral propagation. International flights, urban density, and large gatherings allowed the pathogen to move rapidly across geographic boundaries. Once the gradient was established, containment became progressively more difficult.

ATT — Alignment Threshold Time

For a period, the global system remained within its tolerance window. Health authorities monitored the situation, and many governments assumed that the outbreak could be contained locally.

However, as transmission accelerated and cases began appearing across multiple countries, the system approached its Alignment Threshold Time. When healthcare systems started facing surges beyond their capacity, gradual correction was no longer possible.

At this point, governments were forced to implement abrupt interventions—lockdowns, travel bans, and emergency measures. These actions were not gradual adjustments but threshold responses triggered by the system's inability to absorb further pressure.

Coupling

Modern societies are deeply coupled across economic and social networks. Supply chains connect manufacturers across continents, labor markets depend on mobility, and information spreads instantly through digital platforms.

This high coupling amplified the consequences of the pandemic. Local outbreaks quickly disrupted global logistics, halted industrial production, and destabilized financial markets. At the same time, the information environment became saturated with conflicting narratives, making it difficult for individuals and institutions to distinguish signal from noise.

In UED terms, both biological transmission and informational confusion propagated through the same interconnected field.

Consequence

The visible outcomes of the pandemic were therefore multidimensional. Healthcare systems faced intense strain, millions of lives were lost, and economic activity contracted sharply across the world. Education systems, workplaces, and social institutions were forced to reorganize almost overnight.

These consequences were not merely the effects of a virus. They were the structural response of a tightly coupled global system resolving accumulated delta under severe constraints.

Correction

Over time, several mechanisms helped restore partial equilibrium. Vaccination programs reduced biological pressure, governments expanded healthcare capacity, and institutions adapted to new patterns of work and interaction.

However, the pandemic also revealed structural weaknesses in global preparedness. Many systems had optimized efficiency at the expense of resilience, leaving little tolerance for unexpected shocks.

Structural Insight

From a UED perspective, the pandemic highlights a critical lesson about modern systems:

- Early signals are often weak and easily dismissed.
- Noise can overwhelm accurate information.
- Delayed response allows delta to accumulate rapidly.
- High coupling spreads pressure across domains.
- Once thresholds are crossed, correction becomes abrupt and costly.

In a highly interconnected world, stability depends not only on controlling immediate threats but also on maintaining the capacity to detect and resolve pressure before thresholds are reached.

Case Study 3 — Organizational Collapse

Loss of Coherence Inside a Complex System (The Boeing 737 MAX Crisis)

Organizations, like physical systems, function through alignment between internal processes, constraints, and external realities. When this alignment weakens, pressure accumulates inside the system long before visible failure appears. The crisis surrounding the Boeing 737 MAX illustrates how organizational instability often arises not from a single mistake but from a gradual loss of coherence within a highly coupled institutional field.

Constraint

In the years preceding the development of the 737 MAX, Boeing faced intense competitive pressure from the European aerospace company Airbus. Airlines demanded fuel-efficient aircraft that could be introduced quickly without requiring extensive pilot retraining. To respond to these market conditions, Boeing chose to modify the existing 737 platform rather than design an entirely new aircraft.

This decision created a series of structural constraints. Engineers were required to integrate larger engines into an airframe originally designed decades earlier. Certification processes were accelerated, and design decisions were shaped by regulatory timelines and market competition. These constraints did not immediately produce failure, but they limited the system's flexibility to address emerging design complexities.

Delta

Within these constraints, technical and organizational pressure began to accumulate. Adjustments to the aircraft's aerodynamic behavior required additional software systems to maintain stability. The Maneuvering Characteristics Augmentation System (MCAS) was introduced to compensate for changes in aircraft handling.

Each design modification resolved part of the problem while introducing new dependencies. The system relied heavily on sensor input and automated correction mechanisms. Meanwhile, internal communication between engineering teams, management, and regulatory bodies became increasingly fragmented.

This fragmentation allowed unresolved pressure to remain embedded in the design and certification process. Instead of being fully addressed through redesign or extended testing, several issues were absorbed into procedural and software-based solutions.

Gradient

The dominant gradient influencing the organization during this period was the pressure to deliver a competitive aircraft quickly. Market expectations, investor confidence, and airline contracts created a powerful directional force toward speed and efficiency.

When such gradients intensify, systems tend to prioritize short-term alignment with external demands over deeper structural correction. Decisions that reduce immediate friction appear rational, even if they increase long-term instability.

Within Boeing, this gradient gradually shifted attention away from systemic coherence toward operational deadlines.

ATT — Alignment Threshold Time

For a period, the system remained within its tolerance window. The aircraft entered service, and early operations appeared routine. However, the underlying design dependencies had not been fully resolved.

The Alignment Threshold Time was reached when the interaction between software automation and sensor failure produced catastrophic outcomes. Two aircraft accidents occurred within a short period, revealing the accumulated pressure embedded within the system.

At this point, gradual correction was no longer possible. The aircraft fleet was grounded worldwide, regulatory investigations intensified, and the organization faced severe operational and reputational consequences.

Coupling

Modern aerospace systems operate within a dense network of coupling. Aircraft manufacturers interact with regulators, airlines, suppliers, maintenance providers, and global passengers. When instability appears in one component, it propagates quickly through the entire system.

The grounding of the 737 MAX affected not only Boeing but also airline operations, supply chains, pilot training programs, and international aviation policy. This widespread impact illustrates how tightly coupled systems amplify the consequences of localized instability.

Consequence

The immediate consequences included loss of life, global grounding of the aircraft, regulatory scrutiny, financial losses, and a significant decline in public trust. Beyond these direct outcomes, the crisis triggered broader discussions about safety culture, regulatory independence, and the balance between engineering integrity and market pressure.

These consequences were not isolated events. They were the systemic release of accumulated delta within the organizational field.

Correction

In response, Boeing and regulatory agencies initiated extensive corrective measures. Software systems were redesigned, pilot training protocols were revised, and certification procedures underwent deeper scrutiny. Organizational structures were also reexamined in an attempt to restore coherence between engineering, management, and regulatory oversight.

These actions represent efforts to reopen resolution pathways that had previously been constrained by institutional pressure.

Structural Insight

From a UED perspective, the Boeing 737 MAX crisis illustrates how organizational systems fail when internal coherence erodes under external gradients:

- Constraints limit the range of acceptable solutions.
- Delta accumulates when structural tensions remain unresolved.
- Gradients push decisions toward short-term alignment.
- High coupling spreads instability across the system.
- Crossing the threshold eliminates gradual correction.
- Consequences emerge suddenly but originate from long-term imbalance.

Organizations therefore remain stable not merely through efficiency or technological capability, but through sustained coherence between design, decision-making, and accountability.

Case Study 4 — Climate Change

Slow Delta in a Long-ATT Planetary System

Climate systems evolve slowly, often across decades or centuries. Because of this timescale, pressure within them can accumulate quietly while appearing stable at the surface. Climate change offers a clear example of how long-duration systems behave when delta builds gradually under persistent constraints. From a Universal Energy Dynamics perspective, the current environmental crisis is not a sudden planetary failure. It is the delayed resolution of pressure created by sustained imbalance between human activity and the Earth's energy equilibrium.

Constraint

The modern industrial economy operates within structural constraints that strongly influence energy use. Global infrastructure—transportation networks, manufacturing systems, urban development, and agriculture—has been built around fossil fuel energy sources for more than a century. These systems are deeply embedded in economic and political structures.

Such constraints limit the speed at which societies can shift to alternative energy systems. Even when awareness of environmental risk grows, existing infrastructure, financial investments, and institutional inertia restrict rapid structural change. As a result, corrective adjustments occur more slowly than the forces generating the imbalance.

Delta

Within these constraints, pressure accumulates in the planetary energy balance. Human activities release large quantities of greenhouse gases into the atmosphere, altering the Earth's ability to retain and distribute heat. Each additional emission contributes incrementally to this imbalance.

Because climate systems respond slowly, the immediate consequences of these emissions are not always visible. Temperature shifts, ocean warming, and atmospheric changes occur gradually. This delay allows the accumulation of delta to continue for long periods without triggering immediate corrective action.

In effect, the system stores unresolved pressure across decades.

Gradient

The gradient driving this accumulation is the global demand for energy, economic growth, and material development. Industrial societies rely on continuous energy consumption to sustain production, transportation, and technological progress.

This gradient encourages further extraction and combustion of fossil fuels. As long as these energy sources remain economically efficient and widely available, the system continues moving in the same direction, reinforcing the imbalance in atmospheric composition.

ATT — Alignment Threshold Time

Planetary systems possess long tolerance windows compared with human institutions. The Earth's climate can absorb gradual changes for extended periods before major shifts become visible. This long Alignment Threshold Time has historically masked the structural pressure building within the system.

However, as atmospheric concentrations increase and global temperatures rise, the system approaches thresholds where gradual adjustment becomes more difficult. Once these thresholds are crossed—such as large-scale ice loss, ocean circulation changes, or ecosystem collapse—correction may occur through abrupt environmental transitions rather than slow stabilization.

Coupling

The climate system is highly coupled across atmospheric, oceanic, ecological, and human domains. Changes in one component propagate through many others. Rising temperatures influence weather patterns, agriculture, water availability, and biodiversity. Economic systems, migration patterns, and geopolitical stability can also be affected.

This coupling means that climate-related pressure does not remain confined to environmental systems. It spreads through social, economic, and political structures, increasing the complexity of resolution.

Consequence

The consequences of accumulated climate delta are already emerging in the form of rising global temperatures, intensified weather events, shifting ecosystems, and increasing environmental stress on human societies. These developments are not isolated anomalies but indicators of the system gradually resolving a long-standing imbalance.

Because the pressure accumulated slowly, the resulting consequences also unfold across extended timescales. However, the closer the system moves toward critical thresholds, the more rapidly these effects may intensify.

Correction

Efforts to restore equilibrium involve reducing emissions, transitioning to renewable energy sources, improving energy efficiency, and developing adaptive strategies for changing environmental conditions. These measures aim to reopen resolution pathways and prevent the system from crossing irreversible thresholds.

Yet correction in such long-duration systems requires coordinated action across nations, industries, and institutions. The longer corrective action is delayed, the greater the quantity of accumulated delta that must eventually resolve.

Structural Insight

From a UED perspective, climate change illustrates the dynamics of systems with long tolerance windows and slow feedback cycles:

- Constraints embedded in infrastructure limit rapid adjustment.
- Delta accumulates gradually across decades.
- Economic gradients reinforce the direction of imbalance.
- Long ATT windows mask growing pressure.
- Coupling spreads consequences across environmental and social systems.
- Correction becomes more difficult as thresholds approach.

Planetary stability therefore depends not only on technological solutions but also on the ability of societies to recognize structural pressure early—before long-duration systems reach points where gradual resolution is no longer possible.

Case Study 5 — Social Media Amplification

When Information Systems Accumulate Delta Faster Than They Can Resolve

Modern information systems operate at speeds far beyond those of traditional social structures. Platforms designed to connect people and distribute information now function as global amplification networks. While these systems appear stable on the surface—processing billions of interactions daily—they often accumulate pressure in ways that are difficult to detect. Social media ecosystems illustrate how informational fields can become unstable when signal, noise, and incentives interact within highly coupled networks.

Constraint

Social media platforms are built around algorithms designed to maximize engagement. These systems prioritize content that captures attention, encourages sharing, and keeps users active on the platform. The underlying business model—advertising driven by user activity—creates structural constraints that reward visibility rather than accuracy.

These constraints do not intentionally promote misinformation or conflict. However, they limit the system's ability to distinguish between content that stabilizes understanding and content that merely intensifies reaction. As a result, the mechanisms that filter information are often weaker than the mechanisms that amplify it.

Delta

Within this environment, informational pressure begins to accumulate. Large volumes of content circulate continuously, mixing verified knowledge with speculation, emotion, and narrative framing. When misleading or incomplete information spreads widely, it generates confusion and disagreement among users.

Because digital platforms move information faster than traditional verification systems can respond, unresolved informational pressure grows rapidly. Competing narratives reinforce each other, and users form clusters around particular interpretations of events.

In UED terms, the system begins accumulating delta in the informational field—pressure created by unresolved differences between signal and perception.

Gradient

The gradient driving this process is attention. Content that provokes strong emotional responses—fear, outrage, surprise, or identity affirmation—tends to spread more quickly than neutral or analytical information. Algorithms that optimize for engagement naturally align with this gradient.

As a result, emotionally charged narratives move rapidly across networks, while quieter signals that might clarify the situation struggle to gain similar visibility. This directional pressure pushes the system toward amplification rather than resolution.

ATT — Alignment Threshold Time

For extended periods, informational systems can tolerate high levels of noise. Individuals adjust by relying on trusted sources or familiar communities. However, when the volume and speed of conflicting narratives exceed the system's capacity to filter them, a threshold is reached.

At this point, the informational environment becomes unstable. Public trust erodes, consensus becomes difficult to establish, and coordinated responses to shared problems weaken. Once this threshold is crossed, restoring informational equilibrium becomes significantly more difficult.

Coupling

Social media platforms are among the most tightly coupled systems in human history. Information spreads across geographic, cultural, and political boundaries almost instantly. A message created in one location can influence public perception in another within minutes.

This coupling allows informational pressure to propagate globally. Political events, public health messages, economic rumors, or social movements can all be amplified rapidly through interconnected networks. Because the same platforms distribute both reliable information and noise, distinguishing between them becomes increasingly challenging.

Consequence

The consequences of this dynamic are visible in many areas of public life: polarization in political discourse, rapid spread of misinformation during crises, erosion of institutional trust, and difficulty achieving collective agreement on complex issues.

These outcomes do not arise from a single actor or event. They emerge from the structural behavior of a system designed to amplify interaction faster than it can resolve informational imbalance.

Correction

Efforts to stabilize the informational field involve improving content moderation systems, strengthening verification mechanisms, and encouraging digital literacy among users. Some platforms are also experimenting with algorithmic adjustments that reduce the visibility of misleading or harmful content.

However, lasting correction requires acknowledging the structural gradients driving amplification. If engagement remains the dominant metric guiding platform behavior, informational delta will continue to accumulate.

Structural Insight

Through the UED lens, social media ecosystems reveal how informational systems behave under extreme connectivity:

- Engagement constraints favor amplification over verification.
- Informational delta accumulates when signal and noise intermingle.
- Attention gradients drive rapid narrative spread.
- High coupling distributes pressure globally.
- Thresholds appear as polarization, distrust, and instability.
- Correction requires restoring pathways that favor resolution rather than amplification.

In an age where information moves faster than reflection, the stability of societies increasingly depends on the ability to manage informational pressure before it reaches destabilizing thresholds.

Appendices

Systems rarely collapse because of a single event. They collapse because unresolved delta silently crosses its threshold.

Appendix A — Canonical Definitions (UED Reference)

1. Delta

Delta is accumulated, unresolved pressure within a system caused by a mismatch between available resolution pathways and imposed constraints.

Delta is not disturbance, activity, or conflict. It is **stored imbalance**—energy that is prevented from resolving along natural gradients. Delta can remain latent for long periods, often masked by buffers, efficiency, or delay.

Delta always seeks resolution. If resolution is blocked locally, delta migrates across domains, scales, or time until a viable pathway opens. Sudden events are usually late-stage delta release, not origins.

Delta is neutral. It is neither good nor bad. Systems require delta to adapt, but excessive or misallocated delta leads to instability.

2. Equilibrium

Equilibrium is a temporary state in which opposing gradients within a system are sufficiently balanced to prevent net movement or rapid change.

Equilibrium is not rest, stillness, or completion. It is a **dynamic holding pattern** maintained through continuous micro-adjustments. In living systems, equilibrium is always provisional.

Mistaking equilibrium for permanence leads to rigidity. When equilibrium is defended instead of recalibrated, tolerance margins shrink and correction becomes abrupt.

Equilibrium does not eliminate delta; it redistributes it.

3. Consequence

Consequence is the inevitable feedback produced when delta resolves through available pathways.

Consequence is not punishment or reward. It is **structural response.** Every action, inaction, or constraint modification produces consequence, whether immediate or delayed, visible or displaced.

Consequence often returns through indirect channels and at different scales than the original action. Delay and migration obscure causality, creating the illusion of randomness or injustice.

Consequence cannot be avoided—only postponed, redirected, or intensified.

4. ATT (Alignment Threshold Time)

Alignment Threshold Time (ATT) is the maximum duration a system can tolerate unresolved delta before forced correction occurs.

ATT is not a fixed value. It varies with system capacity, coupling density, buffering mechanisms, and feedback delay. As tolerance is consumed, ATT shortens.

Crossing ATT does not necessarily cause collapse, but it eliminates gradual resolution options. Correction after ATT is typically nonlinear, abrupt, and costly.

Most crises are ATT failures, not sudden breakdowns.

5. Field

A **field** is a continuous medium through which energy, pressure, information, and consequence propagate.

In UED, a field is not abstract space. It is **the environment of interaction itself**, including physical, biological, psychological, social, economic, and informational domains.

Fields are defined by coupling, not boundaries. What affects one region of a field can propagate to others depending on connectivity and constraint.

All systems exist *within* fields and cannot act independently of them.

6. Gradient

A **gradient** is a directional difference in pressure, potential, or capacity that drives movement or resolution within a field.

Gradients determine **where and how delta can resolve**. Steeper gradients accelerate movement; shallow gradients slow it. Artificially steepened gradients increase risk; suppressed gradients accumulate delta.

Gradients are descriptive, not intentional. They exist regardless of awareness or desire.

Effective action aligns with gradients rather than opposing them.

7. Coherence

Coherence is the degree to which a system's internal processes align with each other and with external field conditions.

A coherent system resolves delta efficiently with minimal dissipation. Incoherent systems waste energy through internal conflict, redundancy, or misdirected effort.

Coherence is not harmony, agreement, or calm. It can exist under stress and tension as long as resolution pathways remain clear.

Loss of coherence precedes visible failure.

8. Dissipation

Dissipation is the diffusion of energy during resolution that does not contribute to useful alignment or work.

Dissipation is not loss in a moral or functional sense; it is **inevitable leakage** in real systems. Excessive dissipation indicates misalignment, poor coupling, or blocked gradients.

Healthy systems minimize dissipation relative to resolution, not by eliminating it, but by maintaining coherence.

Chronic dissipation without resolution signals structural failure.

9. Constraint

A **constraint** is any limit—structural, temporal, energetic, cognitive, institutional, or environmental—that restricts resolution pathways.

Constraints are necessary. Without them, systems cannot form structure. However, rigid or misaligned constraints cause delta accumulation.

Constraints shape outcomes more powerfully than intent. Most failures arise not from wrong goals, but from **ignored or misunderstood constraints**.

Effective systems adapt constraints over time.

10. Signal / Noise

Signal is information that accurately reflects underlying delta, gradient, or consequence.

Noise is information that obscures, distorts, or distracts from structural reality.

Noise increases under high coupling, narrative amplification, emotional load, and delayed feedback. Signal is often quiet, early, and easily dismissed.

Intelligence without awareness amplifies noise. Awareness distinguishes signal.

Misreading noise as signal accelerates misalignment.

Appendix A Integrity Note

These definitions are **non-negotiable** within UED. They are not metaphors, interpretations, or philosophical positions. They exist to preserve consistency across scale, domain, and application.

No extension or usage should contradict these definitions.

Appendix B — Crosswalk: UED vs Mainstream Frameworks

This appendix positions Universal Energy Dynamics (UED) relative to—not against—mainstream scientific frameworks. It clarifies what remains valid, what is reinterpreted, and where each framework structurally fails under scale, coupling, or delayed consequence.

UED does not negate historical theories. It absorbs them at their valid resolution scale and removes assumptions that break under extension.

1. Classical Mechanics — Isaac Newton

Primary Concept: Force acting on mass

Core Assumption: Objects move because forces act upon them

What Stays

- Quantitative accuracy at human-scale interactions
- Predictive usefulness in low-coupling, low-delay systems
- Engineering applicability where boundaries are rigid and local

What Is Reinterpreted

- Force is not fundamental; it is a derived effect of gradient resolution
- Motion is not caused by push/pull, but by energy seeking equilibrium under constraint
- Action–reaction reflects field redistribution, not object interaction

Where It Breaks

- At high coupling (fluids, fields, crowds, markets)
- At delayed consequence (friction, dissipation, fatigue)
- When force language hides constraint and gradient dynamics

UED Position: Newtonian mechanics remains valid as a local approximation, not a universal ontology.

2. Relativity — Albert Einstein

Primary Concept: Geometry of spacetime governs motion

Core Assumption: Mass–energy curves spacetime; objects follow geodesics

What Stays

- Accurate description of large-scale gravitational phenomena
- Predictive success for orbital dynamics, lensing, time dilation
- Insight that force-based gravity is insufficient

What Is Reinterpreted

- Geometry is not causal; it is a measurement artifact of field equilibrium
- Curvature reflects stable gradient configuration, not an active agent
- Spacetime is descriptive scaffolding, not a physical substance

Where It Breaks

- Inside black holes (singularities, information paradox)
- At quantum coupling boundaries
- When geometry is treated as fundamental rather than emergent

UED Position: Relativity maps how resolution appears when equilibrium stabilizes, not why it occurs.

3. Quantum Mechanics

Primary Concept: Probabilistic state resolution

Core Assumption: Reality exists as superposition until observed

What Stays

- Statistical accuracy of predictions
- Recognition that determinism fails at small scales
- Measurement as a critical event

What Is Reinterpreted

- Probability reflects unresolved field states, not intrinsic randomness
- Wavefunctions describe resolution potential, not physical waves
- Measurement is alignment collapse, not observer magic

Where It Breaks

- Interpretation dependence (observer problem)
- Non-locality paradoxes
- Separation between measurement and physical process

UED Position: Quantum behavior is resolution in-progress, not indeterminacy at the foundation.

4. Thermodynamics

Primary Concept: Entropy increase governs direction of processes

Core Assumption: Systems evolve toward disorder

What Stays

- Directionality of irreversible processes
- Energy conservation
- Statistical behavior of large systems

What Is Reinterpreted

- Entropy is not disorder; it is resolution dispersion
- Dissipation reflects alignment loss, not chaos
- Irreversibility arises from delayed consequence and coupling

Where It Breaks

- When entropy is moralized as decay
- When applied blindly to living or adaptive systems
- When disorder is treated as fundamental

UED Position Entropy tracks how resolution spreads, not the purpose or fate of systems.

Cross-Framework Summary Table (Conceptual)

Question	Mainstream Answer	UED Answer
What causes motion?	Force / curvature	Gradient resolution
Why do systems change?	External action	Internal imbalance
Why does prediction fail?	Randomness / complexity	Delayed consequence
What is stability?	Equilibrium state	Managed resolution
What breaks theories?	Extreme conditions	Wrong primitives

Why This Crosswalk Matters

UED does not compete with existing frameworks. It reorders their primitives.

Mainstream theories describe *what happens* with great precision. UED explains *why those descriptions stop working* when:

- scale changes
- coupling increases
- feedback delays grow
- interpretation replaces structure

This crosswalk exists to prevent false opposition. UED stands downstream, not above.

Appendix B Integrity Note

This appendix is not a critique of science. It is a structural reconciliation.

All mainstream frameworks remain valid within their proper resolution domain. UED exists only to explain what happens when those domains are exceeded.

Appendix C — UED Diagnostic Templates

Each template follows the same invariant sequence:

Constraint → Delta → Gradient → Threshold → Coupling → Consequence

This order matters. Skipping steps produces false clarity.

C1. Physical Systems (Engineering / Applied Physics)

Purpose: Diagnose failure, inefficiency, instability, or unexpected behavior in mechanical, thermal, fluid, electrical, or structural systems.

1. Constraints

- Physical limits (material strength, capacity, geometry)
- Boundary conditions (fixed, free, insulated, open)
- Time constraints (response rate, damping, delay)
- Energy input/output limits

2. Delta

- Where energy is accumulating without resolution
- Stress concentration points
- Heat buildup, pressure pockets, charge accumulation
- Repeated micro-failures or oscillations

3. Gradients

- Pressure, temperature, voltage, velocity differentials
- Artificially steepened gradients (overdriving)
- Suppressed gradients (over-control)

4. Thresholds

- Yield limits, resonance points, fatigue limits
- Nonlinear transitions (stall, cavitation, buckling)
- Loss of reversibility

5. Coupling

- Mechanical → thermal
- Electrical → thermal
- Fluid → structural
- Local → system-wide propagation paths

6. Consequence

- Where failure actually appears (often not origin)
- Delayed damage vs immediate failure
- Dissipation vs catastrophic release

Diagnostic Signal: If behavior looks "sudden," the delta was ignored.

C2. Personal Psyche (Behavioral / Cognitive)

Purpose: Diagnose stress, indecision, burnout, emotional looping, loss of clarity, or chronic dissatisfaction.

1. Constraints

- Cognitive load
- Emotional tolerance
- Time, energy, social expectations
- Identity commitments and self-image

2. Delta

- Unresolved pressure (avoidance, guilt, tension)
- Effort without progress

- Repeated rumination
- Numbness masking load

3. Gradients

- Desire vs capacity
- Expectation vs reality
- Responsibility vs autonomy
- Internal conflict steepening gradients

4. Thresholds

- Burnout onset
- Emotional collapse
- Behavioral shutdown
- Sudden impulsive decisions

5. Coupling

- Emotional → behavioral
- Personal → relational
- Work → identity
- Stress displacement into health or relationships

6. Consequence

- Anxiety, depression, withdrawal, irritability
- Delayed regret
- Narrowing options
- Loss of resilience

Diagnostic Signal: If clarity feels urgent, the system is near ATT.

C3. Organizations (Leadership / Management)

Purpose: Diagnose dysfunction, loss of morale, execution failure, politics, stagnation, or sudden collapse.

1. Constraints

- Authority structure
- Incentives and metrics
- Communication bandwidth
- Cultural rigidity
- Resource ceilings

2. Delta

- Unspoken tension
- Misaligned incentives
- Decision avoidance
- Chronic firefighting

3. Gradients

- Power vs responsibility
- Accountability vs authority
- Vision vs execution capacity
- Growth pressure vs stability

4. Thresholds

- Talent exit points
- Trust collapse
- Coordination breakdown
- Escalation into control culture

5. Coupling

- Leadership → culture

- Culture → performance
- Performance → market response
- Internal → external reputation

6. Consequence

- Attrition
- Bureaucratization
- Political infighting
- Strategic failure appearing "unexpected"

Diagnostic Signal: If performance metrics improve while trust degrades, delta is migrating.

C4. Societies (Policy / Geopolitics)

Purpose: Diagnose instability, polarization, economic stress, governance failure, or geopolitical escalation.

1. Constraints

- Institutions and legal frameworks
- Demographics
- Resource access
- Technological dependency
- External pressures

2. Delta

- Inequality
- Suppressed dissent
- Debt (financial or social)
- Dependency accumulation
- Narrative contradiction

3. Gradients

- Wealth concentration
- Power asymmetry
- Opportunity gaps
- Regional disparities

4. Thresholds

- Social unrest
- Market collapse
- Political legitimacy loss
- Sudden radicalization

5. Coupling

- Economic → social
- Social → political
- Political → geopolitical
- Local → global propagation

6. Consequence

- Protests
- Policy overreach
- External conflict
- Institutional breakdown

Diagnostic Signal: If stability relies on narrative rather than correction, ATT is shortening.

Canonical Usage Rule (Applies to All Templates)

- Do not jump to solutions
- Do not moralize
- Do not optimize prematurely
- Do not confuse signal with noise

Diagnosis precedes action. Action without diagnosis multiplies consequence.

Appendix C Integrity Note

These templates are scale-invariant but scale-sensitive. They must be applied with correct boundaries and timing.

They are not predictive tools. They are error-reduction instruments.

Appendix D — Common Misinterpretations

(And Why They Are Incorrect)

This appendix exists because powerful frameworks attract incorrect extensions. What follows are the *most frequent misreadings* of Universal Energy Dynamics and their precise correction.

1. "UED rejects experiments" — No

UED does *not* reject experimentation. It rejects misinterpretation of experimental results when structural assumptions are wrong.

Experiments measure outcomes under constrained conditions. UED fully accepts empirical data as real and binding. What it challenges is the habit of elevating measurement artifacts into fundamental causes without examining constraints, coupling, and delayed consequence.

In UED:

- Experiments validate *descriptions*
- They do not automatically validate *ontologies*

An experiment can be correct and still incomplete in explanation. UED uses experiments as boundary checks, not as final arbiters of causality.

Rejecting experiments would be anti-scientific. Reinterpreting what experiments actually confirm is not.

2. "UED is spirituality" — No

UED is not spiritual, metaphysical, religious, or belief-based.

This confusion arises because UED uses words such as *energy*, *alignment*, and *equilibrium*, which are often misused in spiritual contexts. In UED, these terms have strict operational definitions tied to constraints, gradients, and resolution.

UED makes no claims about meaning, purpose, transcendence, or salvation. It does not address faith, belief, or inner fulfillment.

If UED appears "calming" or "clarifying," that is a consequence of reduced misalignment, not spirituality. Structural clarity often reduces anxiety—but that does not convert explanation into belief.

Any use of UED as comfort doctrine or spiritual narrative is misuse.

3. "UED explains everything, therefore predicts everything" — No

UED explains why prediction fails, not how to bypass uncertainty.

UED identifies:

- Where pressure accumulates
- Which resolution paths are blocked
- Why systems oscillate or collapse
- Why delayed consequences appear sudden

It does **not** predict:

- Exact events
- Specific timings
- Final outcomes
- Winners, forms, or identities

Explanation is not prediction. Orientation is not foresight.

Claiming predictive authority from UED is a violation of its core discipline and usually signals overconfidence or ideology formation.

UED reduces avoidable surprise, not uncertainty itself.

4. "UED replaces mathematics" — No

UED does not replace mathematics. It reorders its role.

Mathematics remains indispensable for:

- Local modeling
- Engineering precision
- Measurement consistency
- Predictive bounds within valid domains

What UED rejects is the assumption that mathematical elegance implies physical primacy. Mathematical models describe behavior *after* constraints stabilize. They do not define what is fundamental.

In UED:

- Mathematics is descriptive, not causal
- Equations map resolved states, not resolution itself
- Math fails when primitives are wrong, not when nature is wrong

UED explains *why* mathematical models break at boundaries—it does not discard them.

Appendix D Integrity Statement

UED is:

- Empirical-compatible
- Non-spiritual
- Non-predictive
- Non-anti-mathematical

It is a structural interpretive framework, not a belief system, prophecy engine, or replacement for science.

Any claim beyond this scope is not UED.

Appendix E — Red Flags & Misuse Patterns

(Quick Recognition Guide)

This appendix lists *early-warning signals* that Universal Energy Dynamics is being misunderstood, overstretched, or turned into belief, authority, or identity.

These red flags apply *regardless of intelligence, intention, or expertise.*

Red Flag 1 — Absolutist Language

Pattern

- "UED explains everything."
- "This always happens."
- "Once you understand UED, you see the truth."

Why This Is a Problem: UED is conditional by design. Absolutes indicate closure, not clarity. Field-based systems do not permit total explanation.

Correction: Reintroduce boundaries, scale, and uncertainty. Replace certainty with orientation.

Red Flag 2 — Predictive Claims

Pattern

- "UED predicted this event."
- "According to UED, this will happen next."
- "This outcome proves UED is correct."

Why This Is a Problem: UED explains *why outcomes become likely*, not *what must occur*. Predictive framing converts explanation into prophecy.

Correction: Shift from outcome claims to pressure, constraint, and timing analysis.

Red Flag 3 — Spiritual Substitution

Pattern

- Energy described as feeling, vibration, or positivity
- Alignment equated with peace or bliss
- Language becomes soothing but non-operational

Why This Is a Problem: Structural concepts are being replaced by emotional metaphors. Explanation is dissolving into suggestion.

Correction: Return to operational definitions. If it cannot be mapped, constrained, or tested, it does not belong.

Red Flag 4 — Identity Attachment

Pattern

- "As someone who understands UED..."
- Using UED to signal superiority or detachment
- Defensive reaction when UED is questioned

Why This Is a Problem: Identity demands consistency. UED requires revision. Once identity forms, feedback is filtered.

Correction: Detach understanding from self-description. Let models change without personal cost.

Red Flag 5 — Over-Application

Pattern

- Applying UED to every situation regardless of scale
- Using UED where simpler reasoning suffices

- Modeling replaces judgment

Why This Is a Problem: Universality does not mean uniformity. Over-application flattens reality and delays action.

Correction: Ask: *Does UED add clarity here—or am I using it out of habit?*

Red Flag 6 — Moralization of Structure

Pattern

- Treating misalignment as failure or flaw
- Using UED to judge people rather than diagnose systems
- Framing consequence as deserved

Why This Is a Problem: UED is non-moral. Moral overlays distort feedback and block correction.

Correction: Separate diagnosis from judgment. Focus on structure, not blame.

Red Flag 7 — Language Inflation

Pattern

- Excessive terminology
- Concepts repeated without added precision
- Talking more, seeing less

Why This Is a Problem: Language is compensating for missing structure. Noise is increasing.

Correction: Reduce vocabulary. Increase specificity. Stop when structure is clear.

Red Flag 8 — Resistance to Revision

Pattern

- Defending old models despite new evidence
- Explaining away consequence instead of updating boundaries
- Treating surprise as error in reality, not in modeling

Why This Is a Problem: This is the precise moment modeling becomes belief.

Correction: Redraw boundaries. Discard models without regret.

Quick Self-Check (One-Minute Test)

If using UED has made you:

- Quieter → ✔ aligned
- Slower → ✔ aligned
- Less certain → ✔ aligned
- More patient with ambiguity → ✔ aligned

If it has made you:

- More confident → ⚠ check
- More explanatory → ⚠ check
- More detached → ⚠ check
- More corrective toward others → ⟐ misuse

Appendix E Integrity Note

UED does not fail loudly. It fails *silently through misuse.*

These red flags exist to prevent that failure.

Recognition is enough. Correction follows naturally.

Appendix F — One-Page UED Reference Sheet

A compact orientation aid for diagnosing systems, making decisions, and avoiding misuse—without carrying the framework cognitively.

UED Core Primitive

Resolution, not force. Systems change because pressure seeks release under constraint, not because entities push each other.

Canonical Chain (Always in Order)

Constraint → Delta → Gradient → Threshold (ATT) → Coupling → Consequence

If analysis skips a step, expect error.

Key Definitions (Ultra-Brief)

- **Constraint:** Limits that shape possible resolution paths
- **Delta:** Unresolved, accumulated pressure
- **Gradient:** Directional difference driving movement
- **ATT:** Time before forced correction replaces gradual resolution
- **Coupling:** How effects propagate across domains or scales
- **Consequence:** Structural feedback after resolution
- **Coherence:** Internal alignment enabling efficient resolution
- **Dissipation:** Energy spent without contributing to alignment
- **Signal / Noise:** Structural information vs distracting amplification

Three Diagnostic Questions

1. Where is delta accumulating?
2. Which constraints are blocking resolution?
3. What consequence is being delayed, not avoided?

If answers are unclear, do not act.

Decision Filter (Personal / Organizational)

- Will this action reduce delta or displace it?
- Does it align with gradients or fight them?
- Does it shorten ATT or buy alignment time?
- Does it increase coherence or export pressure?

If uncertain → reduce intervention.

Common Failure Signals

- Sudden collapse after long stability
- Escalation despite "fixes"
- Rising effort with falling clarity
- Narratives replacing diagnostics
- Control increasing as trust declines

These indicate missed delta, not bad luck.

What UED Does NOT Do

- Predict specific events
- Replace mathematics
- Offer certainty
- Justify outcomes
- Provide moral judgment

If any of these appear, stop and reassess.

Correct Usage Posture

- Explain less
- Observe more
- Act later, but cleaner
- Revise without defense
- Let resolution finish when possible

When UED works, it becomes invisible.

End Condition (Self-Check)

You are using UED correctly if:

- Decisions feel lighter, not heavier
- Intervention decreases over time
- Surprise shortens, not disappears
- Understanding no longer seeks closure

If not, return to constraints.
Print. Fold. Keep nearby. Forget the rest.

Appendix G — Short Glossary (General Reader Edition)

This glossary exists to make the book readable without requiring technical background. Terms are defined *only as they are used in UED*, not as they appear elsewhere.

Alignment

A condition in which actions, constraints, and available capacity are consistent enough to allow resolution without excessive friction.

Alignment does not mean comfort, agreement, or success. It means *pressure can move without obstruction.*

ATT (Alignment Threshold Time)

The time window a system has before unresolved pressure forces correction.

When ATT is crossed, systems lose the option of gradual adjustment. Outcomes become abrupt and costly.

Coherence

The degree to which parts of a system work together without internal conflict.

High coherence allows systems to operate under stress. Low coherence causes energy to be wasted resolving internal contradictions.

Constraint

Any limit that restricts what a system can do.

Constraints include physical limits, time, rules, resources, expectations, attention, and tolerance. Constraints are necessary—but must be adapted as conditions change.

Consequence

What follows when pressure finally resolves.

Consequence is not punishment or reward. It is *structural feedback*. It may appear delayed, indirect, or displaced from its original cause.

Coupling

How effects in one part of a system spread to others.

Strong coupling spreads pressure quickly. Weak coupling delays feedback. Hidden coupling causes surprises.

Delta

Unresolved pressure inside a system.

Delta builds when resolution is blocked. It can remain invisible for long periods before releasing suddenly.

Dissipation

Energy spent without contributing to meaningful resolution.

Dissipation is normal. Excessive dissipation indicates misalignment or poor structure.

Equilibrium

A temporary balance where opposing pressures cancel enough to prevent change.

Equilibrium is never permanent. Treating it as stable leads to rigidity and later failure.

Field

The environment in which interactions occur.

A field includes physical space, social context, economic systems, psychological states, or information environments. Systems cannot act outside their fields.

Gradient

A directional difference that drives movement or change.

Examples include pressure differences, incentives, demand gaps, or emotional tension. Gradients move systems whether they are acknowledged or not.

Noise

Information that distracts from what actually matters structurally.

Noise increases with emotion, narrative amplification, and delayed feedback.

Resolution

The process by which pressure finds a path to release under constraint.

Resolution is continuous. Systems do not "solve" problems permanently—they adjust.

Signal

Information that accurately reflects underlying pressure, constraints, or consequence.

Signal is often quiet, early, and easy to ignore.

Stability

The ability of a system to continue functioning without escalating correction.

Stability comes from **managed resolution**, not control or rigidity.

System

Any set of interacting elements that influence each other's behavior.

Systems can be physical, personal, organizational, or societal. All systems generate consequences.

Threshold

A point beyond which behavior changes qualitatively.

Crossing a threshold often turns gradual pressure into sudden events.

Appendix G Note

If a word feels comforting, moral, or inspirational, it is being misread. UED language exists to *reduce distortion*, not to motivate or console.

Appendix H — Reader Guide

This book is not meant to be read as a linear argument or a body of knowledge to be accumulated. It is structured to *change how attention is placed*, not what conclusions are held.

If read correctly, it should feel less impressive over time.

1. Read Once for Structure, Not Agreement

The first reading is only to recognize the structure:

- What is treated as fundamental
- What is treated as derived
- What is deliberately left unresolved

Agreement is irrelevant at this stage. Disagreement is acceptable. What matters is whether the internal logic remains consistent across scale.

If you find yourself trying to "accept" or "reject" the framework, pause. That impulse signals identity formation, not understanding.

2. Do Not Read It as a Philosophy

This book does not offer values, meaning, or direction. It does not answer questions of purpose or belief.

If it begins to feel philosophical, something is being projected onto it. Return to constraints, gradients, and consequences. Those are the only anchors.

3. Use Chapters as Reference, Not Sequence

Later readings—if any—should be selective.

Revisit chapters only when:

- A system behaves unexpectedly
- Pressure feels delayed or displaced
- Intervention increases instability

Do not reread for reassurance or clarity. That converts the framework into comfort, which guarantees misuse.

4. Stop Reading When Behavior Changes

The correct stopping point is not the last page.

If you notice:

- Reduced urgency to act
- Increased tolerance for ambiguity
- Earlier detection of misalignment
- Fewer corrective interventions

Stop reading. The framework has already transitioned into use.

Continuing past this point risks over-application.

5. Do Not Teach It Before It Disappears

Explaining UED to others too early is a common misuse.

If you still need to reference terms, defend definitions, or correct misunderstandings, do not teach it. Teaching freezes understanding and converts structure into doctrine.

Only what no longer needs explanation can be shared safely.

6. Let Appendices Remain Appendices

The appendices exist to prevent drift, not to expand authority.

They are safeguards, not extensions. Do not build theories on top of them. Do not argue from them.

If you find yourself citing appendices to justify a position, return to diagnosis instead.

7. When to Put the Book Down Permanently

This book should eventually become unnecessary.

Put it down permanently when:

- You notice pressure before collapse
- You intervene less but more effectively
- You revise models without defending them
- You no longer feel the need to explain outcomes

At that point, rereading adds no value.

Appendix H Closing Note

This book does not reward loyalty. It does not improve with repetition. It does not scale through advocacy.

Its usefulness ends quietly.

If it has worked, it will leave no trace— except fewer avoidable consequences.

References

The references below are provided for **contextual grounding**, not authority transfer. Universal Energy Dynamics does not derive its validity from these works; they are cited because they represent the **best-known formal descriptions of domains UED intersects with, reinterprets, or places boundary conditions upon**.

Classical & Relativistic Physics

- **Isaac Newton** — *Philosophiæ Naturalis Principia Mathematica.* Royal Society, 1687.
- **Albert Einstein** — *Relativity: The Special and the General Theory.* Henry Holt, 1916.
- **Hermann Minkowski** — "Space and Time," *Physikalische Zeitschrift*, 1908.

Quantum Theory & Foundations

- **Max Planck** — "On the Law of Distribution of Energy in the Normal Spectrum," 1901.
- **Erwin Schrödinger** — "Quantisation as an Eigenvalue Problem," *Annalen der Physik*, 1926.
- **Werner Heisenberg** — "Über den anschaulichen Inhalt der quantentheoretischen Kinematik," 1927.
- **Niels Bohr** — *Atomic Theory and the Description of Nature.* Cambridge University Press, 1934.

Thermodynamics & Systems

- **Rudolf Clausius** — "On the Mechanical Theory of Heat," 1850–1865.
- **Ludwig Boltzmann** — *Lectures on Gas Theory*. 1896–1898.
- **Ilya Prigogine** — *Order Out of Chaos*. Bantam Books, 1984.

Complex Systems & Cybernetics

- **Norbert Wiener** — *Cybernetics: Or Control and Communication in the Animal and the Machine*. MIT Press, 1948.
- **Herbert A. Simon** — *The Sciences of the Artificial.* MIT Press, 1969.
- **Donella Meadows** — *Thinking in Systems*. Chelsea Green, 2008.

Information, Signal, and Noise

- **Claude Shannon** — "A Mathematical Theory of Communication," *Bell System Technical Journal*, 1948.
- **Norbert Wiener** — *The Human Use of Human Beings*. Houghton Mifflin, 1950.

Human Behavior & Organizations

- **Daniel Kahneman** — *Thinking, Fast and Slow*. Farrar, Straus and Giroux, 2011.
- **Karl E. Weick** — *Sensemaking in Organizations*. Sage Publications, 1995.
- **W. Edwards Deming** — *Out of the Crisis*. MIT Press, 1986.

Methodological Boundaries

- **Thomas S. Kuhn** — *The Structure of Scientific Revolutions*. University of Chicago Press, 1962.
- **Karl Popper** — *The Logic of Scientific Discovery*. Routledge, 1959.

Reference Note

These works are **not prerequisites** for understanding this book. They are listed to indicate where existing frameworks are respected, reinterpreted, or structurally limited.

Universal Energy Dynamics stands or falls on its **internal consistency, cross-domain coherence, and consequence alignment**—not on citation authority.

No reference above is treated as foundational. None are rejected. All are bounded.

— **Sandeep J. Chavan**

Further Reading — Selected Works by the Author

The works listed below extend specific aspects of **Universal Energy Dynamics (UED)**. They are optional. None are prerequisites for understanding this book.

Readers are encouraged to choose by intent, not by completion.

I. Core UED Foundations (Physics-First)

For readers who want deeper structural grounding.

1. **Universal Energy Dynamics (UED): A Consequence-Based Framework for Reality in a Still Universe**
2. **UED Layered Architecture: From Ripple–Resolution Ontology to Alignment Completion**
3. **UED at a Glance – The Ripple Framework that Replaces the Particle Paradigm**
4. **The Ripple Age Report: Universal Energy Dynamics and the Science of Consequence**
5. **Universal Energy Dynamics vs Established Theories**

These works formalize the field-first, resolution-centric ontology introduced in this book.

II. Resolution, Fields, and Physical Phenomena

For readers interested in how UED reframes physics without metaphysics.

1. **Structured Vacuum Energy: Reinterpreting Physical Reality**

Beyond the Wave–Particle Paradigm

2. **The Still Universe: Resolving the Hubble Tension**
3. **Cosmic Ray Showers as Multi-Layer Field Resolution Phenomena**
4. **The Myth of Wave-Particle Duality**
5. **Formation of Black Holes: A UED Perspective**
6. **Chavan's Law: A Universal Energy Equilibrium Model for Black Holes**

These works show how force, particles, and spacetime emerge as consequences, not primitives.

III. Alignment, Awareness, and the Human System

For readers exploring mind, behavior, and consciousness through UED.

1. **Consciousness as an Emergent Alignment**
2. **Chavanian Dynamism: A UED Interface Between Awareness and Intelligence**
3. **The Active Mind**
4. **There Is No Subconscious Mind**
5. **Layers of Consciousness**
6. **Why It Feels Like "Me"**

These works extend UED into psychology without spirituality.

IV. Consequence, Ethics, and Human Systems

For readers interested in morality, responsibility, and social coherence.

1. **The Law of Consequences (LOC)**
2. **Ethics & Morale: Shades of Consequence**
3. **Universal Skill: Navigating Consequences**
4. **The Equation of Life**
5. **The Clear View: Living Through Universal Lenses**

These texts translate consequence-based logic into human decision systems.

V. Organizations, Power, and Geopolitics

For readers working in leadership, policy, or strategic domains.

1. **The Philosophy of Power**
2. **The Unified Power Architecture (UPA)**
3. **The Globalization of Consequences**
4. **Sovereignty 2035: The India Equation**
5. **The War of Thoughts**
6. **The Human Game**

These works apply UED-consistent reasoning to large-scale coupled systems.

VI. Interpretive & Boundary Texts (Read With Restraint)

For readers who enjoy conceptual stress-testing.

1. **Chavan vs Einstein: The Universe Was Never Curved**
2. **Chavan vs Hawking**
3. **Quantum Courtroom: The Trial of a Century**
4. **Myth Wrapped in Math**
5. **The End of Particle Age**

These are diagnostic provocations, not foundational texts.

How to Use This List

- Do *not* read sequentially
- Do *not* attempt completeness
- Stop reading when behavior changes
- If explanation becomes comfort, pause

UED works best when reading decreases over time.

Further Reading Note

This list exists to support continuity, not expansion. Universal Energy Dynamics does not require reinforcement through volume.

If curiosity fades, stop. If clarity increases, stop sooner.

That is the intended outcome.

Final Insights from UED

The most powerful insights from Universal Energy Dynamics (UED)

- What appears sudden in the world is rarely sudden. It is the visible moment when accumulated pressure finally finds a path to resolve.
- Equilibrium is not the absence of movement. It is the quiet, continuous resolution of imbalance before it grows dangerous.
- Systems rarely fail because of external shocks. They fail because unresolved delta silently crosses the limits of alignment.
- Constraints do not imprison systems. They define the safe pathways through which change can occur.
- If pressure is resolved early, change appears gradual. If it is delayed, correction appears dramatic.
- Power does not lie in controlling events. Power lies in shaping gradients and constraints before consequences arrive.
- The most fragile systems are not the chaotic ones. They are the systems that appear stable while imbalance quietly accumulates.
- Consequences rarely remain local. Through coupling, every resolution becomes part of a larger field of effects.
- Intelligence does not eliminate consequences. It reduces the cost of encountering them.
- Time does not heal imbalance. Time only allows pressure to accumulate until resolution becomes unavoidable.
- The universe does not operate through intention or punishment. It operates through consequence and correction.
- Universal Energy Dynamics does not promise control over reality. It offers something more practical: the ability to recognize when systems are approaching the limits of their stability.

Acknowledgements

This book exists because many influences converged—some visible, many not.

I acknowledge the traditions of science that taught me discipline, precision, and respect for evidence, even when questioning their assumptions became unavoidable. Every framework I examined, whether I agreed with it or not, contributed by revealing both its strength and its limits. Understanding grows as much from where models fail as from where they succeed.

I am grateful to the thinkers, teachers, and students—known and unknown—whose questions exposed gaps that explanations alone could not fill. Repeated confusion, recurring paradoxes, and stubborn inconsistencies were not obstacles; they were signals. Without them, this work would never have taken shape.

I also acknowledge the quieter influences: everyday systems, ordinary conversations, classroom experiences, organizational behavior, social tensions, and personal moments where consequence arrived later than expected. These were not anecdotes; they were data. They kept the framework grounded when abstraction tempted it to drift.

My family deserves special acknowledgment—not for agreement, but for tolerance. Living alongside someone who continuously questions fundamentals is not easy. Their presence provided the friction necessary to test whether ideas remained functional beyond the page.

Finally, I acknowledge restraint itself. Much that could have been written was intentionally left out. Many claims that could have been made were deliberately avoided. This book owes its coherence as much to omission as to inclusion.

If **Universal Energy Dynamics** proves useful, it will not be because of what is acknowledged here, but because of what quietly works when nothing else needs to be said.

— **Sandeep J. Chavan**

About the Author

Sandeep J. Chavan is an independent researcher, educator, and author whose work focuses on understanding how complex systems behave across physics, psychology, organizations, and societies. Trained as an engineer and long engaged in teaching, his approach is shaped by sustained observation of where established models succeed—and where they quietly fail under scale, coupling, and delayed consequence.

He is the originator of **Universal Energy Dynamics (UED)**, a framework developed to explain systemic behavior without resorting to force-centric, belief-based, or purely mathematical abstractions. His work emphasizes structural clarity, operational restraint, and consistency across domains rather than prediction or ideology.

Chavan's writing avoids motivational, spiritual, and doctrinal framing. Instead, it aims to reduce misinterpretation, over-intervention, and premature certainty—both in scientific reasoning and everyday decision-making. His broader body of work spans foundational physics, human behavior, leadership dynamics, and geopolitical systems, unified by a single question: *why systems fail when intelligence alone is not enough.*

He lives and works in India, continuing his research and teaching independently, with a focus on building frameworks that remain useful even after they are no longer consciously referenced.

Visit: https://www.sandeepjchavan.com/

ORCID Profile: https://orcid.org/0009-0005-4190-7164

Disclaimer and Statement of Originality

Disclaimer

This book is a work of independent research and interpretation. It is not intended to replace established scientific theories, mathematical formalisms, professional advice, or domain-specific expertise. Universal Energy Dynamics (UED) is presented as an **interpretive and diagnostic framework**, designed to clarify structural behavior across systems—not as a predictive engine, belief system, or universal solution.

Nothing in this book should be construed as medical, legal, financial, psychological, or policy advice. Any application of the ideas discussed—whether personal, organizational, technical, or societal—remains the sole responsibility of the reader. The author assumes no liability for outcomes arising from interpretation, application, or misuse of the framework.

UED does not claim certainty, completeness, or final authority. Where ambiguity remains, it is acknowledged intentionally. Where boundaries are drawn, they are drawn to prevent overreach rather than to assert dominance over existing knowledge.

Readers are encouraged to maintain critical judgment, contextual awareness, and domain sensitivity at all times.

Statement of Originality

The framework known as **Universal Energy Dynamics (UED)**, including its core primitives, definitions, structural logic, diagnostic templates, and interpretive posture, is the original work of the author.

While this book references and respects established scientific, philosophical, and systems-theory traditions, UED does not derive its central constructs from any single existing theory. Concepts such as **delta, resolution, alignment threshold time (ATT), field-based causality, constraint-first modeling, and consequence-driven interpretation** are developed here as a coherent, internally consistent system unique to this work.

All reinterpretations of mainstream frameworks—classical mechanics, relativity, quantum theory, thermodynamics, systems thinking, and behavioral models—are offered as **structural reframings**, not reproductions or summaries. Any similarity in terminology reflects shared language, not borrowed doctrine.

This book has been written without reliance on proprietary sources, confidential material, or undisclosed collaborations. Where ideas overlap with known concepts, they are acknowledged openly and bounded clearly. Where departures occur, they are stated explicitly.

The originality of this work lies not in novelty for its own sake, but in **the disciplined refusal to extend beyond what can be structurally supported**, and in maintaining coherence across scale without collapsing into belief, ideology, or abstraction.

— Sandeep J. Chavan

Don't miss out!

Visit the website below and you can sign up to receive emails whenever Sandeep Chavan publishes a new book. There's no charge and no obligation.

https://books2read.com/r/B-A-EPGPC-TYHCJ

BOOKS 2 READ

Connecting independent readers to independent writers.

Did you love *Universal Energy Dynamics*? Then you should read *Participation – From Identity to Awareness*[1] by Sandeep Chavan!

[2]

Why do systems behave the way they do?Why do individuals struggle to control outcomes in organizations, societies, and personal life?Why do narratives spread faster than understanding?

In *Participation*, Sandeep J. Chavan offers a powerful new lens for understanding these questions.

Rather than viewing individuals, institutions, and events as isolated entities, Chavan reveals a deeper structural reality: everything exists and evolves through participation within a shared field of interaction.

From classrooms and workplaces to global systems and social movements, human life unfolds through networks of relationships, signals, incentives, and feedback. Yet most people interpret these dynamics through simplified stories about individuals, intentions, or isolated decisions. As a result, many complex patterns remain misunderstood.

---

1. https://books2read.com/u/4AZego

2. https://books2read.com/u/4AZego

In this book, Chavan introduces the **Participation Framework**, a way of seeing how identity, intelligence, behavior, and awareness emerge from ongoing interaction within dynamic systems.

Through clear explanations and conceptual models, Chavan explores:

• How identities form as anchors within larger systems• Why narratives amplify faster than understanding• How misalignment in participation creates conflict and instability• Why control is often an illusion in complex environments• How awareness allows more intelligent coordination within systems

At its core, **Chavan argues that human beings are not separate observers of the world but active participants embedded within the systems they attempt to understand.**

When this shift in perspective occurs, many familiar problems begin to look different. Leadership, responsibility, conflict, cooperation, and even personal awareness take on new meaning once we recognize the deeper dynamics shaping them.

Written in a clear and reflective style, Chavan bridges ideas from systems thinking, philosophy, psychology, and complexity science to offer a structural view of human interaction. The book is designed for readers interested in understanding how modern societies, organizations, and human behavior function beneath the surface.

Participation does not offer motivational advice or simple solutions. Instead, Chavan provides a conceptual lens through which readers can observe the patterns that quietly shape everyday life.

Once seen, these patterns are difficult to ignore.

Because participation is not something we occasionally choose.

It is the condition through which reality continuously unfolds.

Read more at https://www.sandeepjchavan.com/.

www.ingramcontent.com/pod-product-compliance
Lightning Source LLC
LaVergne TN
LVHW100246210926
848480LV00003B/838
9798233193606